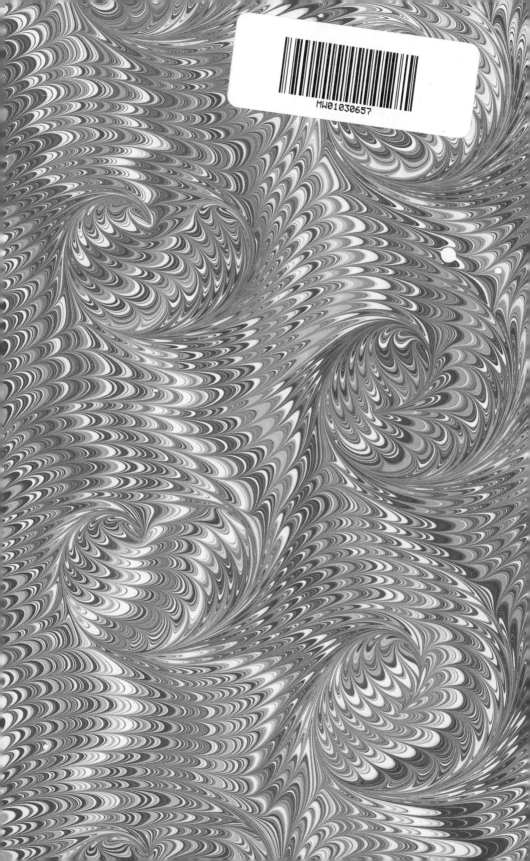

The Legendary Secret Norden Bombsight

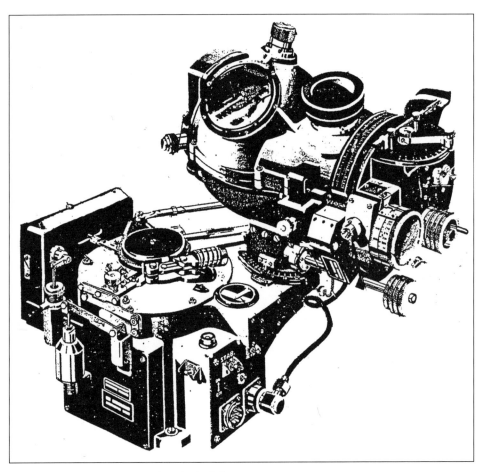

M-9 Norden Bombsight mounted on the Stabilizer.

The Legendary Secret Norden Bombsight

Albert L. Pardini

Schiffer Military History
Atglen, PA

Book Design by Ian Robertson.
Copyright © 1999 by Albert Pardini.
Library of Congress Control Number: 98-87635

All rights reserved. No part of this work may be reproduced or used in any form or by any means—graphic, electronic, or mechanical, including photocopying or information storage and retrieval systems—without written permission from the publisher.
The scanning, uploading and distribution of this book or any part thereof via the Internet or via any other means without the permission of the publisher is illegal and punishable by law. Please purchase only authorized editions and do not participate in or encourage the electronic piracy of copyrighted materials.
"Schiffer," "Schiffer Publishing Ltd. & Design," and the "Design of pen and inkwell" are registered trademarks of Schiffer Publishing Ltd.

ISBN: 978-0-7643-0723-2
Printed in China

Schiffer Books are available at special discounts for bulk purchases for sales promotions or premiums. Special editions, including personalized covers, corporate imprints, and excerpts can be created in large quantities for special needs. For more information contact the publisher:

Published by Schiffer Publishing Ltd.
4880 Lower Valley Road
Atglen, PA 19310
Phone: (610) 593-1777; Fax: (610) 593-2002
E-mail: Info@schifferbooks.com

For the largest selection of fine reference books on this and related subjects, please visit our website at
www.schifferbooks.com
We are always looking for people to write books on new and related subjects. If you have an idea for a book, please contact us at proposals@schifferbooks.com

This book may be purchased from the publisher.
Include $5.00 for shipping.
Please try your bookstore first.
You may write for a free catalog.

In Europe, Schiffer books are distributed by
Bushwood Books
6 Marksbury Ave.
Kew Gardens
Surrey TW9 4JF England
Phone: 44 (0) 20 8392 8585; Fax: 44 (0) 20 8392 9876
E-mail: info@bushwoodbooks.co.uk
Website: www.bushwoodbooks.co.uk

Contents

Dedication 4
Preface 7
Introduction 8
Acknowledgements 12

CHAPTER		
I	Theory of Bombing and the Bombing Problem	15
II	Aerial Bombardment	24
III	Development of Bombsights	28
IV	Gyroscopes	35
V	The Navy's Norden Bombsight	40
VI	The Navy's Norden Mark XI Bombsight	43
VII	The Navy's Norden Mark XV Bombsight	67
VIII	The AAF and the Navy Mark XV Bombsight	87
IX	Production Increases and Problems Escalate	100
X	Pearl Harbor. The Navy's Mark XV Goes to War.	108
XI	AAF Expansion and the Norden Mark XV Bombsight	120
XII	Preparing for Expansion and Production	132
XIII	The Navy's Norden Mark XV, M-9 Bombsight	140
XIV	Augmenting the Mark XV, M-9 Bombsight	163
	Glide Bombing Attachment	164
	Automatic Gyro Leveling Device	167
	Reflex Open Optic Sight	170
	Super Altitude Attachment and High Altitude Bombsight	172
	Norden Super Altitude Attachment	173
	Low Altitude Bombing Attachment	174
	Photo Gallery	182
XV	Norden Bombsight Ancillary Equipment	200
	Disc Speed Tachometer	200
	Bombsight Stop Watch	201
	Anti-Glare Lens	202
	Modification of the C-1 Auto Pilot Steering/Turn Control	203
	Formation Stick	204
	Intervolometer	205
	Automatic Release Lock	206
	Ballistic Coefficient of Bombs	206
	Wooden Tweezers	207
	Illustrated Manuals or Pamphlets	208
	Data Books for Bombsight and SBAE/AFCE	209
	Electrically Operated Heating Blanket for the Bombsight	211
XVI	Security	217
	Repair Shops & Storage Facilities	224
	Typical AAF Bombsight Repair Building	226
	Milton Stomsvick's Story	228
	Field Repair Box-Airborne	231
XVII	Norden Mark XV — From Visual to Radar/Radio	236

XVIII	AZON/RAZON Radio Guided Bombs	241
	Unknown Facts About the Norden Bombsight Mark XV	249
XIX	Mark XV Bombsight Training — Mobile and Stationary Devices	260
	7A-3 Trainer	263
	A-6 Photographic Trainer	264
	Visual Link Trainer	264
	Mock Up Training Devices	265
	Plexiglass Bombsight	266
	Classroom Training	268
	Radar Classroom Training	270
	Field Services Training	271
	Mobile Trainers	271
XX	Myths of Spider Webs, Human Hair, Etc., Etc.	274
XXI	The Honeywell C-1 Automatic Pilot	280
	Mr. Covington's Story	281
	Mr. Booen's Story	283
XXII	Cancellation of The Sperry Sight	286
XXIII	Foreign Interests	294
	Letter from Prime Minister of Great Britain	298
	President Roosevelt's Answer to Mr. Chamberlain	299
XXIV	Compromise	300
	Norden Bombsight Spy Case	302
	The Mellilla Incident	302
XXV	Counting the Cost	307
XXVI	A Bomb Drops on The Norden Company	317
XXVII	Norden Mark XV Bombsight Bombing Accuracy — The Results	321
XXVIII	The Navy's Mark XV Bombsight Last Combat Use	327
	Epilogue	330
	Bibliographical Note	332
	The Navy "E" Award	335
	In Memorium	336
	Aviation Ordnance, Mark XV	337
	Norden Bombsight Contractors	339
	Norden Bombsight Production	342
	Supporting the Norden Bombsight	339
	Definitions and Nomenclature of Norden Bombsight Terms	343
	Glossary	347
	Government Publications	349
	Sources of Information	351

PREFACE

A large part of the difficulty lay in the bombsight itself. The sight was not ready for production line type of work. It had not been designed with an eye to mass production. At first the Norden Company and the selected manufacturers found that they did not speak the same language. Norden craftsman had been working to extremely close tolerances for many years, and many were from the Old World. The contractors had never worked under such conditions.

Extreme difficulty was initially encountered in supplying these new manufacturers with tooling, blue prints, technical and engineering data and the availability of skilled mechanics. Dies, jigs and fixtures were used wherever possible. The idea of using gang drilling was alluring but could not be used in the machining process.

To the readers of this book the author wishes to impress the fact that some of the information or data: *may* or *may not* be accurate. We are at the mercy of files. Most of the time, correspondence referring to other letters for action were never found. At times, some assumptions were made since the author had used much of the equipment produced for the NBS, and knew that it had been manufactured. Many of these items are now being sold in military surplus outlets at a cost more than the original. A good example is an original copy of the Bombardier's Information File sells for about $275-$350 depending on it's condition. This book is not intended to be controversial as only the facts found in the examined files are reported here.

If some of the information appears incomplete or disjointed, it is, as we had no other source. In some areas, a Boeing B-17 can be flown through it. This is the first attempt to try and record the untold story of the NBS. It is a compilation of events beginning in 1922 and ending in 1967-68.

INTRODUCTION

The information in this book is based almost entirely on government documents in the National Archives and Records Centers in Washington, D.C., Suitland, Maryland, Navy Bureau of Ordnance, Bureau of Aeronautics, Naval activities and Air Force organizations. Very little data was obtained from private companies, research centers and universities.

Many years have passed since the beginning of the development of the Norden Bombsight (NBS). Critical documents such as engineering, testing, technical and administrative documents have been destroyed thereby forestalling any attempt to accurately portray a concise history. This history on the NBS has been done to show how the sight progressed from its initial experimentation with the Mark XI to the famous Mark XV. When we started our research for documentation, we were dismayed to discover that very little data remained from a vast source, was incomplete and was not readily available. None of the data was in one repository or office, but scattered over a wide area of the Eastern United States. To compound the problem, there were very few people to interview. Those who were at the decision making level were no longer living, addresses unknown, refused to be interviewed or furnish any information. On the other hand, there were many repair people who, having worked on the sight, gave freely of their time and experiences in direct contrast to either General or Flag Officers who refused to answer letters or telephone calls.

The complete story of the NBS will never be completely written as it is too enormous to be fully mastered by one or more researchers. For nine years we have been searching for any relevant primary data. A very small amount of documentation covering the period of 1921-1945 was available for review. The growing interest in the sight contributed to the reason for this book. The lack of any definitive book or books on such an important subject was not found. When our bombsight, Directional Stabilizer and many parts of the Automatic Pilot was placed on display at the Museum of Flight at the Santa Maria Airport, we were amazed at the interest shown. Of course, this display triggered many questions by visitors concerning the sight. We could not intelligently answer these questions nor could we refer them to any printed material. Since information was not readily available that chronicled this weapon, we embarked on a journey in an attempt to write a history. Little did we realize that very little data was available. We felt that such an important weapon that was used by the Army Air Forces, Navy and our Allies there would be vast amounts of material preserved for future research.

When we started out in an in-depth and intensive research, we were appalled at the lack of documentation. After nine years of writing letters, telephone calls and travel, we were able to find some pertinent data. What we did find was very disappointing. There were no spectacular finds. No boxes hidden away. No bonanzas. The National Archives and Smithsonian Institution personnel tried to find any material that possibly could have been misfiled. Even under this intense scrutiny, nothing more was discovered. There has been and are still being published many magazine articles on the NBS. Some of these articles are very well written and contain good information. Many of the articles written go as far back as 1942. Some of them give an excellent description on how the Norden bombsight worked and what its main job was. Although no specific data was given to these magazine publishers by the government, they apparently knew enough about the bombing problems but some publications such as Popular Science, Flying, etc., printed interesting facts on the use of aerial bombing. These articles did not use any classified information, but their guess was pretty accurate.

During research, many thousands of letters, reports, technical data and engineering reports were reviewed and thousands of copies were made for future references. The files in the National Archives and Records Administration (NARA) in Washington, D.C. and Suitland, Maryland, did not reveal very much technical data. In most instances attachments or enclosures to letters consisting of studies, reports, etc. had been withdrawn from the basic letter so that only the letter remained. Any action or disposition on a certain subject will never be known. A constant stream of letters, telephone calls and visits were made to the various companies, government offices, universities that manufactured the sight or had done contract work for the Army Air Forces and Navy Bureau of Ordnance that were still in business. With a few exceptions, this effort did not provide any significant or extraordinary data.

The task of assembling this documentation and material from various sources presented a formidable task. For some inexplicable reason, no key documents were around relating to the formation of the companies that would eventually produce bombsights. Except for a few, no contracts or purchase orders for equipping new plants were found. There were a few documents detailing the amount of funds appropriated by the government for new plants. Exact costs on the amount of machine tools to equip new plants, converting from peace time to war production was not discovered. In order to determine if additional data existed that we had inadvertently overlooked in government offices that might have information on the BS such as the Library of Congress, Navy Bureau of Ordnance/Aeronautics, NARA, the services of a professional researcher in the Washington, D.C. area was obtained. Again, very little information on the sight was uncovered and the disappointing results did not warrant the expenditure of further funds. Also, the services of a graduate student at the Columbia University was obtained, but the results were the same. Columbia University did considerable work for the government on the sight and attachments, as did the University of Michigan and the Massachusetts Institute of Technology.

    This book contains information from official government records, company narratives, personal accounts, universities and from the author's recollection as supervisor of a bombsight machine shop from 1941 - 1949. since we do not wish to overwhelm the readers of the book with footnotes, we have used only the most important source notes so that it would not be cluttered with explanations. Complete and unequivocal credit is given to all of our sources. Without the data that they provided and their assistance, this book could not be written. The following pages is the story of the NBS before, during and after WWII. It is a story of the perfection of mechanism in which bombing success was dependent on the inherent precision of the sight. It solved the problems of synchronization and gyro stabilization. This book is written so that the average person can easily understand the contents. It is also a story of one of the U.S. Government's most secret devices up to the time of our entry into WWII. It is a tragedy that no action was taken to write a comprehensive history after WWII when much of the data was still available, however, the old saying of "half a loaf is better than none".

    We have not been able to cover every aspect of the sight in detail. Every effort was made to find reliable material or any documentation that pertained to the sight. The task of determining fact from fiction or myth constantly created a problem. Greatest care has been taken to make sure that the reported information is authentic and not just hearsay. Any data that we felt was questionable is labeled as such. We are relying on the documents that were published by the Army Air Forces, Bureau of Ordnance/Aeronautics, Naval Proving Ground, Air Material Command (AAF), etc. We feel that they are authentic and were the only ones that were still available for research. We have accepted the fact that this book will never be a best seller. At best, we will be fortunate to break even financially. With the sophisticated weapons that were used during the Persian Gulf War, many people will not be interested in a weapon that was used some 50 years ago. However, it was the state of the art at the time and the best that our government could produce. Attempts were made to contact all of the many contractors and sub-contractors that were furnishing material to the various factories that were manufacturing the NBS. For example, in trying to obtain information on ball bearings, optics, etc., used in the sight, we were not able to trace any of these sub-contractors. We wrote to these firms that had an address. Invariably our letters were not answered but they also were not returned. We felt that they did not want to be bothered or assist us in our project.

The concept behind this book is a simple one. It is to document a very important weapon in the conduct of the air war over Germany, Japan and Italy. There is no question that this weapon changed the course of the war, and it effected its outcome. It is a fact that the continuing air bombardment of enemy targets caused grave problems by the use of daylight precision bombing. With the systematic destruction of their industrial, transportation, fuel, and chemical facilities caused them to devise alternate methods to survive and to keep destruction to a minimum. It is readily apparent that the daylight bombing phase over Germany and Japan had a devastating effect. After the end of hostilities, the German Generals and Albert Speer (Minister of Production), were unanimous in stating that the precision daylight bombing conducted by the Americans had a very disastrous, if not a catastrophic effect on their capability to continue to produce war material.

ACKNOWLEDGMENTS

We would like to acknowledge those who assisted us in obtaining material for this book. One of the biggest problems was in locating the data. None of it was in a centralized depository. We attempted to contact any government agency that might have participated in the Norden Bombsight program (NBS) that are still operating. Some offices had fragmentary data, others were defunct and some did not know what we wanted. We are sure there is more information stored away in some obscure office or file. These bits and pieces of material, if found, would add to the story. We owe a great debt to the organizations that are listed in the "Sources of Information". Out thanks also goes to the untold number of civilian and military personnel who took the time from their daily duties to answer our letters or telephone calls, to provide what information they had on the sight and ancillary equipment.

Contrary to popular belief about service from government offices and personnel, we had *excellent results*. Many went out of their way to assist us. Without exception, our requests for assistance was either answered by letter, telephone call or both. If their office or agency did not have this information they would suggest another source of which many we were able to obtain needed data. On occasion, it may have taken a period of time to receive an answer but it was worth it as we invariably received a reply. This was in direct contrast with private companies who rarely answered. The exception was Honeywell, Inc. and Barden Company who made available any information they had. It is not feasible to mention all the persons that we came in contact with as there were so many. If they read this book,

Acknowledgments

they will know that they were a part of its formation, and their contributions are certainly appreciated. Due to the ever growing interest in the NBS, bombsight collectors, first time owners of the sight and military surplus outlets selling NBS memorabilia, it is not possible to list all of these people. Information was also found to be available under the Freedom of Information Act. The data was excellent and it was readily made available to us. The material was mostly technical in nature and it dealt with the sight before, during and after World War II.

We are greatly indebted to the following:

Congressman Robert J. Lagomarsino, 19th District, California, House of Representatives, who gave his time and personnel in assisting us in locating difficult, obscure and defunct organizations.

Col. Ross Whistler, USAF (Ret) who unselfishly furnished material, data and sources of information.

Mr. Carroll J. Watkins who, over the years, gave us much information and first hand data.

Mr. James Archer, Colchester, England, who provided information on the NBS concerning its use by the RAF.

Col. Paulo J. Pinto (Ret) Brazilian Air Force gave his time and furnished information on the NBS used in Brazil.

Mrs. Nancy Meddings, Marcia O'Neill and Linda Maneses of the Santa Maria Library who suffered with me in obtaining, locating and researching for information on government agencies, companies, universities, defunct organizations, etc. relating to the NBS. To them we owe a debt of gratitude.

Mr. Ben K. Weed with whom I established a close relationship over the years concerning the sight.

Gratitude is also given to the following:
Mr. Ross Loucks, Carmicheal, California
Mr. Don Bodie, in Rememberance
Mr. Mike Polansky, New Baltimore, Ohio
Mr. Sherman Boen, Richfield, Minnesota
Mr. Mike Stapp, Honeywell, Inc., Minneapolis, Minnesota
Mrs. Carol Church, Motion Picture, Sound and Video Branch, Washington, D.C.
Mrs. Diane McClurkin, Motion Picture, Sound and Video Branch, Washington, D.C.
Mr. Robert M. Valentine, Spokane, Washington
Mr. Fred Miller, Folsom, California
Mrs. Elise Maguire, Folsom, California
Mrs. Marge (Gorman) Eiseman, Scottsdale, Arizona

Mr. John E. Torbert, III, whose life long association was an inspiration and made valuable bombsight material available for display in the local air museum.

To my wife, Frances, who patiently and faithfully accompanied us in our quest for data. She became adept in screening the thousands of documents that we found in the various offices, libraries and government agencies.

Our son, Robert, whose persistence in forcing me to do something with the Norden bombsight. He kept after me until I started research, gathering information and documentation. His help was invaluable in tracing and locating bombsight attachments, bombing tables, bombardier equipment, computers and other ancillary equipment and in the process became an extremely competent researcher.

To Mrs. Helen Bright who took the time from her busy schedule to proof read the entire manuscript.

I

Theory of Bombing and the Bombing Problem

The theory of mechanics and physical laws of an explosive or shell by the use of a cannon is well known and has been around for hundreds of years. By simply elevating the gun barrel to higher angle, a greater distance for the shell to be hurled can be obtained. The same principle applies to dropping a bomb from an airplane. The difference being that it is carried aloft by an airplane hundreds of miles deep into enemy territory and then released at the highest point in the trajectory. The path that the bomb travels through the air is the bomb's trajectory. The bomb's trajectory is subtended by the line of sight at the instance of release, forming in space a right triangle. The sides are represented by the altitude and the horizontal distance along the ground the bomb travels from the point of release to the point of impact. The distance is actual range. The third side of the triangle is the line of sight, or the hypotenuse. From the point of release, it follows the same trajectory as that of an artillery shell (Plate #1). So in essence, aerial bombardment is a long range extension of artillery, although on the surface this does not seem probable, as most people do not associate artillery with airplanes. [1]

In order to present the theory of bombing and the bombing problem as simply as possible only the parts required to understand the use of bombsights will be used. The action of a bomb in the air must be started in an environment that we all are familiar with. To determine this, it is necessary to study the action of a bomb dropped in a vacuum. Many people think that a bombing plane must be directly over the target before releasing the bomb. This is entirely erroneous. The men who invented the bombsights had a thorough knowledge of the bombing problem and mathematics before actually constructing the sight. In a vacuum there is no air or gas to restrain the motion of a body; regardless of size, weight, or shape, it will fall with exactly the same speed as any other body. The laws of physics provide a formula for the speed of a falling body, and it is expressed in terms of the attraction of gravity and of the amount of time or the distance through which it acts. Since gravity is essentially constant and is acting on the bomb continuously until the target is hit, the bomb will accelerate at a constant rate. This picking up of speed is known as the acceleration of gravity. At sea level, this acceleration is approximately 32.2 feet per second. This means that during each second of fall, the body will increase its speed by 32.2 feet per second.[2] In the use of bombing mathematics, distances along the ground can be measured in feet or in angular measure. In the latter case, measurements are made from the bombing airplane in units called "mils" instead of degrees. A mil is the angle subtended by an arc whose length is one thousandths of the distance from the observer to the object, or the angle tan - 1 = .001; 17.4 mils equals one degree and 3.43 minutes equals one mil. At a distance of 10,000 feet, one mil = 10 feet; or 100 feet at a distance of 10,000 feet would be 10 mils.[3]

The solution of the bombing problem by the bombsight is done in terms of angles instead of distances, and for this reason all values except altitude put into the bombsight by the bombardier must be in terms of angles. Trail is always used in terms of mils and is inserted as such in the bombsight. The mil system can also be used to compare bombing scores. For example, a bombardier whose average error from 10,000 feet is only 100 feet should be considered as good as one whose average from 5,000 feet is only 50 feet. In each case, the average mil error is 10 mils. The mil system expresses distances on the ground and must be qualified by a statement of bombing altitudes.[4]

In order to release a bomb so that it will hit the target, the exact point in space must be determined. To find this critical point, an accurate sighting device is required. It would take volumes to fully explain this phenomenon. In order to keep this so that it can be easily understood, the various bombing terms and explanations have been reduced to fairly simple terms. A comprehensive explanation in the development and solving the bombing problems cannot be detailed in this book, as many years of experimentation, testing, building of factories, and great sums of money—government and private—with many inventors laboring were applied in trying to develop a high altitude precision instrument which started before World War I.

The bombing problem consists of two parts. They are the COURSE and RANGE. The course problem is well known and is fairly simple. However, the range problem is a different matter. A falling bomb, in order to hit the target, must be released at the correct distance back from the target so that it will not fall short or over.[5] When a bomb is released from an airplane, various forces immediately act upon it. These forces determine the path which the bomb follows and the point of impact. They are Gravity, True Airspeed, Air Resistance, and Wind (Plate #2). The following is an explanation of the above and some of the bombing terms that need to be explained.

GRAVITY
Gravity pulls the bomb toward the earth at a continuously increasing speed. This acceleration, due to gravity, is the same for all bodies, regardless of size, shape, or weight.

TRUE AIRSPEED
The second force that affects a bomb is that at the same time that gravity is pulling the bomb downward, velocity is pulling it forward. The airplane is traveling at definite speed with respect to the air. Since the bomb is part of the airplane up to the moment of release, it leaves the airplane with the same forward velocity. In bombing, this forward velocity of the airplane and the bomb relative to the air is called true airspeed. The time between release and impact the bomb follows a path between the direction of these two forces. The time element between release and impact is called actual time of fall.

AIR RESISTANCE
The third force affecting the bomb in its flight is one which acts against the first two—air resistance. Airspeed is driving the bomb forward, and the air through which the bomb moves is resisting this motion, causing the bomb to lag behind the airplane. The distance on the ground resulting from this resistance to the forward motion of the bomb is horizontal lag.[6] This resistance tends to keep the bomb in flight longer. During the additional time required for the bomb to fall, the airplane continues to move forward. The distance on the ground over which the airplane travels during this additional time is vertical lag.[7]

WIND
Before considering the fourth force, some fundamental bombing criteria should be explained. Because of the complex nature of high altitude precision bombing, these terms will, in some way, explain how the bombsight works to solve these problems. The following terms are required to be solved before a sighting instrument could be developed with any degree of accuracy.

TRAIL
Trail is the horizontal distance measure on the ground from the point of impact to a point directly beneath the airplane at the instant of impact. Actually, trail is the sum of the two distances—horizontal and vertical lag. Trail is the result of several forces which are acting on the bomb. While true airspeed is driving the bomb forward, air resistance is tending to hold it back; while gravity is pulling it down, air resistance is tending to hold it back; while gravity is pulling it down, air resistance is tending to hold it up. If true airspeed increases, the resistance of the air increases; thus, the horizontal lag is greater. As true airspeed increases, trail increases. If the downward

velocity increases, the resistance of the air to that force increases and the vertical lag is greater. The resistance which the air offers to the bomb depends on the size and shape of the bomb. Each type of bomb has been classified into different types according to its own **BALLISTIC COEFFICIENT**. This means the relative amount of resistance of the air offers to it. A bomb with a high ballistic coefficient will fall faster with less than a bomb with a low ballistic coefficient, therefore, as ballistic coefficient increases, trail decreases.[8]

CROSSTRAIL
The distance between the true course of the airplane and the collision course is known as crosstrail. Crosstrail is measured from the point of impact to the true course. The bombsight automatically measures the crosstrail. Also, it depends on the trail and drift. To make this computation, the bombsight uses the SINE of the drift angle. The sine of an angle in a right triangle is the number obtained when you divide the opposite side of the hypotenuse. The side opposite the drift angle is the crosstrail; the hypotenuse is the trail. Therefore, the sine of the drift angle is the crosstrail divided by trail. If there is no drift, there will be no crosstrail. Accordingly, if trail is not inserted into the bombsight, it cannot compute crosstrail. The bomb will fall not only short, but also downwind of the target.[9]

DROPPING ANGLE
The dropping angle is the angle formed between the line of sight and the vertical reference at the instant the bomb drops from the airplane. This particular sighting angle set up by the bombsight at the instant of release is called the dropping angle.[10]

LINE OF SIGHT
In looking at the target through the bombsight, you are looking along a line from the bombsight to the target—this is called the line of sight. As the airplane moves forward toward the target, the line of sight changes. When the proper course toward the target has been set, the actual range needs to be determined; that is, the correct distance back from the target that the bomb must be released in order to score a hit. If the proper data has been programmed into the bombsight it will automatically solve this problem. It measures an angle which subtends actual range, thereby locating the proper point in space for the bomb's release.[11]

SIGHTING ANGLE
Actual range is the linear distance measured along the ground from a point vertically beneath the airplane at the time of release to the point of impact. This distance may be computed by subtracting trail distance from whole range. The bomb will lag a certain distance behind the airplane (trail).[12]

TRAJECTORY
The path the bomb follows through the air mass, from the point of release to the point of impact, but more particularly fired from a gun or a bomb dropped from an airplane.[13]

VACUUM TRAJECTORY
The path described by any object moving in a vacuum.

TIME OF FALL
Time of fall is the time it takes for the bomb to fall to earth from the time of release to the time of impact with the target.[14]

ACTUAL TIME OF FALL
Actual time of fall depends primarily on the exact height of the airplane over the target, or the vertical distance which the bomb must fall, but it is affected by true airspeed and bomb ballistics. The actual time of fall for each bombing altitude, true airspeed, and type or bomb has been determined by trail and error and are shown in various bombing tables.

DRIFT ANGLE
The angle between the airplane heading and the ground track. In other words, the angle formed between true heading and the true course is called the drift angle.

GROUND SPEED
Ground speed is the speed at which the airplane is traveling over the ground. It is calculated by adding or subtracting wind effect from true airspeed.[15]

WHOLE RANGE
Whole range is the horizontal that is measured on the ground that the airplane travels during the time of fall of the bomb, or from the time of release to the time it strikes the earth, providing the airplane maintains its speed and course. The whole range is measured on the ground. This distance may be computed by multiplying the ground speed by the time of fall. In computing the whole range, ground speed must be in feet per second. The time used is actual time of fall, and it is given in seconds.[16]

VELOCITY of CLOSURE
This is the velocity at which the airplane is closing with the target. If the target is stationary, the velocity of closure is equal to the ground speed of the airplane. If the target is moving, it is the ground speed plus or minus the ground speed of the

moving target. For example, if the target is moving at a ground speed of 30 mph in the same direction as the airplane, and the airplane speed is 250 mph, the velocity of closure would be 220 mph. If the target is moving towards the airplane at a ground speed of 300 mph and the airplane has a speed of 250 mph, the velocity of closure is 280 mph.[17]

Plates #2, 3 and 4 graphically show the various relationships of the forces that are encountered when a bomb is dropped from an airplane and the effect of these motions on the target.[18]

CROSSWINDS

The fourth force acting on the bomb is wind. If winds are from any direction except from dead ahead or directly behind, drift enters the bombing problem. Wind is the movement of the entire body of air surrounding the airplane. When the wind moves to the right, the airplane moves to the right. This is called right drift. If the body of air is moving to the left, the movement of the airplane is to the left. When there are crosswinds, the pilot "crabs" the airplane into the wind, that is, he heads the airplane upwind sufficiently to compensate for the effect of drift. The angle formed between true heading and true course is called the "drift angle." The speed of the wind causes both airplane and bomb to drift the same distance away from true heading. The bomb will strike the ground behind the airplane, along the longitudinal axis of the airplane and downwind of true course.[19]

HEADWINDS AND TAILWINDS

Headwind or tailwind does not effect the true airspeed of the airplane. Trail depends only on true airspeed, bombing altitude, and type of bomb, and none of these has any effect on the amount of trail. Groundspeed is the factor which is affected by a head or tail wind. If the airplane is flying at a true airspeed of 150 mph, with a tail wind which pushes the air forward at 10 mph, then the airplane's speed over the ground is 160 mph, or vice versa. Since the whole range is found by multiplying ground speed (in feet per second) by the actual time of fall, an increase in groundspeed causes an increase in the whole range.[20]

Chapter I: Theory of Bombing and the Bombing Problem

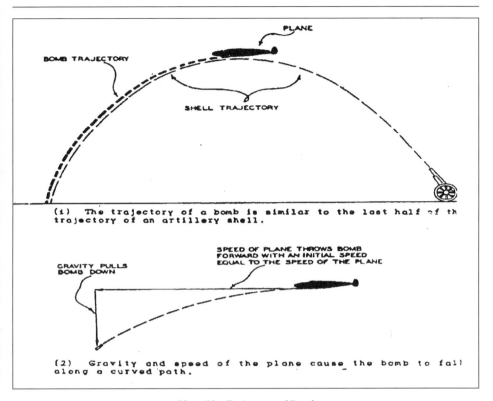

Plate #1 - Trajectory of Bomb

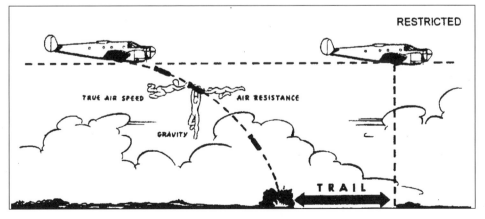

Plate #2 - Forces Acting on Bombs

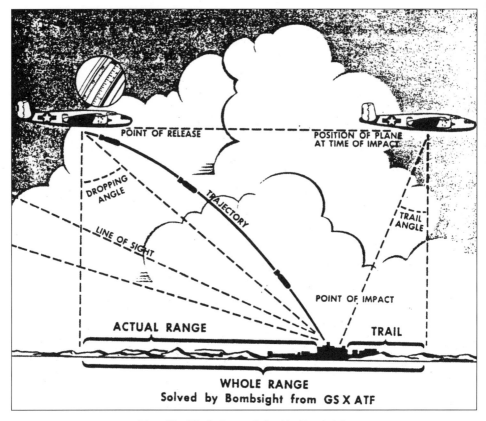

Plate #3 - Whole Range Solved by Bombsight

Notes:

[1]   The Norden Bombsight, Maintenance and Calibration, Department of Fire Control, Lowry Field, Denver, CO, September 1943, page 92.
[2]   Aviation Ordnance, Ordnance Pamphlet No. 649, Bomb Sight Mark XV, Mods 4 and 5. Equivalent Air Corps Designation M-4, 5, 6, and 7. Volume 1, 1943, page 1.
[3]   Ibid, page 9 and 10.
[4]   Ibid, page 9.
[5]   Student's Manual, Bombing Air Training Command, AFTRC Manual 51-346 1, dated September 1, 1949, page 1-1.
[6]   Student's Manual, Bombing, Army Air Forces Training Command, no date, page 1-2-2 and 1-2-3.
[7]   Student's Manual, Bombing, Air Training Command, AFTRC Manual 51-346-1, dated September 1, 1949, page 1-2.
[8]   Ibid, page 1-2.
[9]   Student's Manual, Bombing, Army Air Forces Training Command, 51-346-1, dated September 1, 1949, page 1-2 and 1-3.
[10]   Student's Manual, Bombing, Army Air Forces Training Command, no date, page 1-2-10 and page 1-2-11.
[11]   Ibid, page 1-2-5.

[12] Ibid, Page 1-2-4, 5.
[13] Ibid, page 1-2-5.
[14] M-Series Bombsight, Maintenance and Calibration, Bombsight Maintenance Lowry Field, CO, dated September 1944, page 101.
[15] Ibid, page 102.
[16] Ibid, page 103.
[17] Student's Manual, Bombing, Army Air Forces Training Command, 51-346-1, dated September 1, 1949, page 1-3, M-Series Bombsight, Maintenance and Calibration, Lowry Field, CO, dated September 1944, page 102.
[18] M-Series Bombsight, Maintenance and Calibration, Bombsight Maintenance Division, Maintenance and Calibration, Lowry Field, CO, dated September 1944, page 103.
[19] Ibid, page 92 for plate #1 and #3.
[20] Student's Manual, Bombing, Army Air Forces Training Command, 51-346-1, dated September 1, 1944, for plate #1, #3 and #4.
[21] Student's Manual, Bombing, Army Air Forces Training Command, no date, page 1-2-1.
Plate #1 Ibid, page 1-2-9.
Plate #2 Maintenance and Calibration, BSM Division, Department of Armament Lowry Field, CO, September 1, 1943, page 92.
Plate #3 Student's Manual, Bombing, Army Air Forces Training Command, no date, page 1-2-3.
Plate #4 Student's Manual, Bombing, Air Training Command, AFTRC 51-340-1, September 1, 1949.

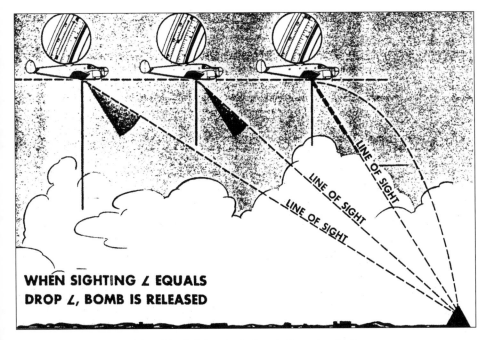

Plate #4 - Sighting Angle Equals Dropping Angle

II

Aerial Bombardment

Some form of aerial bombing has been attempted throughout mankind. Prehistoric man went to great lengths to find higher ground in order to drop rocks on their enemies. In the early 1800s, many of the European countries used balloons in warfare to drop explosives or other demoralizing firearms on the enemy. Mostly used in those early days were hot air balloons, and not much faith was put in their use.

With the successful flight of an airplane at Kitty Hawk, North Carolina, in 1903, a new weapon of war was ushered in. Although this was not immediately accepted at the time, the potential for such a new weapon did not become readily apparent. Man now had realized the age old goal of getting above the enemy was now within range. It became possible for man to go higher and further than any known artillery being used at the time. It did not take the military planners very long to determine the value of this new flying machine.

Some experimentation on a limited basis was done, and some bombing missions were attempted in the years before World War I. These bombing missions created more panic on the ground than damaging the enemie's ability to carry on a war. Prior to WWI, bombers were being produced at a slow rate. At the time, emphasis was on the fighter, pursuit, and scout airplanes. After hostilities started in Germany, the Allies' real importance was placed on individual air combat between the opposing pilots. Also, another form of entertainment was balloon busting, in which Frank Luke was one of the leaders. The advantage of aerial bombing and its effect on enemy troops was not fully understood at that time. The military realized that it was a terror weapon against the civilian population which could be used to a good advantage. The major military powers did not have much faith in aerial bombardment as an offensive weapon. At that time they felt that nothing could be more effective than the fire power of land based armies and naval gunfire. They felt that the most useful role of the airplane would be in scouting to ferret out the enemie's troop concentrations, industrial capability, etc. It readily became evident during World War I that the airplane was a new and useful military weapon that had to be reckoned with. Here was now a chance where explosives could be carried beyond enemy lines and strike the rear echelons by destroying troop areas, marshaling yards, industrial plants, etc.

Prior to the beginning of WWI, military strategy did not include the airplane as a basic part of the war machine. At the time aircraft were small, light, and underpowered. After WWI started, more thought was given to the role of the bomber. Bombing missions were attempted by these aircraft carrying only small loads of bombs, which were dropped by the pilot or observer. These planes were not equipped with any type of bombing devices. They were forced to bomb from low altitudes, which exposed the aircraft and crew to ground fire. In order to escape from ground fire the planes were forced to fly higher, which in turn made accurate bombing impossible. The aiming methods used at the time were very primitive, and it was generally hit or miss—mostly misses. The pilots and observers, in trying to obtain better bombing results, tried to devise their own type of sighting devices. Some used parts of the airplanes as a reference, and others used fixed points on the wings, such as nails, dowels, etc., for releasing the bombs as the target came into sight. Towards the end of the war, larger and more powerful bombing planes were starting to be developed so that heavier payloads could be carried to be dropped on the target. However, the basic bombing problems had not yet been solved.

Prior to WWI and after the end of hostilities, there were many proponents for air power. One of the foremost and vocal of this group was a former artillery officer, General Guilo Douhet of the Italian Army. In the late 1910s to the early 1930s his ideas were beginning to be noticed by the major military powers, but a cautious approach was taken. The aeronautical use of aircraft for war purposes was used by the French in the Moroccan War in the 1920s. Also, the Italians used aircraft to bomb the rebels in Tripolitana in the 1920s. Most of these bombings were ineffective, as the targets were so widely dispersed. In WWI the Germans used zeppelins to bomb enemy targets, including London. As usual, the results were about the same—poor. Douhet also felt that if a large bomber force was used to bomb enemy industrial, transportation, fuel facilities, etc., the enemy would be forced to sue for peace. The most vocal American counterpart of Douhet's theories was General Billy Mitchell. He labored long and hard for a large striking air force, but the result of his efforts was that he was court marshaled and was forced to leave the service in the 1930s.

There were many theories on how effective air power would be during wartime conditions. One of Douhet's concepts was to develop a strategic air force capable of delivering devastating raids. The old guard military mind was not yet able to accept this type of warfare where hundreds of planes would be used in a single raid.

It would not be for another 20 years before the idea of using a strategic air force was developed and used. This concept came into being during World War II when the bomber stream consisted of hundreds of planes, and at times it would be a thousand in one mission. Douhet's idea of a separate air force came in 1947 with the formation of the U.S. Air Force. Prior to that time, it was part of the U.S. Army. Most major military nations now have a separate department for the air force. Both Douhet and Mitchell were far ahead of their time.

The fallacy of Douhet's theory of bombing a nation into submission was clearly demonstrated during WWII and the Persian Gulf War. However, there is no question that air power was a significant factor in the defeat of Germany and Japan. It remained for the ground soldier to complete the job.

After the end of hostilities in 1918, the role of the bomber was relegated to the back burner and no significant development of a bomber force was undertaken. The generals and admirals based their war plans and strategy on massive land armies and huge flotillas of naval warships, and to some extent the use of aircraft carriers. Usually after the end of a major conflict there is not too much interest in developing new weapons. Although the airmen were in the minority, they were not to be denied their concern for an adequate modern air force. The United States was to learn a bitter lesson for neglecting air power as part of their defensive and offensive military planning. Fortunately, the United States was planning to dramatically increase the Army Air Forces. President Franklin D. Roosevelt asked Congress to authorize the expansion of the air forces to 50,000 airplanes. It was quite evident by 1939 that the war in Europe, and with the Japanese occupying much of southeast Asia, that the United States would eventually be drawn into the conflict.

The day that the Japanese struck Pearl Harbor on December 7, 1941, the aircraft program was well underway and was increasing daily in strength. The rearmament program appeal by President Roosevelt would soon bear fruit.

It became readily evident that if the bombardment arm was to do its job, then development of an accurate fire control system was imperative. Before, during, and after WWI, many attempts were made to develop accurate bombing systems. One of the first types of bombsights used in airplanes with success was the drift type. This type of sight was fairly accurate, and bombing could be accomplished from various altitudes. Many of the bombing problems were incorporated in these early sights. To make this calculation, the drift of the aircraft over the ground was necessary. The idea of a level bombing platform during this time was also thought to be required. The concept of a level bombing platform had not yet been developed, and many years would pass before a workable one would be installed in airplanes. Many of the bombing problems were known to inventors working on bombsights. These early pioneers knew that the bomb lagged when it was released from the airplane. They also knew that the ballistic coefficient of a bomb was a very important element in a successful bombing mission. As is normal with any new type of weapon development was slow, as basic design data was not available, such as ballistic coefficient of bombs. Basic principles of bombing, testing, and engineering also lagged behind the development of the bombing plane. Bombing airplanes were built to carry a load of ordnance to the target. The bombsight was not included in the specifications when the airplane was built, and the bombsight became an accessory to the airplane. It soon became apparent that concept was in error, and that the airplane had to be designed around the bombsight. This makeshift operation compromised the effectiveness of the bombardier. Despite inadequate knowledge of aerodynamics, instrumentation, and testing means, these handicaps were overcome by the early bombsight inventors, and they were able to produce a fairly accurate sighting device.

The development and procurement of bombsights and fire control equipment in the late 1910s and after WWI was in its infancy. The Army Air Service by 1921 became serious about procuring bombardment airplanes, and the development and testing of larger type aircraft was started. In a memo dated September 15, 1921, submitted by the Engineering Division for auxiliary equipment, shows a total of $33,000.00, of which $10,000 was for bombsights. The advancement of larger airplanes brought on the development of more accurate bombsights. In 1926, the Air Service was phased out and replaced by the Army Air Forces. The Air Corps was interested in developing a bombing force and was investigating the feasibility of using precision bombing equipment.

III

Development of Bombsights

The United States did not enter WWI until nearly the end of the war. The Allied Powers and Germany had been fighting since 1914. It was not until April 1917 that the United States became an active participant in the conflict. During this period, it became evident that precision bombing equipment was a must if any degree of accuracy was to be obtained in destroying enemy targets. Experimentation on this phase of the air war had been going on in Europe for several years—before and during the war. The British and French had already made strides in developing some form of fire control equipment. The war ended in 1918, and interest lagged for a time in further exploiting the use of bombsights; as is always the case after a major conflict, there was not much interest in developing new weapons, plus the financial strain on the warring nations curtailed the expenditure and appropriation of large sums of funds for new weapons.

There are two types of bombsights that can be used. They are synchronous and fixed angle.[1] In this book we will discuss only the synchronous type. Errors are the same for each type of sight. In synchronous bombing, the bombsight solves the ground speed and the dropping angle. The word synchronous means that one thing will maintain the same relationship to another under constantly changing conditions. In this type of sight, the rate or angular drive of the optics keeps pace with the rate of angular change of the sighting angle. For example, if the sighting angle is 50 and the transverse hair rests synchronized upon the target, then when the sight angle reaches 40, the transverse hair will also automatically be at 40 from the vertical, and in that configuration will rest on the target. When sighting through the optics, there are two points: one above and one below the transverse hair. The first is the position of the eye, which sights to the transverse hair. The second is the point of projection to the ground of this line of sight. As the plane approaches the target,

Chapter III: Development of Bombsights

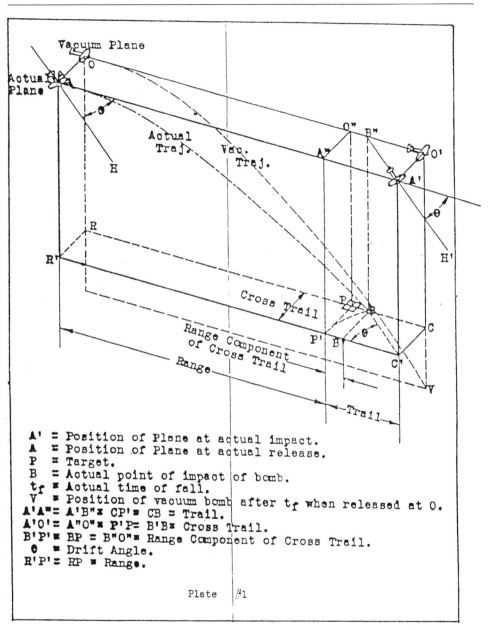

A' = Position of Plane at actual impact.
A = Position of Plane at actual release.
P = Target.
B = Actual point of impact of bomb.
t_f = Actual time of fall.
V = Position of vacuum bomb after t_f when released at O.
A'A"= A'B"= CP'= CB = Trail.
A'O'= A"O"= P'P= B'B= Cross Trail.
B'P'= BP = B"O"= Range Component of Cross Trail.
θ = Drift Angle.
R'P'= RP = Range.

Plate #1

Plate #1 - Relationship of Trajectories, Trail, Draft, and Cross Trail with the sight synchronized, the point of projection would move along the surface of the earth at a speed equal to velocity of closure. The fact is that the transverse hair (therefore, this point of projection) remains on the target and does not move relative to it. This is because the "forward" velocity of closure of the airplane is equaled exactly by the "rearward" velocity of the point of projection relative to the target. All this means the rate of angular closure of the transverse hair is exactly equal to the rate of change of the sighting angle.[2] No bombsight can synchronize itself. It must be properly adjusted and operated by a trained bombardier in order that synchronization can be obtained so that a mission can be carried out.[3] The manual act of synchronizing causes the optics to drive at the correct speed for synchronization and sets up the correct actual range angle for the given altitude, velocity of closure, and trail characteristic of the bomb. With this correct actual range angle set up it would cause the bomb to be released at the correct actual range distance away from the target. From tables, trail and actual time of fall are determined for a particular type of bomb, for a given altitude and calibrated air speed.[4]

Many years of constant experimentation would pass before a true synchronous sight would be developed so that it could perform as a precision high altitude bombsight. In order to calculate and compensate for the forces acting on a bomb, some form of sighting instrument was mandatory if a target was to be accurately hit. First of all, tremendous amounts of engineering, mathematics, and testing would be conducted to determine if a bombsight could actually drop a bomb from a great height with any degree of pinpoint accuracy. To begin with, airplanes prior to and during WWI did not have a very high service ceiling. The maximum altitude for airplanes in those days was about 15,000 feet. Most operated at a much lower altitude. Also, the power plant was not conducive to lifting a great load of ordnance. It was fortunate if some got off the ground. The main weakness in bombing was the human element. In using a bombsight, it was imperative that the bomb dropper or bombardier know accurately the bombing altitude, true air speed, type of bomb, correct trail, actual time of fall, and the velocity of closure. When these items are inserted in the bombsight it will automatically solve these problems. The sight will find the correct point in space for the bomb to be released, and it will release the bomb at the proper instant.[5] If the bombardier did a poor job of using the bombsight and missed the target because of an error, then the work of all of the crew that had anything to do with getting the airplane to the target goes for naught. The bombing airplane is built to deliver a load of bombs on the target, but if the bombardier fails in his tasks, then not only has the mission failed, but the enemy was not damaged and the considerable cost of men and equipment is wasted. To prevent this from happening it was necessary that a bombsight be engineered so that errors of any type could be kept to an absolute minimum. In addition, the bombing airplane could not hover over a target for any length of time because of enemy fighters and anti-aircraft fire.

Chapter III: Development of Bombsights

It was imperative that a sight be designed so that the very minimum time in seconds be spent over a target.

The use of synchronous sights went a long way in solving this bombing problem. Plate #1 graphically shows the various relationships of the trajectories, trail, drift, and cross trail. In this figure, the airplane is heading in the direction of AH in the upper plane OO'A'A, with track along AA' the direction of the wind being represented by A'O'. The bomb is released at A vertically over R', the latter in the horizontal plane through the target P. If there was no wind, the trajectory would lie in the vertical plane OO' CR with OO' as the travel of the airplane during the time of fall, P as the point of hit, and CP as the trail distance. The cross wind is represented by A'O'. While the bomb is in the air, this air moves the bomb the distance A'O', carrying with it the vertical plane AA'C'R', so that when the bomb hits, this plane, with the airplane and bomb, has moved to the position OO'CR, and the actual hit is at B. The apparent, or ground speed, direction of the airplane is along the line A'A', but the hit does not lie along this line. The trail lies always in the vertical plane through the airplane axis, and is the same whatever the wind. If we consider only the projection of these lines on to a horizontal plane we have B as the point of hit and the angle B is the "drift angle." At the instant of hit, the vacuum bomb would have reached V. Trail is a function of the type of bomb, air speed, and altitude of release.[6] From tables, trail and actual time of fall are determined for a particular type of bomb for a given altitude and calibrated air speed. The velocity of closure is computed from calibrated indicated airspeed by the use of tables that were developed at the Naval Proving Ground.[7] A truly synchronous sight, if properly calibrated by the bombardier, will solve many of the bombing problems providing that all of the necessary information was fed into it.[8] It would be many years before a sight of this type would be available for use in aircraft.

One of the first early sights to be used by airplanes to carry out a bombing mission was the High Altitude Drift Sight, Mark 1-A. This sight had the capability of bombing from 1,000 to over 15,000 feet and at speeds of up to over 120 miles per hour. It also incorporated such features as trail, which was part of the bombing problem. There were many such features as trail that were part of the bombing problem. There were many other types of sights, such as Tachometric bombsights, Negative Lens, Low Altitude, Vector, etc.[9] The Royal Air Force (RAF) was well ahead of other nations in experimenting and developing fire control and bombing equipment. The British began experimenting with sights as far back as 1916 and were far ahead of all other nations in this effort. During this period, many types of devices were introduced. One of the most promising of these was the Course Setting bombsight. This sight was designated by H.E. Wimperis of the Air Ministry Laboratories in London, England, and was in use by the RAF by 1917. It was by far the most accurate sight at the time, and was used by the RAF with many modifications over the years, but the basic principles remained the same. This particular sight measured airspeed, wind speed, and wind direction, which in turn could measure ground speed. It also provided altitude and bomb ballistics, and it incorporated many of the elements of the bombing problem. During the intervening years, many modifications were made, starting with the Mark 1 through 10.[10] At this time, many other inventors were developing their own sights. Some showed promise, and others did not.

The early types of bombsights incorporated the design of the High Altitude Drift models. At the time it was felt that this was the only way to go. These sights measured the velocity of the wind present at the time of bombing, and in order to make this calculation it was necessary to measure the drift over the ground. These sights became known as "Drift Sights."[11]

World War I ended without any great significant tests conducted against stationary and moving targets or surface ships. Beginning in 1919 and 1920 there was much interest in developing a refined sight capable of hitting surface ships or any moving target. Ballistic data for land weapons were adapted to bombing theories and problems, because there was no other information available to the bombsight inventors. It was felt by a few that aerial bombardment was an extension of artillery fire. There were those who also thought that high altitude precision bombing was not possible and was a waste of time and funds. As new, faster, and higher flying airplanes were being built, they were compounding the problem of fire control and bomb equipment. From the very first of there few inventors working on sights had enough faith in the future of high altitude bombing to continue exploring the feasibility of finding a solution of the vexing bombing problem.

Developing a bombsight or sighting instrument to accurately determine in space the exact point of release on the surface appears to be a fairly simple task. However, just the opposite is true. This task required a tremendous amount of complex mathematical computations, engineering, experimental testing, etc. It took man many years of hard labor and financial resources to build a sight that would have such pinpoint precision quality. At about this time the Americans were also showing signs of interest in developing an accurate bombing system. One of the early attempts to expand the use of a gyro stabilizer in airplanes was Elmer Sperry and Henry Tanner of the Sperry Gyroscope Company of New York. An application for a "Gyroscopic Stabilizer" was filed July 14, 1914, and a patent was issued to them in August 14, 1917.[12] By the use of this invention a reference line, or plane, could be held at a fixed angle in respect to the earth's surface under any conditions. In other words, it was a means of stabilizing a reference point or line on an unstable vehicle, which would then create a stabilized platform. This invention was further refined by an application for a patent in June 1917 for a "Dampening Means for Gyroscopic Pendulums." A patent was issued to Henry L. Tanner in December 1919, assignor the Sperry Gyroscope Company of New York. The gyroscopic pendulums were to be used to maintain a reference plane, or platform, for an airplane. Through much testing it was found that small oscillations of a gyro pendulum would rise, which caused trouble if not dampened (deadened). This devise was to be used in airplanes for the purpose of obtaining a stabilized platform from which a bomb could be dropped. [13]

One of the first to file an application for a stabilized bombsight was Morris M. Titterington. An application was filed on September 9, 1916, for a stabilized bomb sight. A patent was issued in February 1923.[14] Mr. Titterington worked for the Sperry Gyroscope Company, and the patent rights were assigned to the Sperry Company. Since Elmer Sperry had invented the gyroscope for use in ocean going ships it was a natural outgrowth to go to the airplane. He accurately theorized that in order to be able to conduct any type of precision bombing it was necessary to have a true bombing platform. With the use of this stabilizer, this platform could be maintained regardless of the changing position of the aircraft. This invention was used in conjunction with the pendulum, which also was gyro stabilized and which the Sperry Company had already patented. The U.S. Navy Bureau of Ordnance wanted to develop an accurate bombing system that could be used to attack moving ships or targets. The type used by the Navy toward the end of WWI was the Wimperis sight, developed by the British. The Navy Bureau of Ordnance, in keeping with their objective in obtaining a better sight, issued a contract in 1916 to the Sperry Gyroscope Company to develop such a sight. Considerable testing of the sight showed that the degree of accuracy wanted by the Bureau of Ordnance could not be achieved. Sperry declared that he could not accomplish this requirement. Since there were no other sights available at the time, and the Wimperis, which was a course setting device that had a good track record with the RAF, the U.S. Navy decided to also use them. The sight had been redesigned, which permitted greater flexibility during a bombing run, resulting with fairly good results. The Bureau of Ordnance, not being satisfied with the Wimperis sight in its present form, modified it by incorporating the Pilot Director mechanism. After tests were conducted, the Navy was satisfied with the performance on their redesigned sight and issued a contract to the International Signal Company for 3,000 units in April 1918. This sight was labeled the "Pilot Directing Bombsight Mark III." However, before all sights could be delivered, the Armistice was signed on November 11, 1918, ending WWI. The contract was terminated shortly afterwards.[15] The Navy Bureau of Ordnance (Bur Ord) was still not satisfied with the redesigned Mark III sight. They still wanted a high altitude precision bombsight capable of hitting ships or any moving target. The Mark III sight did not have these attributes, but it did contain elements to the solution of the bombing problem. This was the state of the art on bombsights at the beginning of the 1920s. Several companies and inventors were working and experimenting on various methods of solving this annoying bombing problem. Up to this time, most of the sights were using a clock and pendulum as a means to determine the correct instant to release a bomb so that it would hit the target.

Notes:

1. Student's Manual, Bombing, Army Air Forces Training Command, no date, page 1- 4-1.
2. The Norden Bombsight, Maintenance and Calibration, B.S.M. Division, Department of Armament, Lowry Field, CO, dated September 1, 1943, page 104.
3. Ibid, page 104.
4. Ibid, page 108.
5. Student's Manual, Bombing, Army Air Forces training Command, no date, page 1-2- 3.
6. Aviation Ordnance, Ordnance Pamphlet Number 649, Bombsight Mark XV, Mods 4 and 5. Equivalent Air Corps Designation M-4, 5, 6, and 7, dated 1943, page 7.
7. Student's Manual, Bombing, Army Air Forces Training Command, no date, page 1- 2-2.
8. Ibid, page 1-2-3.
9. Literature on bombsights, no date, and no identification. Received from Colonel Pinto of the Brazilian Air Force, pages 275 to 277.
10. Ibid.
11. Ibid.
12. Patent Number 1,236,993, issued to Elmer A. Sperry and Henry L. Tanner for a Gyroscopic Stabilizer, application filed 14 July 1914, patent issued 14 August 1917, Department of Commerce.
13. Patent Number 1,324,478, for Damping Means for Gyroscopic Pendulums, application filed 8 June 1917, patent issued 9 December 1919, Department of Commerce.
14. Patent Number 1,446,280, for Stabilized Bomb Sight, by M.M. Titterington, filed September 1919, issued February 20, 1923, Department of Commerce.
15. The Navy Mark XV, Norden Bombsight, Its Development and Procurement 1920 - 1945.

Plate #1: Aviation Ordnance, Ordnance Pamphlet No. 649. Bombsight, Mark XV, Mods 4 and 5. Equivalent Air Corps Designation M: 4, 5, 6, and 7, page 7.

IV

Gyroscopes

The Norden bombsight (NBS) used two spinning gyroscopes to hold the optical system of the bombsight in a stable position, regardless of movement of the airplane (pitch, roll, turn). The cross-hairs of the telescope could not be synchronized on the target if it were not for this means of holding the telescope in a steady fixed position relative to the earth. The vertical gyroscope in the bombsight unit also established the vertical reference, which is absolutely necessary in solving the bombing problem. It is apparent that a spinning gyroscope can be used only in planes containing its spin axis, and for complete stabilization in every place, two gyroscopes, having their spin axis at right angles to each other, are required. This is the reason both a horizontal gyro (in the stabilizer unit) and a vertical gyro (in the bombsight unit) were required to give complete stabilization to the optical system.

Since gyroscopes and gyroscopic principles were used in designing and developing a truly synchronous bombsight such as the NBS, it is appropriate that a few words be devoted to this instrument. The following is a very, very short explanation on how it behaves in relation to the earth. [1]

A gyroscope is a flywheel so mounted that only one point, its center of gravity, is in a fixed position, the wheel being free to turn in any direction about this point. Such a wheel has three angular degrees of freedom.

The wheel or rotor revolves in bearings in a concentric inner ring. This ring is free to revolve in pivot bearings in an outer ring about an axis, which is always at right angles to the axis of rotation of the wheel. The outer ring likewise is free to revolve in pivot bearings in a supporting frame about an axis which is always at right angles to the axis of rotation of the inner ring. With a universal mounting such as this, the axis of the wheel may be pointed in any direction by the touch of the finger.

In the bombsight the movement of the gyro is limited by the construction of the sight. The reason for the limitation of gyro movements is because the purpose of the sight is accomplished without entire freedom of movement of the gyro. The inner gimbal ring is the housing which encloses the rotor; the second gimbal ring is the cardan, and the third gimbal ring is the sight case.

All of the practical applications of gyroscopes are based upon two fundamental characteristics, namely, "Gyroscopic Inertia" and "Precession."

When a gyroscope is subjected to a couple, or a force, about an axis at right angles to its axis of rotation it resists that couple, and the degree of resistance which the gyro turns, or precesses, about a third axis, which is called the axis of precession. The resistance continues until the gyro has precessed so that its plane of rotation coincides with the plane of applied couple, or force. Then precession ceases, and with it all resistance to the applied couple.

Rigidity of plane and fixity in space are also characteristics of gyros. The laws of precession apply to the gyroscope. If a force is applied perpendicular to the plane of rotation of a spinning gyro, its spin axis will precess ninety degrees in the direction of rotation from the point of applied pressure. If a torque is applied about one of the axis of a spinning gyro, its spin axis will tend to align itself with that axis, and in the direction of rotation.

A convenient way to remember the direction of precession is to regard the applied couple or torque as a push acting at a single point on the rim of a wheel. This point will not move in response to the push, but a point ninety degrees beyond in the direction of the wheel's rotation will move away instead.

Technically, there are two kinds of precession. They are induced precession and apparent easterly precession. Induced precession is caused by shifting of the weight mass of the gyro in relation to the spin axis of that gyro. A second type of induced precession, and precession over which the maintenance man has no control, is north hook, which is caused by bearing drag and flexible lead drag.

As it applies to the bombsight, only natural or apparent easterly precession will be considered. Just so long as the gyro behaves properly and gives a true precession in relation to the turning of the earth on its axis, it will be satisfactory. In order to have normal apparent easterly precession it must be made certain that the gyro is so balanced as to eliminate all induces precession excepting bearing drag and flexible lead drag. It will be necessary to shift movable weights on the gyro housing in an effort to do this and leave only apparent easterly precession.

Easterly precession can be understood if one's thinking is based on the following three statements:

a) The earth rotates from West to East, and requires 24 hours for a complete revolution of its 360 degree circumference.

b) It is characteristic of a gyro for its spin axis to be parallel to its original position at any given moment of time.

c) Gravity operates at all times to hold all bodies and things perpendicular to the earth.

In Plate #1, the earth is shown rotating from west to east. AB represents the spin axis of the gyro at the beginning of a precession run. It has been erected to the vertical. CD represents the axis about two hours later. Notice that the spin axis illustrated by CD is still parallel to the original starting position at AB, but is not in the true vertical in relation to the earth.

AB represents the spin axis in the vertical position at the start of the precession run. Two hours later it is represented by CD. CD is no longer in the vertical. EF represents the vertical. If the outer gimbal ring is horizontal at the beginning of the run, it will naturally remain that way two hours later, when it is designated as G'H'. FI represents the amount of apparent easterly precession.

In the bombsight, the case and the carden are not stabilized for lateral tilt, i.e. roll. Therefore, they are free under any force to move independently of the gyro housing. Gravity is such a force, and it will force the right end of the case and cardan to move downward as the earth rotates (sight on North heading). This means the tilting will take place on the gyro gudgeon bearings. Since no bearing is absolutely frictionless, this tilting will exert a slight force upon the right face of the gyro housing, tending to depress the right face downward. Under the laws of the precession, the rear face will precess downward, thus moving the bottom of the spin to the north, i.e. north hook.

North hook is or can be caused by (1) flexible lead drag, (2) bearing drag, and (3) pendulousness of the vertical gyro. Notice Plate #1 position of the bearings.

A gyro whose axis is vertical at six o'clock in the morning appears to be horizontal at noon and upside down at six o'clock in the evening. This creates the illusion that the gyro has turned over. Actually, the gyro has maintained its position in space, and the earth has moved under it. This movement in relation to the gyro is called apparent precession, which is at the equator, decreasing as the gyro is moved toward the north or south poles where apparent precession is zero. Apparent precession of a gyro makes it unfit for use as a reference over an extended period of time without some sort of compensating or correction mechanism. Over a relatively short period of time, however, a gyro can be used to establish a satisfactory reference, as is the case in the gyro of the bombsight unit. In apparent precession, the earth moves in relation to the gyro. However, the gyro may be used to move in relation to the earth. This is called induced precession.

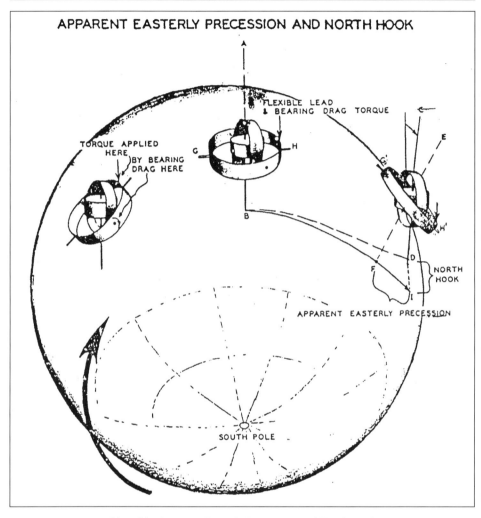

Plate #1 - Apparent Easterly Precession and North Hook

Notes:

[1] Bombsight Maintenance Division, Department of Armament, Lowry Field, CO, Sep 1, 1943, page 115.

Plate #1 Bombsight Maintenance Division, Department of Armament, Lowry Field, CO, September 1944, page 114.

Plate #2 Handbook Operation, Service, and Overhaul Instructions, Bombsight Type M-9 11B41-2-2-1, 15 Aug, 1958, page 55.

Plate #3 Ibid.

Chapter IV: Gyroscopes

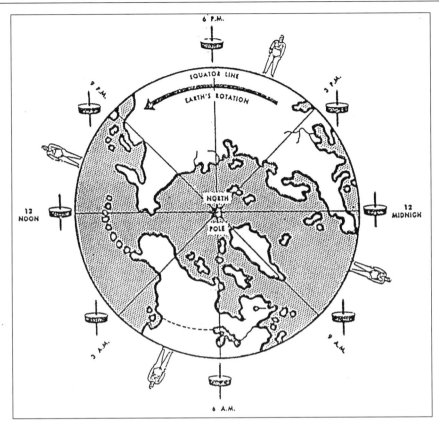

Plate #2 - Apparent Gyro Precession

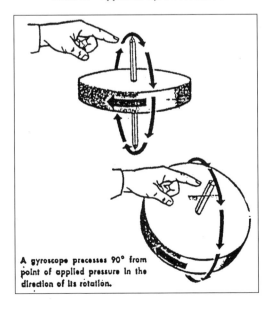

Plate #3 - Induced Gyro Precession

V

The Navy's Norden Bombsight

The Norden Bombsight (NBS) was one of the few weapon systems that was not developed during WWII. It had been in the development, testing, and redesigning phases almost 20 years previously. It was one of the most secret weapons in the Navy Department and U.S. Army Air Forces arsenal before and during the war. At one time, it was considered impossible to conduct accurate high altitude bombing missions. As bombers were being built to fly faster and higher to evade anti-aircraft artillery, it compounded the problem tremendously. It became very evident that, with the introduction of new high performance bombers (B-17, B-24, etc.), the pilot could not fly the airplane sufficiently accurate enough and conduct a bombing run at the same time. It became mandatory that some form of fire control system was needed that would not be too complex to operate, require a very short bomb run, ease in maintenance, and be extremely accurate at high altitudes. This was a tall order to accomplish at the time.

The NBS was born in the early 1920s involving fundamental and applied research, such as unprecedented bombsight engineering, design, complex mathematics, unheard of machining of metal parts to tolerances in the range of one thousandth of an inch (0.001) to one or two ten thousanths of an inch (0.0002) on massed produced parts manufactured to watch-like measurements, development of precision anti-friction ball bearings, optical equipment refined to new tolerances, and the industrial ability to produce delicate instruments in mass quantities never before attempted. All of this was accomplished long before the discovery of high speed computers, calculators, and the event of the micro chip. This was a testimony to the ability of the American industrial complex to react to a very critical time during the first part of WWII. This led the way for new, exotic, and sophisticated weapons that were used in the Persian Gulf War.

Chapter V: The Navy's Norden Bombsight

The basic bombing problems that had plagued bombsight inventors over the years were finally solved and incorporated in the sight by the use of gyroscopic principles. Much of the early engineering data is missing, but some correspondence, letters, reports, and sketches covering almost 20 years that were available for research revealed the many problems, defects, and deficiencies in the early NBS models. Due to the precision accuracy of the sight at high altitudes over other sights, both national and international, it was classified to be treated with an "unusual degree of secrecy." Very few knew of its existence. The only personnel that were allowed to be associated with it was severely limited to those directly involved with its manufacture, use, maintenance, and repair. The personnel allowed on this program was limited to absolute minimum. No photos from airplanes were taken of the area where the bombsight was located. The censors made sure that it was completely blocked out. In transporting the NBS, it was always under wraps in a canvas case and escorted by a heavily armed military or civilian guard. Many foreign nations were very much aware of its existence and made many attempts to obtain the plans through, in, and out of diplomatic channels. Among the nations that were vitally interested were France, Russia, Japan, and especially Great Britain. It was in direct contrast to the Manhattan Project (Atomic Bomb), where only a few very top military and civilian leaders knew about it. Little or no information was leaked to the press.

The only information that was released on the sight stated that it was a highly accurate device for pin point bombing at high altitudes. Any of the performance specifications that were released to the public were in very vague terms shedding no light on the sight. To the average person, the NBS is not a spectacular looking weapon. It is not a large instrument, weighing about 40 lbs., but it was one of the most versatile visual/optical sighting devices ever developed up to that time. All engineering and testing was done by laborious hand work. This was many years before the advent of high speed computers that can solve many thousands of problems in a very relatively short time. As WWII progressed, many attachments and modifications were made. Strangely, the NBS was not developed by the Army Air Forces or its predecessors (Signal Corps or Army Air Service), who would eventually be allocated almost 80% of all the sights produced during WWII. It was and remained a Navy weapon throughout the war. In the 1940s, many articles on the NBS appeared in various magazines (Popular Science, Aero Digest, Flying, etc.) trying to explain how it worked. Some of the write ups were fairly accurate. As previously mentioned, the bombing principles and theory had been around for many years. Some of the articles also brought out the rivalry between Navy Bureau of Ordnance and the Army Air Forces.

The concept of sending a thousand airplane raids over Germany was an appealing one to the British. The RAF had become proficient in bombing in an attempt to destroy targets, such as factories, ball bearing facilities, airframe plants, etc., that had the potential to produce war material. Arthur Harris of the RAF Bomber Command thought that by bombing Berlin and reducing it to rubble, the Germans would be forced to surrender. The result was just the opposite. They dug out of the devastation and continued to produce war material to pre-bombing levels. It soon became evident that area bombing was not doing the job. From the beginning of the development of a bombardment force and bombsights, it had always been the American goal to conduct daylight precision bombing of industrial targets, and not residential areas not devoted to the war effort. The potential for destruction of a ball bearing plant could cause severe or catastrophic problems in the production of all types of war materiel. These were the ideas expounded by General Mitchell and General Douhet. In other words, the aim of precision bombing was to destroy the enemie's internal ability to such an extent that it could not conduct war.

There is no question that the NBS was one of the outstanding weapons of WWII, aside from the atomic bomb. Although the sight was maligned by people such as the press, news commentators, reporters, etc., issuing statements to the contrary that it was not an accurate system. If properly used by a trained bombardier, it could strike a target within acceptable limits. There are records where ships at anchor have been sunk on the first bomb run, specific targets destroyed, etc. Admittedly, with anti-aircraft fire and fighter planes attacking a bomber would be sufficient to rattle anyone's nerves. This was not the fault of the bomb sight, but in not having sufficient fighter protection. Regrettably, vital engineering and source data is fragmentary, but a determined attempt was made to locate sufficient facts to document a coherent evolution of the NBS.

The state-of-the-art in the bombsight industry in the early 1920s was practically non-existent. There were only a handful of inventors working on this complex subject. The most important single element in this market was that very few, if any, companies were either willing or able to spend great sums of their own money in research and development and then have the government reject their product since the only customer would be the military. The other alternative was to sell their ideas to foreign countries. This, of course, was frowned upon by the government and the military.

The market for bombsights was in the hands of an extremely few manufacturers. Although bombsights had been developed and used during WWI, they were not reliable or very accurate. The research centers for bombsights were extremely limited, and were located in either the U.S. Army Air Forces and Navy facilities. These facilities were in their infancy and primitive to today's standards. These were the conditions that existed from 1920-1941.

VI

The Navy's Norden Mark XI Bombsight

Through the persistence, determination, farsightedness, and in the face of many failures, obstacles, and setbacks, the U.S. Navy Bureau of Ordnance continued with the concept of the NBS. The BurOrd wanted a bombsight that could bomb both moving and stationary targets. During and after WWI, the bombsights in use were not reliable. The initial steps to develop an accurate sighting instrument were undertaken by the BurOrd shortly after WWI. Mr. Carl L. Norden had been working on Navy projects for a number of years when he was employed by the famed Elmer Sperry of the Sperry Gyroscope Company of Brooklyn, New York. The Navy was to be his only customer over the years. It was in this manner that Carl Norden designed, engineered, developed, and tested the first successful high altitude precision bombsight. The U.S. Navy had spent great sums of money and time in this development—would this high stake "gamble" pay off? The answer to that question is an unequivocal yes, as evidenced in the conduct of the air war over Germany, Japan, and Italy. To trace the beginning of this formidable weapon, it is necessary to start with the development of the Mark XI in the 1920s, which was the precursor of the famous Mark XV that was used in the unprecedented aerial bombardment of enemy targets. This is the story behind the efforts and events that led to the development and evolution of the legendary Norden Mark XV bombsight.

Mr. Carl L. Norden was born in Semarang, Java, Dutch East Indies on April 23, 1880, to parents who were citizens of the Netherlands. He was five years old when his father passed away. The widow returned to Holland with her children. He attended various mechanical schools in Holland and Germany, and in 1900, he entered the prestigious Federal Institute of Technology in Zurich, Switzerland. He graduated from the Institute in 1904 with a degree in mechanical engineering. After graduation, he soon migrated to the United States. Old world engineers were much sought after for their expertise. He worked for several engineering firms, including the H.R. Worthington Pump and Machine Works. Later, he worked for the Lidgerwood Manufacturing Company. It is not clear where and how he met Elmer Sperry. Sperry, at the time, was a leader in developing gyroscopic equipment for ocean going vessels. In 1911, he was hired by the Sperry Company to design gyrostabilizing equipment for the U.S. Navy. While employed by the Sperry Company, he spent most of his time in the capacity of improving gyroscopes. The relationship between the two men was anything but harmonious. He left Sperry in a huff several times, but always returned to work. However, after another dispute in 1915, he decided to leave Sperry for good. He opened his own business as a consulting engineer in Brooklyn, NY. With his departure from Sperry, the question about patents arose on the gyroscopic stabilization equipment that he had worked on and developed while he was employed. The question arose concerning as to who should be entitled to the patents on the equipment that Norden had worked on. The application of gyroscopic application was not a secret, as other companies were also developing their own type of equipment. Because of this, Sperry dropped the issue and ended up with Carl Norden working an additional two years for Sperry as a paid consultant on work that he had contracted with the Navy. Apparently, Sperry had sufficient confidence in Norden's ability to have him work on his projects. There was no question that Norden was one of the world's leading experts on gyroscopes and gyroscopic application at that time. The Navy, and for that matter, the U.S. Army (Signal Corps) had not developed a sight of their own. They were using the Wimperis Sight developed by the British, which was a course setting type. The Navy was not satisfied with the Wimperis sight, and decided to try and improve it. In 1920, the BurOrd worked on the incorporation of a telescope for a high altitude version of the Mark III and added a mechanism that could set the speed and course in the new version. However, with all these improvements on the Mark III, it still was not satisfactory. Because of this, the BurOrd asked Mr. Norden to investigate the possibility of developing a gyroscopic stabilized bombsight.[1] These redesigned sights ended up with the addition of the Pilot Director Indicator feature on the Wimperis, and after exhaustive testing and redesigning of the sight, it was accepted. Preliminary testing revealed the redesigned sight worked well, and the Navy figured it would only need further refinement, so that an accurate sight could be developed. The testing revealed that this sight was not what the Navy really wanted. Due to the unsatisfactory sight, Mr. Norden was asked to look into this situation. By 1920, the bombsight problem was the most important item that required a solution. The development of a satisfactory stabilized bombsight was the first step in this program.[2]

Between 1917 and 1920, Mr. Norden was involved with the Navy on several projects. Some he was directly involved with, and the for others he only prepared engineering designs. One of these was the blueprints for a flywheel type of launching device. This device was to be part of the flying bomb project. Mr. Norden's ideas on the project vitally interested the Bureau of Ordnance. They wanted Mr. Norden to supervise the development of a flywheel type catapult so that the flying bomb airplane could be operated automatically. The project was stopped in 1919 because the aircraft that were to be used were not safe to be flown by the pilots during testing. This concept, as envisioned by the Navy, was to have an airplane launched from a surface ship by means of a catapult system that would fly the airplane to a predetermined target, which would fall to earth as soon as the airplane reached the target. Mr. Norden was one of the chief consultants on this project. The design included gyro controls. It became evident at this early stage that the use of gyroscopic principles would be very much in the forefront in the designing of bombsights. This concept was similar to the V-1 weapon developed by the Germans that indiscriminately bombed English cities before and during WWII. The idea was the same, but only refined. As soon as the weapon ran out of fuel, it would drop to the earth, releasing a charge of explosives. It was not a tactical weapon, because it did not pinpoint any specific target. It did cause considerable damage to residential areas and also to the general population. It has also been reported that he designed for the Navy the arresting gear for the aircraft carriers U.S.S. *Lexington* and *Saratoga*. However, no firm data was found during extensive research to actually document these inventions credited to him. These landing devices used on aircraft carriers have been in use for many years. Of course, many modifications have been made over the years, especially during WWII. In the years 1917—1920, Mr. Norden was involved with the Navy Bureau of Ordnance in various capacities. It readily became evident to the Navy that Mr. Norden had an exceptional ability and knowledge in the field of gyroscopic application to the various weapon systems that the Navy was interested in.

In 1920, the Bureau of Ordnance discussed with Mr. Norden the possibility of developing a precision high altitude bombsight capable of attacking moving ships. He felt that such a bombsight was possible, providing that a considerable amount of research was devoted to the project. In January 1920 Mr. Norden was asked to submit a preliminary design on a gyroscopically stabilized bombsight. He also brought out and explained to the Navy that many of the bombing problems had not been resolved. The BurOrd was extremely interested in Mr. Norden's idea of developing a new sight. Although he did not receive a contract, he had done considerable work on this on his own time. He discussed the possible improvements on the Mark III. He explained that it was a finished designed and extensive redesigning could not be made, however, it could be altered to some extent.[3] Since the Bureau of Ordnance had done considerable modifications previously on sights that were in the Navy inventory, Mr. Norden requested any and all information or data that the Bureau of Ordnance had on file concerning these sights. The Navy had a good supply of these. Neither Mr. Norden nor the BurOrd felt that a completely new bombsight was required at this time. He began work on designing, developing, manufacturing, and testing a stabilized sight which was to be mounted on the Mark III-A.[4] To further this cause, the BurOrd issued a contract to cover this production in June 1920. Problems developed in determining the direction and magnitude of the drift relative to the moving target. Designing a sight for use against moving targets was very complex. At that time there were no answers to this problem, and further research was required.

Mr. Norden speculated that even with the new base, the Mark III-A could still be further improved, however, considerable testing and engineering would be necessary before a workable sight could be produced, and the cost would be very high. After continued and lengthy studies and testing, he came to the conclusion that the Mark III-A could not be efficiently redesigned. The studies conducted by Mr. Norden indicated the best approach would be by the use of a timing device and pendulum as he had previously advised the Bureau of Ordnance.

In the meantime, in 1921 Mr. Norden was hired by the Navy to do experimental work leading to explore the feasibility of designing an accurate sight at a rate of $35.00 per day plus expenses for the time actually spent on sight work. The redesigned Mark III-A sight mounted on a newly built Norden base was sent to the Naval Proving Ground at Dahlgren, Virginia, for testing. The personnel responsible for the testing at this base installed it in a Martin Torpedo plane assigned to the Atlantic Fleet, Torpedo Plane Squadron, who did the testing in July 1920.[5] This was the Navy's first stabilized bomb sight. The results of these tests were very encouraging, and the sight performed well, but trouble was encountered in keeping the gyro functioning properly. Many runs were made, and the results were good, so further testing was highly recommended.

The U.S. Army got wind of the Navy's experimentation on stabilized sights, and not wanting to be left out in the development of bombsights, formed an Army Bomb Board. In 1919, a group of three Army officers from the Air Service Command and two officers from the Bureau of Ordnance made up this board. The only purpose of this organization was to develop a policy for future bombsight development, including bombs and fuses. After conducting in-depth research into the bombing problem, the Board came to the conclusion that the most important item was the development of a satisfactory fire control system. It was the consensus among the military that the Michelin was the best bombsight. The Bureau of Ordnance conducted many exhaustive tests between the Michelin and the Wimperis and arrived at the conclusion that the Michelin was superior. It also felt that, because it was partially stabilized, it gave the edge over the Wimperis. Although this was not the best sight, the Board suggested that gyroscopic stabilization be used in obtaining a solution for a high altitude precision bombsight. It was determined that the sight must be absolutely level at the time the bomb was dropped, as well as in determining the correct setting and other adjustments depended upon the sight being true vertical to the surface of the earth. If the bombsight was not level, all the other elements that make up the sight were of no value.

At about the same time, Brig. General Billy Mitchell of the U.S. Army was for the development of a heavy bombardment force capable of conducting bombing missions. In a speech at McCook Field on October 26, 1921, he explained the different methods of attacking ships and stationary targets and the types of bombs used. He went on to explain the type of training that was done with the 14th Squadron in bombing anchored ships. He went into great detail about how experimentation was made and the use of different types of bombs. He gave a graphic description of the attack on the Ostfriesland. This was a captured German battleship and was one of the strongest battleships in existence at the time, and was well protected by armor against all types of attack. They used 1,100 pound bombs and, when it attacked the ship, three direct hits were made out of five. After the first hits, the bombing was discontinued at the request of the Navy. In the attack on the U.S.S. *Alabama*, again 1,100 pound bombs were used, and in the first drop, the bow was taken off. On another attack on the Ostfriesland, it was sunk. The 1,100 pound bombs turned the ship completely bottom side up, and it started to sink at the stern.[6] General Mitchell was already in the forefront in demanding the development of airplanes and bombsights so that a heavy bombardment force would become a reality. Unfortunately, the higher military hierarchy did not go along with this concept.

The Bureau of Ordnance had not yet decided which avenue they would take in trying to develop an accurate sight. Discussions were continuing with Mr. Norden concerning a timing sight. In using timing sights, it was necessary to have a timing device or clock to determine the speed in which the airplane was closing in on the target.[7] The use of a universal control system would enable the bombsight operator to make the airplane fly in a straight path. From January through June 1922, considerable and lengthy discussions, drawings, sketches, and engineering designs were underway between Mr. Norden and the Navy Bureau of Ordnance. The Navy, having faith in Mr. Norden's ability as an engineer and his expertise in gyroscopic applications, awarded contract number 56166 to Mr. Norden in June 1922 for three experimental models of a timing sight at a fixed price of $10,700, including three Pilot Director Indicators (PDI) Mark I.[8] In order to conduct a bombing run, it was necessary the airplane fly straight flight or any desired curve or course maneuvers and was to be suitable for general pilot directing, including the use with a bombsight. The cost of the equipment required by the Bureau of Ordnance was extremely expensive. To lower the cost, it was decided to simplify engineering and construction and improve on the original design.[9]

With the introduction of this new sight, as it is with all weapons, it was necessary to give this new sight a number, and it was designated as "Bombsight Mark XI." The Navy and U.S. Army Air Service had previously signed an agreement to use the same series of bombsight marks. The Mark I, III, XI, and XV were assigned to the Navy, and IV to X to the Army. The Navy Pilot Directing Bombsight was assigned the Mark III to the new pilot directors. The Pilot Directors left over from WWI were not the best, and were erratic in operation. Mr. Norden submitted various designs for these instruments, and in 1922 the Navy ordered these new units, calling them Pilot Director, Mark I. Thus began the transition of the bombing problem into an instrument that would be capable of precision bombing, but the task would be long and arduous.

Awarding of this contract by the Navy presented Mr. Norden with the task of enlarging his fabrication shop to take care of this added production of three sights and the Pilot Directors. These instruments were mostly handmade by trial and error. There were no previous designs for any tools, necessary moulds, castings, dies, etc., that information could be drawn from. Because of this factor, every item had to be handcrafted, taking much longer to complete their manufacture. The Pilot Directors were furnished to the Navy in 1923, and the three bombsights Mark XI in March 1924.[10] His office and shop was located at 18 East Forty-First Street and 24 Lincoln Street, Larchmont, New York. In a letter to the Bureau of Ordnance, he stated that the three sights were completed with each one being assembled and reassembled many times and run for more than twenty hours. He had suitable hardwood boxes made to transport the sights, including spare parts and drawings. He realized that these instruments would need to be thoroughly tested. He had delayed delivery so that these instruments would be thoroughly tested. He had delayed delivery so that he could fully check their performances. He expected to hand carry these sights in person to Washington, D.C., by April 1, 1924, to the Naval Inspector of Ordnance. He also requested that the Bureau of Ordnance allow him to accompany the sights should they need to be reworked. The new sights were accepted and sent to the Naval Proving Ground, Dahlgren, Virginia, for bench testing before any air testing was done. The sights were tested and retested and many changes were made. Results of actual air tests were very discouraging. Personnel at the test center felt that the sight was too complicated, causing much loss of time in keeping the sight calibrated. Mr. Norden was not upset with these tests and setbacks, as he expected this and knew there would be much rework to obtain a serviceable unit. Although problems were encountered with these first sights, it was felt that this was the only way to go. The BurOrd, in order to continue this program, awarded another contract No. 63111 for the modification of two of the Mark XI sights. The modification of these sights was completed in 1925, and they were again tested at the Naval Proving Ground. A different approach was taken. One of the sights was of a pendulum design, and the other sight was gyroscopically controlled. Again, the results of both sights were disappointing. However, modification, redesigning, and testing continued. Not all was lost during the air testing phase. Much needed information was obtained, such as ballistic coefficient of various bombs, formation of bombing tables, mathematical bombing computation concerning sighting angles, range problems, etc. This early day bombing data proved invaluable to the bombsight program for future use in the development of high altitude bombing instruments, especially during WWII.

Mr. Norden was greatly concerned with the performance of the Mark XI. He felt that the bombsight engineering and design was accurate and that errors were due to Pilot Directing and course changes during approach to the target. There was beginning to be considerable apprehension and doubt that the Mark XI would not work as a high altitude sight. The Bureau of Ordnance still wanted a high altitude precision bombsight and was not about to give up when so much time, engineering, testing, funds, etc., had already been expended. In addition, much of the redesigning and rework on the sights was borne by Mr. Norden at no cost to the Navy. In early 1926, the Bureau of Ordnance asked Mr. Norden and Mr. Barth to develop two new Mark XI bombsights, three Pilot Directors, and two extra bombsight clocks. The Navy awarded them a contract for $21,650 in April 1926, and they immediately started work on the new sights.[11] The price seemed rather high, but the Bureau felt that it was justified. Again, the results of the new sights were very poor. In addition, another two pilot directors were designed and built.

Around 1924 Mr. Norden took on a partner by the name of Theodore H. Barth. There is not much information on him. He took over the Administrative duties for the two, and Mr. Norden referred to him as his business associate. In this manner, he could take his periodic vacations, primarily to Switzerland, and Mr. Barth would take care of the business at hand. While Mr. Norden was abroad, it did not stop him from continuing work on the Mark XI. All drawings and engineering data that was to be sent to the Navy by him was processed by the State Department through the American Consul in Zurich, Switzerland, via diplomatic pouch. At the time, he was staying at 36 Weinbergattrasse, Zurich, Switzerland. Also, data concerning the sight was sent to Mr. Norden in this manner by the U.S. Navy. The Navy had set up this procedure so that Mr. Norden could keep abreast with the on going work, the Bureau of Ordnance, and Mr. Barth, his partner.[12] In order to keep this material secret, it was strictly processed in accordance with State Department policy. During research no reason could be found why Mr. Norden took these periodic trips to Switzerland, unless it was to visit family and friends, or to take care of property that he had there. While Mr. Norden was abroad, air testing continued at the Naval Proving Ground (NPG) on the newly redesigned and modified Mark XIs. In March 1927, the two experimental sights on the above contract had been completed.[13] One of the sights was in the drafting room and in operation. The Bureau of Ordnance was invited to witness the operation of this sight. The two sights were then sent to NPG for further testing. Even with all of the redesigning, engineering, and testing over the years, the results were still not acceptable to Mr. Norden and the Bureau of Ordnance. The results of actual air testing during these years were not encouraging, and bombing of targets was erratic. Mr. Norden was still convinced that the sight was engineered properly. Continued modification, testing, and the development of the automatic pilot would prove him correct. By 1926, it appeared that the Mark XI would not be ready for production because of the many mechanical, electrical, and maintenance problems being encountered.

The Fleet activities were clamoring for a new sight. They had been using the Mark III-A, which was developed in 1918 and to some extent modified. The Fleet Squadrons were aware that the BurOrd was experimenting and developing a new bombsight. They wanted an up to date bombsight so that they could carry out the mission that they were assigned to. In 1926, to appease Fleet requirements, the Bureau of Ordnance ordered 40 D-4 Army type bombsights (Estoppey) as a stop gap measure, and to appease these activities, 15 were shipped directly from the manufacturer (Geartner Scientific Corporation) on completion to the Aircraft Squadron Battle Fleet.[14] It would not be until late 1927 or early 1928 that any thought would be given to mass produce the Mark XI. These new sights still had several "bugs," but they were being corrected as fast as possible. The most vexing problem at the time was parallax. Services of experts in the field were obtained from various organizations and universities, and research centers were consulted, and their consensus was that it was a mechanical, not an optical, problem. A new adjustable fork was developed, and this corrected most of the problem. In addition, the NPG identified some of the causes of this mechanical parallax, and corrective action was taken to remedy this problem. It was suggested that if the fore and aft axis of the airplane be tilted in flight, the parallax would be eliminated by precessing the gyro out of level by the same angle using the fore and aft bubble.

The Bureau of Ordnance experimentation on the Mark XI bombsight had been continuing since 1922, and the result had not been encouraging. Considerable modifications, engineering, and design changes had taken place over the past six years. Tests continued to be conducted at the NPG, and by early 1928 results were beginning to show progress. Comparative accuracy of the old and new Mark XI was conducted in 1926-27 to determine how well they would perform. The two sights to be tested had been redesigned under a separate contract. The tests were to be conducted by fleet activities. The mean radial error did not appear in the Reports of Gunnery Exercises, however, the percentages of hits were found within circles of various radii. From the 1926-27 Gunnery Reports, the approximate values of the mean radial dispersion in bomb droppings was computed from 3,000 and 6,000 feet, against fixed targets. The mean error of the drops made to date at 3,000 feet was 67 feet, and at 6,000 feet was 110 feet. Therefore, the sight seemed to be 3.3 times more accurate at 3,000 feet and 3.6 times more accurate at 6,000 feet. It was also brought out that the bombing of a moving target did not introduce any difficulties in the Mark XI, whereas it would seriously increase the inaccuracy of the other sights presently in use.[15]

Chapter VI: The Navy's Norden Mark XI Bombsight

With the Mark XI starting to prove very good in tests in late 1927, the Bureau of Ordnance put in an order for 25 sights to the Norden Company for the type being tested at the NPG. With the correction of many of the deficiencies, the sight was one of the best in existence. The BurOrd increased this quantity to 80 sights. The unit cost $4,050 per system and 26 spare parts boxes at $950 each, for a grand total of $348,700.[16] Because of such a very large production order, Mr. Norden and Mr. Barth established a new company, called Carl L. Norden, Inc., to perform this contract and also to protect themselves should they not be able to complete the contract and against possible patent infringement, as other companies were also experimenting with bombsights. From review of various notes, correspondence, and letters, it appears that the company was formed in the early part of 1928. No specific date could be found. The corporate headquarters of the new company was located at 70-74 Lafayette Street, Brooklyn, New York. In the formation of the new company, Mr. Carl L. Norden was to be the Consulting Engineer to the company, and Mr. Theodore H. Barth the President. Mr. Norden was a very quiet and retiring man, and preferred to be in the shop working on projects at hand, so he was not involved in the administration and day-to-day operation of the company. Mr. Norden was content to spend his time in the shop, which sometimes was an 18 hour day. Mr. Barth was just the opposite of Mr. Norden. He was an outgoing person and a public relations type. Not much is known of Mr. Barth's background. It was reported that he was a Colonel in the U.S. Army Chemical Warfare Department. He had an exceptional talent when it came to machine work. He took over all of the business aspects of the company. It is not known where and how Mr. Norden met Mr. Barth. However, correspondence indicates that he was associated with Mr. Norden around 1924.

Before signing the contract and production could begin, the question of patents reared its head. As it was in most cases where inventions were involved, a determination was needed as to the propriety rights on patents and patent liability. It was finally agreed that any patent rights on bombsights produced by C.L. Norden, Inc., was to revert to the U.S. Government with assignors to the Untied States of America as represented by the Secretary of the Navy and, in turn, to the Bureau of Ordnance. No specific documentation or data on this subject was available for research, except for some scattered reference in various letters, correspondence, and an opinion from the Judge Advocate General's Office. Mr. Norden and Mr. Barth's work on the two new experimental Mark XI systems made it necessary for them to visit the NPG at Dahlgren, Virginia, 51 times over a period of five and half years to confer with personnel at the test center regarding various mechanical and electrical problems. This amounted to about a trip a month, and considering transportation in those years posed a hardship on them, as they were absent from the shop for extended periods of time. However, they did not complain and continued to work with the Bureau of Ordnance, including Saturdays and Sundays, to perfect the sight that they had under development.[17] Carl L. Norden, Inc., never charged the Bureau of Ordnance for any personal work. They only charged for the actual shop time and materials, plus a small profit on the experimental models being worked on or for those that had already been furnished to the Bureau. However, on various occasions, the Bureau of Ordnance offered them a contract for $50.00 a day each on work they performed for the Bureau. They sustained many losses during the intervening years, but never complained. Both Mr. Norden and Mr. Barth would have preferred to wait until the instruments proved successful and then reap rewards by receiving a contract for their production.

From the beginning, in 1919, Mr. Norden's association with the Bureau of Ordnance was that they were his only customer. Most of his time was spent on designing and experimenting on bombsights. He had no other customers either local, national, or international, and no products were made for sale to the public. Their efforts were limited to services to the U.S. Navy. The bombsights that Norden and Barth produced for the Bureau of Ordnance in the past were for the most part hand tooled. They did not have a large shop, as most of their work was in the designing and engineering phase. This new contract would require them to expand their production facilities dramatically, including new machine tools, floor space, securing trained machinist/mechanics, draftsmen, training assembly workers, etc. With the plant already in production to take care of this existing contract and should the Bureau of Ordnance want additional sights, the price would go down because the plant was already in production and, if expansion of facilities were needed, would result in minimum cost to the Navy. In addition to undergoing shop expansion, it would be necessary to secure the services of sub-contractors to furnish the many parts of sub-assemblies, castings, optics, hardware (machine bolts and screws), electric wiring, precision ball bearings, etc. This would be the first time that such a sophisticated and complicated instrument of this type would be mass produced. The tolerances were very close and required much hand work. It would be many months before any sights would be coming off the production lines. It was not uncommon for machine shop work tolerances to be in the range of plus or minus one or two ten thousandths of an inch (0.0001). Although the Mark XI sight was now in production, Mr. Norden and the Bureau of Ordnance were still not happy with its performance. He advised the Bureau of Ordnance that he would start work immediately on developing a new and more sophisticated sight that would be much less complicated and more accurate. During this period many other companies and inventors were working on gyroscopically stabilized bombsights, hence the concern about patent rights, liability, and infringement. In trying to get a better instrument, the Bureau of Ordnance decided to stay with the Norden Company, who had a proven and an exceptional record over the years with the Bureau of Ordnance in the gyroscopic field.

Now that the Mark XI was a reality and during the intervening years, much work was also done on other things than designing bomb sights. Bombing data was being compiled, such as time of free fall, ballistic coefficient of different types and weights of bombs, date and time of flights, ground speed, ground course, clock readings, name of pilot and bomber (bombardier), characteristics of a falling bomb, etc.[18] To obtain a stabilized horizontal plane, the optical system was directly stabilized, along with two axis by the use of an electrical driven gyroscope. When the range or dropping angle was established, the clock was started, which was set to run about one half of the actual time of fall for the particular altitude and type of bomb being used. Considering that data for ballistic coefficient of bombs had not fully been developed the bombing accuracy was admirable. Mr. Norden was strongly of the opinion that the Pilot Directing device was essential in performing a bombing mission. This opinion would be verified many years later with the invention and use of the auto pilot. He also felt that this type of instrumentation would be beneficial to the flying bomb project if it should be reactivated. He never considered the Mark I developed during WWI as satisfactory or reliable. If it was to be used, this system would require extensive revision and updating before it could be used by fleet activities. Adjustment for trail and other foresight, this movement automatically rotated the back sight geared to the rotation of 2 to 1 in such a position that the clock was stopped. The Pilot Director was used to keep the airplane on course and enabled the bomber (bombardier) to solve the problem of deflection (by keeping the airplane on the correct heading to make it fly over the target). If all of these factors were maintained and when the clock stopped the bomb was dropped at the instant that the target came into view. The clock was so arranged that the timing was for half of the time of fall, therefore, the movement of the backsight was doubled. The electric stop clock, instead of being graduated in seconds, the face was marked with scale height since the time of fall is a function of the sight. This made it more convenient because the clock using the altitude scale could be set directly from the altimeter without stopping to convert these times. In this manner, the bomber or bombardier only needed to turn the needle on the stop clock to a certain reading matching that of the altitude.

Because of the complex nature in the use of a timing clock, it was necessary to develop bombing tables (Plate #2) to make it easier for the bomber in calculating the time of release for the bomb. An engineering study was conducted in January 1929 in an effort to alleviate this problem. The result was the publishing by NPG of a Calibrated Summary for the Mark XI bombsight clock. The Time of Fall Apparatus and Apparatus for Calibrated Observation was developed. The curves representing observation for a 100 lb. bomb differed slightly from that of the 500 lb. bomb. The data on clock settings was obtained by actual releases, averaging pairs with respect both to altitude and range. The final curves are weighted averages for the entire range of altitudes. No compensation was applied to variation in atmospheric factors other than wind. The clock rate was checked by comparison with a standard chronometer, recording both (simultaneously) by an oscillograph. The time delay in release of a bomb, after the radio signal, was measured for two types of release mechanism. The NPG borrowed an ultra-high speed moving picture camera to measure the time of fall. The type of data obtained with the camera was the basis for the charts and tables. This was accomplished by the use of a micro split stop watch (crude but effective) which was used during the time of calibration. The clock calibrations that were obtained did not take into consideration cross winds or disturbed air conditions. The bombsight, by its timing operation, measured ground speed and identified the time of fall. Wind errors were identified by other calculations.

Although some of this data, such as clock settings, was not usable in future bombsight models, the remainder was absolutely an invaluable source of information in the development of the Mark XV sight that was used in WWII. Most of the "spade work" in solving the bombing problem had been incorporated in the Mark XV. Any of the earlier sights would have worked satisfactorily if the true vertical was known. This vertical is most difficult to find in the air. One degree error in the vertical may not appear to be very much, but it produces a far greater error at ground level. It can be calculated that deviation of one degree from the vertical produces a far greater error at the ground than all other errors of similar amount at high altitudes. An airplane traveling 100 miles per hour, actual speed making a turn of 7.6 miles in radius, the complete circle in 28 minutes caused a deviation from the vertical of one degree, causing an error of 176 feet from an altitude of 10,000 feet. The gyroscope appeared to be the only promise of solving the vertical problem. Suitably controlled, the gyro could be kept within two tenths of a degree of the vertical. Because of this, it was Mr. Norden's theory that solving the problem of true vertical was in the use of gyroscopes. The high speed of the gyro augments its inertia, causing it to resist a change in the direction of the axis. If a gyro, properly designed, can be kept within two tenths of a degree of the vertical, then the theoretical accuracy of the Mark XI was one percent of the altitude.[19] These results had been consistently obtained during tests at the NPG. By December 1929, one bombing squadron in the Battle Fleet was completely equipped with one sight per airplane. With the current contract for 80 sights, it was expected that by the end of 1930, all of these sights would be deployed.

The Navy thinking, at this time, was that bombing airplanes in formation would not get direct hits every time, but that the center of the bombing pattern should be close to the target in much the same way that a surface ship (battleship, cruiser, etc.) places a salvo. However, in 1921, General Billy Mitchell proved that individual bombing airplanes could sink a battleship providing the correct size of bombs were used and properly dropped on the target. It was demonstrated through tests that with a bomber (bombardier) entirely new to the Mark XI, the average error was 190 feet. This compared to 286 feet using other types of sights over a considerably longer period (Wimperis, Estoppey, Sperry). By May 1928, the results with the Mark XI sight indicated that every bomb salvo would be a straddle. When this was compared to other bombsights in use before the Mark XI became operational, they could only count on getting one in every four straddles, and this displayed the accuracy of the Mark XI. The Bureau of Ordnance expenditures of funds, time, material, engineering, and testing of the Mark XI was a giant step toward the development of a precision bombsight. The Mark XI sight could be compared as great a relative change in aircraft ordnance as to *battleship gunnery with the revolutionary change from pointer to director fire.* [20]

Feedback was being received from the fleet activities concerning the Mark XI. The U.S. Fleet, Aircraft Squadron, Battle Fleet, Torpedo and Bombing Plane Squadron Two, based at San Diego, California, encountered a host of problems. This Squadron completed Individual Battle Practice on November 11, 1929. Fifteen sights were used to train and fire 36 bombers required by the order for this practice.[21] This subjected the sights to heavy operating conditions with very little time for overhaul. About 30% of the airplanes were forced to return to their base due to bombing problems. Analysis of the reports indicate that many of the problems were caused by dust and sand in the stabilizer main clutch, clock failures, to low voltage, excessive play between the bombsight carden and the case, etc. It was determined that insufficient time and experienced personnel were not available to do overhaul and maintenance duties. The fleet activities estimated that the removal, overhaul, and reinstalling the sights required about three days.

The squadrons were only given two weeks training on the sights for the Battle Practice. The Mark XI required delicate adjustments, both mechanical and electrical, and it was so complicated that under continuous and intensive operations, numerous failures occurred. Personnel assigned to the bombsight section were not properly trained in the maintenance and operation of such a delicate instrument. It became very evident that stringent means had to be taken so that this instrument was accorded the proper maintenance and care. The Mark XI was not just another instrument to be installed in an aircraft and then left to be used without trained personnel. As with all extremely delicate instruments, great care must be taken so that they are in no way abused. To try and alleviate the problem, it was recommended that personnel be trained, either by the fleet or at the manufacturer's plant. In addition, spare sights be supplied to the squadrons to replace sights under overhaul in the ratio of one spare sight to each six airplane group.

In April 1929, the BurOrd issued a Memorandum for Bombsights (a lengthy document) concerning the Navy's requirements for bombsights. It became readily apparent that no one design would take care of all their requirements. It was determined that three major types would be needed. They were: 1) High altitude, high precision sight; 2) High altitude sight of considerable accuracy; 3) A general utility sight. The development of the high precision sight was currently being worked on a contract awarded to the General Electric Company. This type of sight was to evolve very slowly as a result of a thorough study on all of the problems of gyro stabilization, design of the sight, bomb trajectories, etc. It was felt that this sight would not be ready for air tests before the summer of 1930. It was to be a synchronizing type with gyros so precessed as to allow the apparent motion of the sight. The entire sight was expected to weigh about 200 pounds, and the cost would be over $10,000. It was thought that the average error would be about 50 feet at 10,000 feet altitude. The BurOrd was fully aware that they were dependent on only one source for bombsights, C.L. Norden, Inc. The inventor was getting along in years, and it was a small company. They felt that this was a handicap in the development of bombsights and could not be actively pursued from different sources. In view of this situation, the Bureau of Ordnance tried to interest various companies, such as the General Electric Company, Ford Instrument Company, etc. These companies were already skilled in manufacturing fire control instruments. A design contract was awarded to General Electric to study the bombsight problem. However, after working on the project for quite some time, it was abandoned by the company.[22] The Ford Instrument Company did not want to participate. The other source of bombsights was the U.S. Army Air Forces, which was conducting its own research on bombsights.

The Mark XI was already in production, and since it had a good track record, it would fill the requirements for the high altitude sight of considerable accuracy. The average error in theory was 1% of the altitude. The sight could not be used accurately below 2,500 feet. The main drawback of this sight was that it was unduly complicated, required considerable maintenance, and the cost was high. The five year Navy building program contemplated 217 VT airplanes in 1933. It was felt that 150 sights would fill peace time needs. There was an existing contract for 80 Mark XI sights on order, and another 70 would be purchased as soon as the design proved itself under simulated war conditions. The Mark XI was being issued to the Fleet activities as fast as production allowed. The first six sights had already been delivered to the Aircraft Carrier Saratoga.

At the present time, a general utility sight was non-existent. The fleet services were using the Mark III, which was a WWI development of the Wimperis sight, and it was entirely unsatisfactory. It did not have any form of stabilization, and bombing results were erratic. The Fleet also had the D-4 sight, which was a better instrument, but mechanically it was not satisfactory. Both sights required special maneuvering during a bomb approach. To this effect, the Norden Company was requested to investigate the designing of a general utility sight and to submit estimated costs in developing and manufacturing two experimental sights of this type. In April 1929, Norden advised the Bureau of Ordnance that the cost was $20,500 for the two units. Mr. Norden had already spent seven months of his own time and had completed much of the preliminary design, which was almost completed and included some of the parts that were already made. Norden Company stated that it would take an additional six months to complete the first instrument because every part for this model had to be hand made, and there was no precedent for these parts. This attributed to the high cost. Reworking of the optics, special fixtures, and jigs were absolutely needed to machine certain parts. They estimated that, in quantities of 100 or more, the cost would be about $600 per unit.[23]

In early 1930 the first group of sights were returned to the contractor, the Norden Company, for overhaul and modernization. The sights were being issued two or three at a time in numbered groups. Group 1, 2, 3, that had been issued to the USS *Saratoga*, were returned for rework. Inspection revealed that the instructions requiring extreme care be taken in shipment were completely disregarded. In removing the sights from the packing boxes, in many cases pieces of the rack which held the sights in place were broken and pieces of wood as large as 3" x 6" were broken off. Further inspection revealed an extremely dirty condition of the sights. In many cases, a film of course grained dirt at least 1/16" thick was found on top of the pilot director and similar flat surfaces.[24] In removing the clock face, dirt was also found inside the clock and scales. The condition of spare parts boxes was most unsatisfactory. The conditions of the spares indicated that very little care had been given to these accessories. The Chief of the Bureau of Ordnance, in a very strongly worded letter to the fleet activities, requested that all necessary steps be taken to assure that all instructions in the care of these instruments be carried out immediately. To assist the using activities, the contractor offered their techniques involved in the disassembly, overhaul, reassembly, and testing. They felt that it would be of great value to the men on the *Saratoga* who were responsible for the care and upkeep of these extremely delicate instruments. The Norden Company also gave the Bureau of Ordnance the option of ordering several mechanics to the bombsight plant to witness the process and give them thorough instructions on the care, upkeep, and "trouble shooting" in connection with the very sights that were assigned to their care.[25] The maintenance and care of the sights was a continuously serious problem that had to be dealt with if the sights were to operate properly. Strong emphasis was put on this care by higher headquarters.

The story of the Norden bombsight cannot be told without including the part that the U.S. Army Air Forces played in the development of the famous Mark XI sight. The Army Air Corps was aware that the Navy had finally developed a bombsight of considerable accuracy, and that it was one of the best in the world. They were using the Estoppey (D-4) and the Sperry-Inglis (C-1), which were not as accurate as the Mark XI. In early 1929, the AAC requested a complete Mark XI bombsight system from the Navy Department. This one sight would be an addition to the 80 sights being produced under contract. However, since the contract called for 80 sights, either a new contract was needed or the existing contract be increased to 81 sights. The Norden Company quoted the Navy (and the Army) a price of $4,500 for a complete Mark XI. Also, spare parts of sufficient quantity and an assortment would be provided in a standard Mark XI Spare Parts and Tool Box at a cost of $200. The sight would be the same as those being delivered presently under contract. The item could be provided to the Army in 30 days after award of contract.[26] The AAC was to use this sight for test purposes.

In a letter from the Secretary of War to the Secretary of the Navy, the question relative to the secret status of the Mark XI was discussed in detail before delivery of the sight. In the letter, the War Department assured the Navy that all the necessary steps would be taken to insure that the bombsight would be kept secret and that only authorized personnel would be allowed access to it. The AAC also requested that certain high level Army Air Officer personnel be allowed to inspect the sight, including their chief physicist as requested in the Navy letter. The War Department immediately forwarded a letter to the Materiel Division, AAC Wright Field, Dayton, Ohio, instructing them that any personnel, officers or civilians, who would be working with or inspecting the Mark XI be given definite instructions and guidelines to comply with the secret restrictions connected with the sight in order to comply with the policy outlined in the letter from the Navy (the Navy letter was not available for research). Since there was a difference in price between the present contract for 80 sights and one for the AAC, the Navy determined that this purchase should be handled on a new proprietary requisition and not as an amendment to the existing contract. These were the conditions that led the AAC in acquiring a Mark XI sight. This was the beginning of a long and sometimes acrimonious association between the two services that would continue until the end of WWII.

The Mark XI bombsight system had now become part of the standard bombing installation in Naval aircraft. Few publications and directives concerning the redesigning of the airplane for installation of the Mark XI bombing system were found during research. The Fleet activities now had the sight, but needed information on its installation. The fleet activities were urgently requesting the Bureau of Ordnance and the Bureau of Aeronautics for information that would alleviate this problem. They were using makeshift methods instead of a standardized configuration. The only publication that was found relating to the installation of the sight was a publication issued by the Bureau of Aeronautics, dated June 27, 1930, entitled "Armament Specifications for the Mark XI Bombsight Installation, and SR 27, Revision A, dated May 15, 1934." These publications outlined the standards and specifications for the bombsight installation in Naval aircraft. It was not in the form of a Technical Order or Maintenance and Calibration Instructions. It dealt primarily with the placement and installation of the sight in the airplane. It further cautioned that this was of an extremely classified nature, no information was to be revealed to anyone not connected with this project, and great care was required that only those involved would have access to it—military and civilian. The directive also gave the

Mark XI sight the following statistics (in pounds): Sight 18.0, Fork stabilizer 10.5, Timing clock 4.9, Pilot Director 37.0, Pilot Director 1.25, Electrical wiring, switches, etc., 5.5, for a total of 77.15 lbs.

The airplanes coming from the manufacturer were not designed for the placement of bombsights. It would be necessary to redesign the interior of the airplane to accommodate this system. The BurOrd decided that the sight was to be placed in the bow compartment behind a laminated glass window inclined as to more than 45° to the vertical as the lines of the bow of the hull would permit. The sight was to be installed to give unobstructed angles of vision as follows: Side 35° to each side of the centerline when sighting through the sight telescope in its forward maximum position; Forward 90° unobstructed vision when installed in the bow of the airplane; Downward unobstructed vision of 4° to the rear of the vertical when the airplane is in level flight at full speed with fuel load less than 40% of the total fuel for which tankage was provided, and at 90% of the service ceiling for that airplane. In addition, the line of sight of the bombsight would coincide with the center of the bombing window. The bomber (bombardier) was to be given space as to be comfortable and at ease in working the sight. The contractor was to furnish either an adjustable seat or cushions. Such cushions would also serve as life preservers. The safety of the bombardier was also considered. The compartment was to be free of any sharp objects that could cause injury in a rough flight. For a person to maneuver in an area of 2± feet high does not give much room to be comfortable. To obtain the specified angles of vision and to move the installation close to the bombing window authority was given to shorten the fork, but care had to be taken so that it would not emerge from the lower bearing of the fork.

Samples of the bombing windows, drawings of the sight installation, angles of vision, release handles, switches, wiring, and all other items pertaining to this installation were forwarded to the Bureau of Ordnance for approval. In addition, a mock up was required, and a complete set of drawings was furnished to the Trial Board for its use during the trials and investigation. It was readily apparent that bombsight development had outstripped the design of the airplane. Actually, this was a backward approach to aerial bombing. It was necessary to modify the airplane to accommodate this new instrument. The language of the directive decreed that it was up to the fleet activities to obtain a contractor to perform this retrofit. In order to accomplish the terms of SR-7A, a contractor was selected for this modification. Accordingly, the first contract No. 12551 was awarded to the Glen L. Martin Company. The modifications were to be made on 30 PM-1 model airplanes by the middle of the 1930s. On March 6, 1931, a complete Mark XI was taken to the Naval Air Station, Anacostia, D.C., to investigate the suitability of using the sight in a P3M-1 airplane being tested by the Trail Board. The results were favorable. In July 1931, the NPG reported that the sight installation in the X2PM-1 airplane and PM-2 airplane at Anacostia, D.C., were not in accordance with the provisions of

SR-27 specifications, and that engineering change orders would be issued to comply with these specifications. The Bureau of Aeronautics found that some of the deficiencies were major and others were of a minor nature. For example, the port side installation did not have an altimeter, and the two Pilot Director Indicator positions were immediately adjacent to the two magnetic compasses, causing considerable changes in compass readings. Other items were minor in nature, such as cutting the fork stem. The contractor installed corrugated metal for floor covering instead of using a smooth surface as specified. It would be difficult for the bomber to kneel or walk. Outside of these items, the installations were acceptable. The best parts were accepted, and others were rejected. Items that worked in the shop efficiently did not operate that way when actually used in the airplane. For example, the Pilot Director Indicator and air speed indicator were to far apart to be readily seen by the bomber without undue roving of the eyes. Other defects, such as brackets, electrical connections, etc., that were awkward to operate were easily corrected. As it is with all new weapons, these deficiencies showed up only in simulated combat use.

Many Mark XI sights had been in use for quite some time. In January 1931, the Mark XI bombsight No. Exp-1 (Experimental) was returned to the manufacturer for major overhaul. This sight provided a particularly good record during operation in the field. The sight was overhauled by Norden Company before being reissued to fleet activities on May 2, 1931. Considerable failures on the bombsight were noted on the backsight during the period of March 2, 1929, to April 2, 1930. This particular sight was used by 15 students for practice bombing. The sight had been in operation for a total of 249.9 hours. It was also used for assembly and disassembly practice by several bombsight classes.[27]

In April 1931, the Mark XI bombsight No. Exp-2 was returned to the manufacturer for major overhaul. This sight had a better track record than the Exp-1. Although there was some trouble with the bearings, the sight was operated for a total of 269.1 hours, and it was also used as a trainer and for assembly and disassembly practice by bombsight students from March 2, 1929, to March 12, 1930. This was the first time that a comprehensive record was kept on the Mark XI operating under severe conditions. This time would amount to approximately 33 days of continuous running, plus the fact that it was also being used as a training device. Of course, during wartime an instrument would be required to be sufficiently rugged to withstand heavy usage before being returned for major overhaul. Since there was no precedent for this kind of instrument, it was not known if this was a high or low average. Further, tests were required before any conclusion could be arrived at. NPG, in testing the Mark XI received in early 1932, still had many defects that needed to be corrected before they were sent to Fleet activities. However, these defects were not of a critical nature and were corrected by NPG personnel.[28]

The optics in the Mark XI were constantly being upgraded. The optics in Group 18 were much improved over the previous sights. By January 1931, this improvement consisted of clearer images and sharp, clear cross wires. There was still a little "sliding," or displacement, of the image. Later models of the Mark XI were excellent in their sharpness. They were improved by making them slightly heavier. However, some of the bombardiers preferred the thin cross wires (crosshairs). It was recommended that the optics in Group 18 should be used as a standard in future Mark XI sights.[29]

During the Tactical Exercises of November 19-22, 1930, bombing runs were conducted to determine whether the sight could be used at night. Bombing runs were made at an altitude of 4,000 to 6,000 feet on battleship targets illuminated by parachute flares. The ships were clearly visible with the naked eye from altitudes up to 6,000 feet, but could not be picked out with the bombsight. However, the exposed light on the target could be easily picked up with the sight, and bombing could be conducted practically as effectively as in day time when using a light as a point of aim. The inability to pick up an object (other than a point of light) with the bombsight was due to the blinding effect on the bomber of light which was used to illuminate the bombsight gyro. This light caused a dimly lighted target to become invisible. It was felt that additional studies would be required to correct this deficiency. Additional air tests at night would be required on illuminated targets for night bombing. By continuing to study this problem, it was felt by the Bureau of Aeronautics that the use of radium paint would be very effective. Many years would pass before night bombing would be successfully conducted.[30]

The Mark XI bombsight system had proven itself to be a very accurate weapon. Mr. Norden and Mr. Barth filed an application for a bombsight patent on May 27, 1930. The detailed description and drawings consisted of 19 pages. Because it was placed on a secret status, the patent was not issued until October 7, 1947, or seventeen years later (Plate #1).[31] No other names appear on the patent as inventors. Some reports have indicated that there were other co-inventors. Research did not show this fact. There were many early day naval pioneers who worked on the sight, too many to be named, who were involved in the development of the Norden bombsight spanning almost two decades. The patent rights were assigned to the United States of America as represented by the Secretary of the Navy.

Plate #2 displays the bombsight as it was installed in the airplane. Date of the photo is May 4, 1930. The airplane is not identified. Note the knots in the pine boards. It seems that a better grade of lumber could have been used. Plate #3 shows actual bombing conducted by the Naval Proving Ground using the Mark XI bombsight. Notice the ground speed of about 60 knots. These early day records provided much valuable information, as much of it was also used with the Mark XV. (Plate #4)

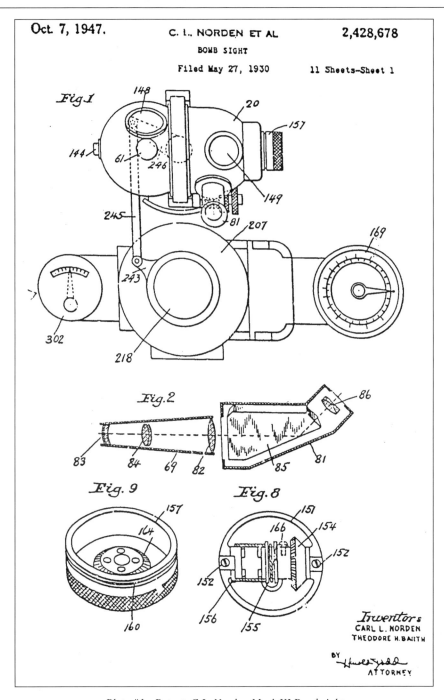

Plate #1 - Patent, C.L. Norden Mark XI Bombsight

Chapter VI: The Navy's Norden Mark XI Bombsight

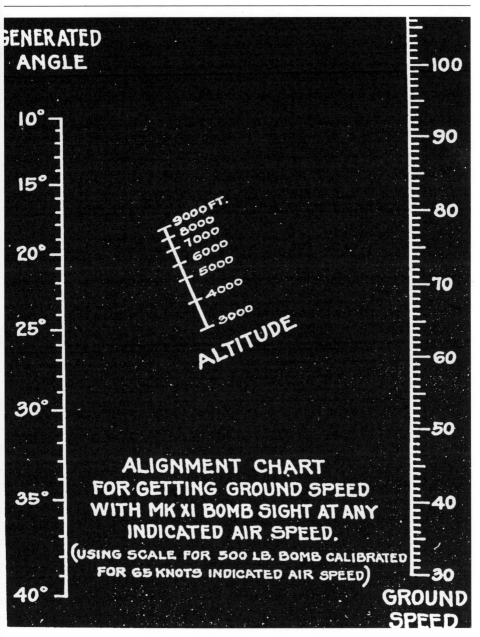

Plate #2 - Alignment Chart, Ground Speed

Plate #3 - Mark XI Installed in Airplane

Plate #4 - Bombing Data, 1927

Chapter VI: The Navy's Norden Mark XI Bombsight

Notes:

1 The Navy's Mark XV (Norden) Bombsight. Its Development and Procurement 1920 - 1945, Page 15, National Archives.
2 Ibid, page 16.
3 Ibid, page 36.
4 Ibid, page 34.
5 Ibid, page 38.
6 Speech delivered by Brig. Gen. William Mitchell at McCook Field, October 26, 1921, National Archives.
7 The Navy's Mark XV (Norden) Bombsight. Its Development and Procurement 1920 - 1945, page 40, National Archives.
8 Ibid, page 45.
9 Ibid, page 42.
10 Letter from Carl L. Norden to Bureau of Ordnance, Subject: Bombsight Mark XI, Contract No. 56166, dated March 25, 1924. National Archives.
11 The Navy's Mark XV (Norden) Bombsight. Its Development and Procurement 1920 - 1945, page 59, National Archives.
12 Letter from the American Consul in Zurich, Switzerland to the Bureau of Ordnance, No subject, dated January 11 & 12, 1932, National Archives.
13 Letter from Commandant, Naval Gun Factory to Bureau of Ordnance, Subject: Manufacturing of Two Experimental Mark XI, dated March 3, 1927, National Archives.
14 Letter from Commander, Aircraft Squadron, Battle Fleet, Subject: D-4 Bombsights, dated July 28, 1927, National Archives.
15 Letter from Bureau of Ordnance, Memorandum, Subject: Comparative Accuracy of Old and New Bombsights, dated July 3, 1928, National Archives.
16 The Navy's Mark XV (Norden) Bombsight. It's Development and Procurement 1920-1945, page 66, National Archives.
17 Letter from Carl L. Norden to Lt. Commander Picking, Subject: Bombsight Requirements, dated April 22, 1929, National Archives.
18 Spread Sheet, Bombing Data, Naval Proving Ground, Dahlgren, Virginia, dated July 11, 1927, National Archives.
19 Memorandum to Chief of Ordnance, Bombsight Mark XI, Subject: General Description, dated December 11, 1929, page 2, National Archives.
20 Ibid, page 5.
21 Letter U.S. Fleet, Aircraft Squadron, Battle Fleet, Torpedo and Bombing Plane to Bureau of Ordnance, dated November 27, 1929, National Archives.
22 Letter to Chief of Bureau of Ordnance, Subject: Bombsight Requirements, dated April 22, 1929, National Archives.
23 Letter from General Electric company to Bureau of Ordnance, Subject: Bombsight Mark XIII, dated March 5, 1931, National Archives.
24 Letter from C.L. Norden to Adm. William D. Leahy, Bureau of Ordnance, Subject: Bombsight Requirements, dated February 28, 1930, National Archives.
25 Letter from Bureau of Ordnance, Inspection of Groups No. 1, 2, and 3 Returned for Modification, Mark XI, dated May 3, 1930, National Archives.
26 Letter from Carl L. Norden, Inc., to Chief of Bureau of Ordnance, Subject: Return of Groups 1, 2, and 3 of Mark XI Bombsights for modernization, dated February 28, 1930, National Archives.
27 Letter from C.L. Norden, Inc., to Bureau of Supplies and Accounts cost of Mark XI Bombsight for the U. S. Army, dated June 12, 1929, National Archives.
28 Armament Specifications for Mark XI Bombsight Installation, SR-27 Revision-A, May 15, 1931, National Archives.

[29] Letter from the Naval Proving Ground, to Bureau of Ordnance, Subject: Mark XI Bombsight No. Exp-1, dated January 27, 1931, National Archives.
[30] Letter from Naval Proving Ground to Bureau of Ordnance, Subject: Operations Record of Mark XI Bombsight, No. Exp-2, National Archives.
[31] Letter from Naval Proving Ground to Bureau of Ordnance, Subject: Special Report on Optics in Group 18 Mark XI Bombsight, dated January 2, 1931, National Archives.
[32] Letter, Base Force, Patrol Plane Squadron Seven, to Chief of Bureau of Ordnance Mark XI Bombsight, Deficiency for Night Use, January 5, 1931, National Archives.
[33]

Plate #1, Patent filed by Carl L. Norden and Theodore H, Barth's on May 27, 1930, issued October 7, 1947, Number 2,428,673, Department of Commerce.

Plate #2, Photo of installed Mark XI bombsight in airplane, National Archives.

Plate #3, Bombing data, Naval Proving Ground, Dahlgren, Virginia, National Archives.

Plate #4, Alignment Chart for Setting Ground Speed with Mark XI bombsight at any indicated airspeed, National Archives.

VII

The Navy's Norden Mark XV Bombsight

With the success of the Mark XI, Mr. Norden, as stated in the previous chapter, advised the Bureau of Ordnance, in 1929, that he was not happy with this sight as was the Bureau of Ordnance. It was determined that a bombsight for level bombing must contain certain characteristics, such as the greatest possible accuracy, speed in operation, simplicity in operation, upkeep and low maintenance cost, along with adaptability to quantity production in the event of war. The most important item was the speed of operation, requiring a minimum of elapsed time over the target between starting the approach and dropping the bomb. During this period, the plane is on a level course, vulnerable, and an easy target for enemy fighters and anti-aircraft fire. A bombsight containing all of these parameters would inherently be somewhat complicated, but the design should be simplified as much as possible. With these guidelines in mind, Mr. Norden started work immediately in developing a completely new sight. By late 1929, the development of the Mark XV was well underway. This new sight under manufacture by C.L. Norden was to be identified by the Bureau of Ordnance as the Mark XV.[1] This sight would not be a timing sight, but a synchronizing type. Other bombsight manufacturers, notably Sperry, Estoppey, and General Electric, were leaning towards this type of sight.[2]

The C.L. Norden Company had been working on the design and engineering of this new sight since 1929, and considerable work had already been done in obtaining new castings, machine tools, jigs and fixtures, electrical parts, etc. The work had progressed very steadily to a point where, on February 9, 1931, the Norden Company delivered the first Mark XV Bombsight, Experimental Model No. 1, to the Naval Proving Ground at Dahlgren, Virginia. The sight did not have a pilot director or fork stabilizer, but it was fitted so that it could be used with the present Mark XI fork bail, stabilizer, and pilot director if the need required it.[3]

In a synchronizing sight, the altitude and air speed are set, and the line of sight takes up automatically a motion which the bomber (bombardier) can regulate. Mr. Norden described this new sight as being able to provide ground speed, angles of drift, and true air wind. It would also hold a true compass course for hours and compensate for earth rotation, all in knots per hour. All these calculations he thought would only take about ten seconds (Plate #1).[4] In actual use it was closer to 20-30 seconds. The introduction of a new constant speed governor went a long way in making these calculations possible, as it was very accurate and sensitive, and it was adaptable to a wide range of revolutions per minute. Also, the introduction of a computer assembly commonly known as the "rate end" contained the synchronous drive transmission in the new sight. With the development of these various mechanical discoveries, Norden went a long way in solving the bombing problem.

The next logical step was the transition of the bombing problem to the sight. The sight determined the correct dropping angle as follows:

Trail for the particular type of bomb is taken from the trail charts. The number of mils of trail is set into the sight by rotating the trail arm from zero mils to, let us say, 50 mils as indicated by the scale on the trail plate. Actual time of fall, which is also taken from charts, is set into the sight. This is done by positioning the altitude knob so that the proper disc speed can be attained. For example, if 20 seconds is the actual time of fall, the disc speed for a sight is computed by the use of the disc speed formula, i.e.: Disc Speed - Sight constant divided by the actual time of fall. In this example, the Disc Speed $5300/20 = 265$ R.P.M. In order to obtain the Disc Speed of 265 R.P.M., it was necessary to use a tachometer and position the altitude knob until the required Disc Speed was obtained. When trail was set in, the roller was positioned from the center of the disc a distance proportional to trail. The correct rate drive for the optics was determined by positioning the roller whole range distance from the center of the disc. This was accomplished by rotating the rate knob and, through a system of gears, the spindle screw was rotated, causing the roller to be moved upward on the spindle screw. When the roller turns, it transmits, through a system of gears and by means of the telescope cable, the drive to the telescope mirror. When the indicator of the sectors (quadrants) coincides, the bomb is released provided the release points were closed. The telescope indicator is positioned at maximum sighting angles at the start of the bombing run, and the rate indicator is positioned at zero. The telescope (quadrant) indicator sector is driven, the driving power being furnished by the telescope motor. The drive is transmitted from the motor to the disc to the roller through intermediate gears and differential to the telescope rack, which translates the telescope (quadrant) indicator sector.[5] Plate #2 shows the train of action which takes place in the rate end as the sight solves the bombing problem. The actual time of fall is set into the sight by positioning the altitude knob. The disc speed varies inversely as the altitude. This is shown by the altitude vector. As the altitude knob is rotated clockwise, the speed of the

Chapter VII: The Navy's Norden Mark XV Bombsight

disc increases and the altitude decreases. When the trail arm is rotated, the roller is raised from the center of the disc a distance proportional to trail.

When the point of synchronization is determined, the roller is driving the telescope quadrant at an angular speed proportional to the ground speed through the intermediate gears, traction gears, the tee head shaft, and telescope rack pinion and telescope rack. The drive was transmitted to the mirror by means of a bronze cable. The cable operating the sheave gear was meshed with the telescope indicator sector (quadrant) through the right carden gudgeon over several sheaves, which were attached to the telescope cradle. The mirror, or telescope, was driven at a speed proportional to the ground speed. The telescope was stabilized by the vertical gyro. If bombing with a cross wind condition and drift is encountered, the rate end solves the problem exactly as in the case of no cross wind or drift. Plate #3 shows how the sight corrects for cross-trail and shows exactly the relationship between the triangles formed by the movement of the crank slide and cross-trail operating shaft and the tilting of the optics and resulting triangle. In order to keep the fore and aft cross hair on the target, the airplane's ground tract is up wind of the collision course cross-trail distance and parallel to the collision course.[6]

Since the Mark XV was of a completely radical design, the primary tests were not intended as an acceptance test, but to determine possible deficiencies and obtain further information/data on this new concept so that any defects could be corrected as quickly as possible, or to make improvements that were required so that it would form the basis for a production model. In many respects the Mark XI and XV were alike, and this made it easier to check during tests. The sight was given a thorough bench check for items such as end play in gyro and carden bearings, electrical defects, gyro balance, optics for sharpness of cross lines (cross-hairs), etc. This was the first time that a comparison was made between a timing sight and one that was gyro stabilized and was of a synchronous type. On completion of bench tests, it was to be air tested. The bombs to be used were the Mark VIII Water filled practice bombs. The sight had previously been calibrated for altitude and disc speeds for the above bombs. A number of bombs were dropped in an attempt to obtain some idea of the accuracy to obtain data for the estimated range, deflection errors, and altitude at which the droppings took place. It appeared that the Mark XV would be relatively simple to operate in comparison with the complex Mark XI. In some quarters, there was some thought that nothing would be as accurate as the Mark XI. Continued air tests would be required to verify this.[7] Excellent results on the Mark XV would soon dispel this misconception. There were also many parts which required very high precision machining measurements that were not in any sight developed up to that time. New production methods in providing these necessary parts for this sight was critically needed if this new weapon was to be mass produced. The completion of Experimental Model Number 1 was the beginning of a *new era of aerial bombardment at very high altitudes.*[8]

During extensive and intensive flight testing, it was found that servo motors were being affected by temperature changes in the range of 40° to 50° creating havoc with the sight caused by the extension of the dural castings. This greater friction also caused axial pressure on the bearings. No further testing could be conducted until this problem was remedied. A partial solution was using a light thin oil and scrapping the bearings to a loose fit at room temperature, while the armature was given axial play by using a gasket. Eventually, a more powerful servo motor with a field and armature of 1 1/8" or 1 1/4" was substituted with excellent results.[9] Another troublesome defect was in determining the best spotting methods for bombing fixed targets. There are two types of errors: 1) Accidental range from inaccurate synchronizing and 2) from an unsteady flight at the time of instant release. Range errors with altitude settings too low puts the bomb over and setting an altitude to high puts the bomb short. Also, any errors from trail angle caused the following: to small a trail angle puts the bomb shorts, and setting to large a trail angle puts it over the target. A gyro tilted athwartship tends to place the bomb off on the side towards which the upper end of the vertical axis established by the gyro inclined.[10] It was also found that the bombs trajectory did not differ excessively from ordinary parabola. In order to alleviate these errors, two spotting doctrines or rules were devised by the use of tables developed by the Naval Proving Ground. Normally, if the range error of the first bomb is less than 15 mils, no spot is made; between 15 and 60 mils, one half of the error is spotted and if it was over 60 mils a practice run was made without dropping the bomb.[11] Rule #2 stated that thereafter, bombs were to be dropped in consecutive pairs on successive runs using identical settings for both bombs of each pair.[12] In keeping with the testing parameters of the Bureau of Ordnance, the Mark XV Exp-#1 and #2 were delivered to the Naval Proving Ground on February 9, 1931 and were run for 54 hours and 90 hours respectively. After 11 hours a new synchronizing unit was installed in Exp-2 and this eliminated previous difficulties of vibration. The sights were then run for 30 minutes at 38° and 12° without developing speed reduction at low temperatures. After 43 hours of running time, this sight had to have carbon dust removed and sticking brushes would be required at 40 hour intervals. This deficiency was later corrected. This and about 30 deficiencies were recorded during this "shake down" test. The letter from Naval Proving Ground to the Bureau of Ordnance consisted of six pages, nine enclosures and 33 pages of deficiencies and suggested what corrective action should be taken.[13]

Even with all of the deficiencies, the Mark XV, Experimental Model #2 was scheduled to be used in the bombing of the U.S.S. Pittsburgh at anchor. In September 1931, an altitude scale was also furnished for that particular test. The tests were conducted in October 1931 using both the Mark XI and XV. During the test, 14 bombs were dropped using the Mark XI scoring three hits for 21 % and using the Mark XV a total of eight bombs were dropped scoring four hits for 50% from an altitude of 5,000 feet. This was dramatic evidence of the accuracy of the Mark XV.[14]

In order to determine the points of weakness and probable life of this new sight and including the life of a representative stabilizer was undertaken in November 1931. The methods employed may have been primitive to our present day standards, however, the desired results were obtained. The sight was mounted on a shop bench with a double acting automatic solenoid installed to pull out the displacement knob as the sight reached the dropping point. The solenoid pushed the displacement knob in again relocking the ring and starting another run. To vary the dropping angle, a second solenoid was mounted to actuate a ratchet mechanism which rotated the rate knob. The solenoid was controlled by a clock which activated it every ten minutes. The altitude scale was varied two or three times a day by hand, being set the greater part of the time around 7,000 feet altitude. The Mark XV bombsight Exp-1 was run 54 hours in the shop and in the air previous to starting the life test. The life test was started September 11, 1931, and ran until November 9, 1931, for a total of 707 hours, which involved about 78,000 cycles of operation. During this test many serious deficiencies were noted. For example, after 415 hours of running the gyro rotor and servo motor, it was necessary to re-pack the bearings with XX grease. This corrected the problem. After 707 hours of running time, the sight was disassembled for inspection and lubrication. It was discovered that the phosphor bronze wire was frayed, but continued to function. The sight was then tested to 1,000 hours. During this testing phase, several defects showed up, and again it was necessary to re-pack the bearings with XX grease. It had been run for 650 hours without bad results. The sight was provided with a rig which made it operate automatically as though dropping about 2 bombs per minute at different ground speeds.[15]

The life test of the stabilizer was started on September 8, 1931. The stabilizer was mounted on a bench in the shop, and the clutch arm was driven with an oscillating motion of an electric fan motor for 1,050 hours. This test involved about 3,600,000 cycles of operation. The most serious defect was bearings. They were running dry after several hundred hours of running time. For example, after 172 hours, rotor bearings needed grease. Also, brush problems showed up after over 600 hours of operation. The bearings were re-packed with XX grease furnished by the manufacturer. At the end of the test about all that was found was carbon dust around the rotor brush holders, field of coils, and brush end case. The rotor brushes wore about 1/4" during the entire test or about .025 per 100 hours of running time. The bearings were removed and cleaned, and all appeared in excellent condition. Also, after 525, 745, and 961 hours of running time, the rotor bearings required grease. In spite of the bearings running dry, they were found to be in excellent condition. There were other miscellaneous deficiencies that surfaced, but they were immediately corrected. These instruments stood up excellently under these rigorous tests and more than fulfilled their expectations.[16] In November 1931, further action on the Mark XV was undertaken for additional testing, investigations and analytical studies other than just the actual data on bombing errors at different altitudes. The report of the studies was issued covering all phases of gyro action detailing almost 40 engineering casualties. Unfortunately, most of the enclosures of this report were not found in the examined files. It was found that when tests in the shop were undertaken by unlocking the gyro by 1° degree it duplicated the gyro performance observed in flight by sighting on the shadow. During these tests the stand was tapped lightly at frequent intervals to simulate transmitted vibration. This tapping would be used in the shops during repair and calibration until the bombsight was phased out of the inventory. Some bombsight repair mechanics swore by it and others scorned it. Plate #3 shows graphically the results of the gyro error. The gyro of Exp-2 exhibited good precessional qualities in all headings except East, on which headings its precession was rapid.

This sight was given more tests than any other sight that the Bureau of Ordnance had ever undertaken. These deficiencies were of great concern to the Naval Proving Ground, Bureau of Aeronautics, Bureau of Ordnance, and the manufacturer. The former tests did not cover the full amount of work which it was intended to undertake. However, it had disclosed a sufficient number of interesting points to warrant further inquiry into these items so that they could be brought to the attention of Mr. Norden while he was still working on the new design. The first and most pressing casualty to be investigated was to determine the actual performance of the gyros with the airplane in flight. After considering various methods the simplest and most direct method was to sight on the airplane's shadow using the Mark XI sight, with the airplane in horizontal flight headed directly away from the sun. The observed depression angle of the shadow was compared with the sun's altitude as determined by navigational methods and corrected for the changing geographical positions. As a result, after a dozen runs, the gyro axis was continually swinging at a rapid rate, and it often wandered as much as 2 or 3 degrees or more. This was worse than expected. It was found that if the gyro was unlocked in the shop with an initial error of about 1 degree (which was typical of what had been observed in flight) that its shop performance practically duplicated the gyro performance which had been observed in flight by sighting on the shadow. Similar precessions also resulted in most cases even when the gyro was released with no initial error. The Mark XV, XP-2 gyro exhibited good precessional qualities on all headings except east, on which heading its precession was rapid. The bombing work done at the NPG was almost entirely on course other than east. With all the testing done, it was felt that the Mark XV had a better gyro than the Mark XI, having determined the magnitude of the total probable bombing error (of 10 to 15 mil) that these gyros might account for (Plate #4). The probable error in athwartships leveling was practically as large as the probable deflection error that was observed in bombing (cor-

Chapter VII: The Navy's Norden Mark XV Bombsight 73

rected for cross-trail), and it was considered that the effects of pilot directing errors at the NPG was exceedingly small. The effect of a fore-and-aft error in the gyro's vertical was more involved. During investigations of it, there were encountered a number of important points that were adjudged to be at complete variance with what the average pilot would imagine unless his attention was specifically directed to them. Making any very late synchronizing or sighting adjustments did not possess any advantages and could result in more harm than good. There was a great desire for improving gyro performance, however, it was equally important to reduce range errors. The large deflection errors also needed to be addressed, as deflection errors could be reduced only in proportion to the improved accuracy of the gyro in athwartships leveling.[17] Many of the deficiencies identified during the analytical studies in 1931 were being worked on by the manufacturer, Bureau of Ordnance, and the Naval Proving Ground before any thought could be given to proceed with production. During testing in January 1932, irregularities showed up in the synchronous end. To correct this problem, C.L. Norden, Inc., provided another synchronous end with various knob ratios or a number of gears being tested. In addition to this, sight XP-1 Mark XV was returned to the manufacturer, which had about 1,100 hours running time, and showed more of this irregularity than the Exp-2.[18] With continued testing and redesigning this problem was corrected.

Great strides had been made since the early days of the Mark XI. It took a little over three years to design, produce, and test the first experimental Mark XV, in comparison to the Mark XI, which took almost eight years. The state of the art on bombsights had advanced dramatically with the introduction of the new sight. The development of the Mark XV was a tremendous leap in high altitude bombing accuracy. It was now possible to hit targets from 10,000-15,000 feet, and at higher altitudes as new planes were being designed to operate at higher service ceilings.

In September 1932, a contract was awarded to C.L. Norden, Inc., to provide for the purchase of bombsights conforming to the recent naval design, a number of which were to be delivered to the War Department (AAF) in accordance with previous arrangements. This production model was almost complete except for a few minor features. The terms and contents of this contract were classified "secret." However, all correspondence and conversations were to be treated in a strictly confidential manner. Every precaution was to be taken in order to prevent information regarding the development of the Mark XV be disclosed to any unauthorized personnel.[19]

By October 1932, the Mark XV had been tested both in the air and on the ground and, in the opinion of the Naval Proving Ground personnel, it was considered that the Bureau of Ordnance could very safely proceed with the necessary order for the manufacture of the sight on a production basis. Tests were continuing during the summer, and the Proving Ground determined the probable error of the Mark XV sight EXP-1 under favorable weather conditions to be 18 mils in range and 13.5 mils in deflection (12 mils if plots were corrected for cross-trail), in the range between 4,000 and 14,700 feet. The probable errors of the production model, 36 bombs having been dropped under favorable and 32 under very unfavorable weather conditions, were 8.5 mils in range and 8.0 mils in deflection. The result was that the production model was considered much superior in accuracy to both the Mark XI and the experimental models of the Mark XV. Drops were made from 2,000 and 3,000 feet with fair results. At these altitudes the amount of forward vision (69.5° forward of the vertical, which figure includes the 60° travel of the telescope plus half the field) was not sufficient to permit long enough runs, except at low ground speeds, to develop a high degree of accuracy. However, this sight could be used for low altitude work, but this form of bombing would require considerably more forward vision. The present altitude scale was graduated from 1,800 to 18,000 feet. This appeared to cover the present range of altitudes from the low point at which the vision of the sight limits one to a high point which the present service bombers did not approximate. However, it was retained as airplanes were starting to be greatly improved with higher service ceilings. There were 17 minor features of the sight that needed to be improved. One of the most notable was the curves of the trail plate. The lines were so crowded that improper settings in poor lighting conditions could occur. The plate was engraved with the trail curves, and the edge of the plate marked in mils, the trail arm having a locking device to prevent being moved accidentally. This feature and many others were incorporated in the later models of the Mark XV. Another item was to relocate the cross-trail setting knob to the after side of the sight and marked to mils to correspond to the markings of the trail angle plate, as well as to mark the left hand synchronous end in degrees and the right in tangents. It was recommended that the tangent scale be provided, and that the left scale of the synchronous end be marked in degrees and the right hand scale in tangents. The telescope cable clamp in the synchronous end required to be locked more firmly. The contractor replaced the disc motor, which was an improvement over the old model. Another troublesome area was screws becoming loose. It was necessary to go over the entire sight to provide positive locking for all screws. There were several different types of pilot director indicators being used. It was found that the round indicators were excellent. Those with the horizontal scales were so affected by vibration that they were of no value. In addition, it was necessary to provide the bombardier's pilot director ventilation to keep moisture from forming. The manufacturer was appraised of these shortcomings, and immediate action was taken to redesign, modify, or engineer new units. With the correction of these deficiencies, the Bureau of Ordnance recommended the manufacture of the bombsight, providing that sufficient time be allowed for the Proving Ground to conduct air testing and calibration with service bombs, and also for life tests be conducted.

The Bureau of Ordnance was also interested in the various bombing equipment

Chapter VII: The Navy's Norden Mark XV Bombsight

that the U.S. Army Air Forces was presently using. The Army Air Forces (AAF) had contracted with the Sperry Company for 28 type C-4 sights that had not yet been delivered. Navy personnel, in examining the sights, found that the AAF sights had many desirable features which they thought could be incorporated in the Navy sights. The Mark XI and XV were undoubtedly superior to the AAF sights, but using some of their features could prove invaluable. The Sperry bombsight (L-1 and C-4) possessed gyro stabilization, but the Pilot Directing System was impossible. In addition, these sights were very heavy and complicated. Even with these deficiencies, the Sperry sights contained detailed engineering and represented an exceedingly serious and intelligent attack on the bombsight problem, and presented many interesting and valuable points, including complete correction for cross-trail.[20] In February 1932, the AAF had sufficient funds for the procurement of 15 Mark XI sights, which were to be delivered immediately, and 25 Mark XV, which would be delivered as soon as production warranted their availability. The purpose of obtaining the Mark XI was to provide for the establishment of a bombsight school which would train special Ordnancemen to care for both types of sights. The AAF was permitted to assign a limited number of Ordnancemen to the Navy's Ordnancemen's School at Hampton Roads, Virginia, for special bombsight training with the view of the AAF establishing their own school. The Bureau of Ordnance immediately honored this request.

By late 1932, it was becoming quite evident that the AAF needed a high altitude precision sight such as the Navy Mark XV, which was superior to any sight then in existence. It was the U.S. Army Air Forces' charter to have a bombardment capability. Admittedly, during these years, the strength in multi-engine bombers was only about 160. It is interesting that all of the developments and redesigning of the Mark XV was accomplished during the Great Depression, when funds were very difficult to obtain for such work. Even in these difficult times, Congress did appropriate funds for the testing and production of the sight. With only about 135 planes in commission at one time, the Air Corps estimated that it would take about 4 months to replace an optical sight due to transportation to and from the manufacturer and re-testing. The Air Corps felt that about 25% of the bombsights should be available for spares. Therefore, for 25% of 135 airplanes in commission, the required amount of spares would be 34.[21] By fiscal 1934, the Air Corps would have a total of 89 sights which would be approximately 10 years old. The Air Corps felt that these sights had outlived their usefulness and should be declared obsolete and replaced with the new precision Mark XV bombsights. By October 1933, the Air Corps had only three Mark XV sights. The Army Air Corps wanted sufficient sights to distribute to the major tactical organization (3 each to March, Rockwell, and Langley Air Bases) for a period of six months. The series of service tests to be conducted by the AAF would not be completed until the end of April 1934, thus seriously impacting their program. Also, due to modification of the sights, it would be August 1934 before any testing could start.[22]

During the early years of the Mark XV, there was a courteous and cooperative spirit between the Bureau of Ordnance and Wright Field (AAF). All of the facilities and personnel were available to each of the services for inspection and research. Only with such cooperation and liaison between the respective counterparts could perfection be obtained in aviation ordnance equipment that could be useful to both services.[23]

In continuing with the development of the Mark XV, Theodore H. Barth filed another patent on July 15, 1931, for a Speed Controlling and Speed Measuring Centrifugal Mechanism. A patent on this invention was issued on November 23, 1933 (No. 1,936,578) with Assignor to the Carl L. Norden, Inc., New York (Plate #5). This invention was used in the Mark XV and was a vital component in the success of this new sight. This invention provided for the manufacture of a control mechanism having a high degree of precision for maintaining rotary motion constant speed. It also provided for and improved construction of a governor mechanism for rotary systems, which insured constant speed without using bulky or complicated parts. In addition, this would also provide a simplified centrifugal governing mechanism in which a control member was shifted under conditions in which variations in speed of rotation tend to occur for maintaining cooperative related rotating mechanism at constant speed. A further object of this invention provided an arrangement of balance weights coordinated with a contractor, which was controlled by a centrifugally operated mechanism with means for selectively controlling the effects of the balance weights upon the centrifugal mechanism for correspondingly controlling the operation of the contactor. It also provided for the construction of a statically balanced governor which could be employed on swiftly moving bodies for controlling rotary mechanism with all moving parts of the governor.

The upshot of the radical changes in the Mark XV triggered a number of patentable items. This was necessary because the number of bombsight inventors was increasing rapidly, and this was the only means that the Norden Company could protect itself. On July 15, 1931, Theodore H. Barth filed Patent No. 1,936,576 and was issued a patent on November 28, 1933, for a Sighting Apparatus, assignor to Carl L. Norden, Inc., New York. This sighting device, or instrument, was used for obtaining the rate of linear or angular motion, or the bearing angles or of some other sight function of a distant moving object in reference to a stabilized plane. This new invention provided a means whereby complex functional mechanism in connection with stabilized, tilting, sight planes could be transferred from the stabilized case to a non-stabilized case without impairing the operation and function of the stabilized system. This invention also decreased the number of parts requiring stabilization and static balancing. This made it possible to use a complete optical system on a stabilized case in a non-tilting non-rotatable sight case, effectively

Chapter VII: The Navy's Norden Mark XV Bombsight

eliminating the troublesome optical and mechanical parallax which had been a problem in the past (Plate #6).

With the success in developing a truly synchronizing bombsight, the Norden Company was now the leader in producing high altitude precision bombsights. They had developed an accurate means for correcting the trail and cross-trail of the bomb, it was simple to operate, and did not take up much space in the airplane. The bombing problem had been finally solved. Accordingly, a patent was taken out in the name of Theodore H. Barth, filed September 28, 1932, Synchronizing Bomb Sight. Patent was issued March 30, 1948, No. 2,438,532, Assignor to United States of America as represented by the Secretary of the Navy (Plate #7). From 1932 to 1935, the Mark XV underwent many modifications. Some of the changes were major, and others were of a minor nature. These did not warrant a patent. However, over the years so many changes were made that in 1935 Theodore H. Barth, on April 5, 1935, filed a patent for a Synchronizing Bomb Sight and was issued in 1950, No. 2,534,397, Assignor to the United States of America as represented by the Secretary of the Navy. A classification of "Military Secret" was applied to all these patents. It still retained the classification of Mark XV. Although many changes were made from 1935 to the end of the war and beyond, it retained the Mark XV designation.

The U.S. government retained the patent rights and patent liability. The introduction of a synchronized bombsight went a long way to completely solving the bombing problem. The movement of the parts of the mechanism were synchronized to move in conformity with the apparent movement of the target, whereby the sight was kept trained on the target. The optical parts of the sight were stabilized so that they would not be affected by the roll or pitch of the airplane. This would also provide for accurate means for correcting trail and cross-trail of the bomb. All of these new inventions would provide for simple and accurate operation of the sight. It was also light in weight and would occupy only a small area in the airplane.[24] Throughout the life of the Mark XV, from 1935 to its final combat use in 1967-68, the configuration of the sight did not change. Many modifications and attachments were developed, but these had no effect on design and use. It was so well engineered that any modification could be incorporated without redesigning the entire sight. For example, the trail spotting device could be attached with a minimum amount of time or disruption on the production line.

It has been reported that a Lt. Entwhistle, United States Navy, was a co-inventor of the Norden bombsight. However, a thorough review of all applicable referenced patents cited in the filing of patents relating to the bombsights developed by Mr. Norden and Mr. Barth, and also various other bombsight inventors (including Sperry, Estoppey, Clark, etc.) did not disclose that there was a co-inventor. Research is continuing in this area to determine if there were indeed co-inventors. Lt. Entwhistle was very much involved in the early engineering and development of the Norden Sight. This may have triggered the notion that he was a co-inventor.

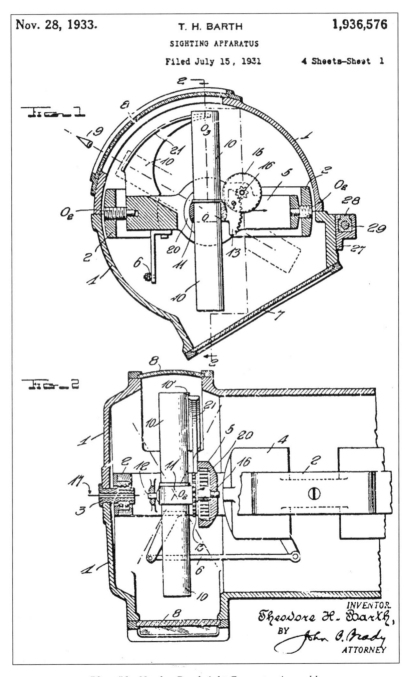

Plate #1 - Norden Bombsight Computer Assembly

Chapter VII: The Navy's Norden Mark XV Bombsight

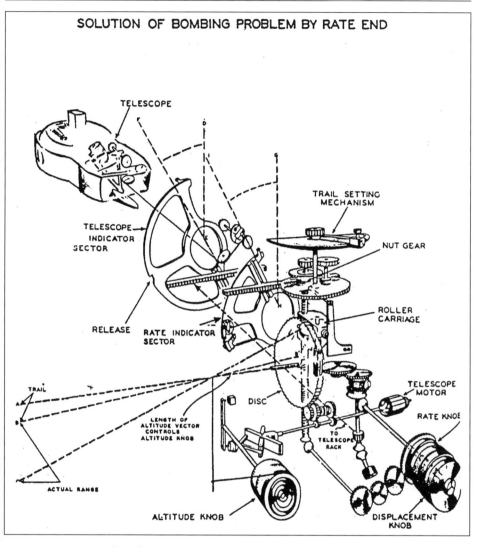

Plate #2 - Solution of the Bombing Problem by Rate End

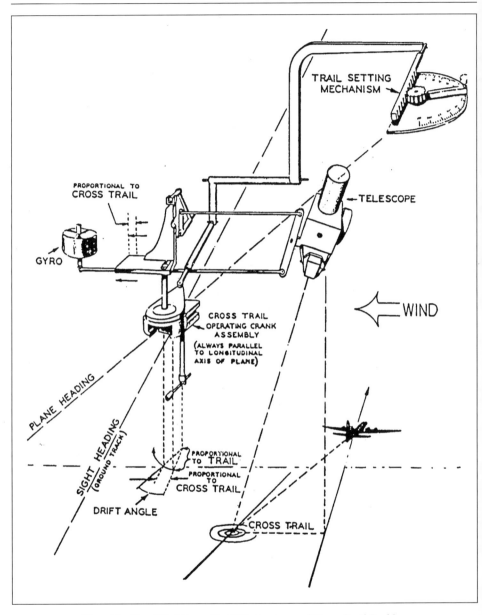

Plate #3 - Simplified Cross Trail Mechanism and Cross Trail Problem

Chapter VII: The Navy's Norden Mark XV Bombsight

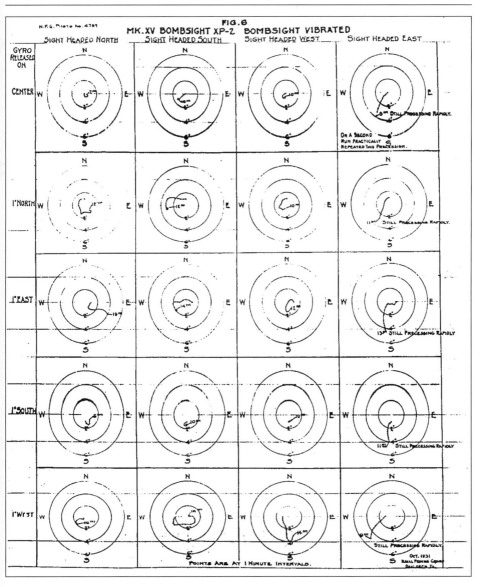

Plate #4 - Gyro Precession in the Shop

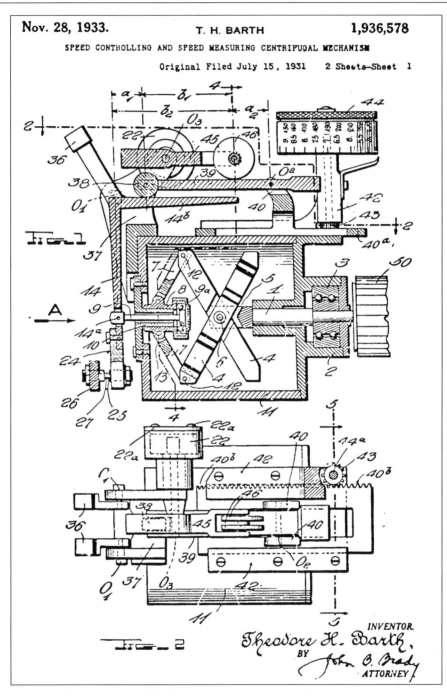

Plate #5 - Patent, T.H. Barth, Speed Controlling and Speed Measuring and Centrifugal Mechanism

Chapter VII: The Navy's Norden Mark XV Bombsight

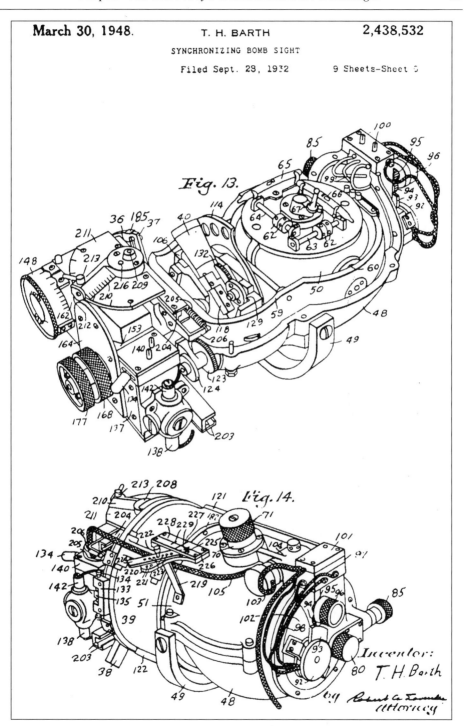

Plate #6 - Patent, T.H. Barth, Synchronizing Bombsight

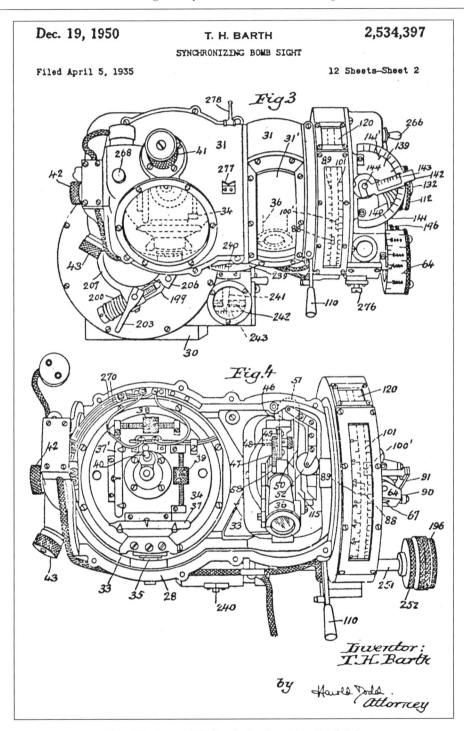

Plate #7 - Patent, T.H. Barth, Synchronizing Bombsight

Chapter VII: The Navy's Norden Mark XV Bombsight

Notes:

[1] Letter from Bureau of Ordnance to Commandant United States Naval Gun Factory, Subject: Assignment of Marks for Bombsights, dated March 10, 1929, National Archives.

[2] Memorandum from Bureau of Ordnance to Chief of Bureau of Ordnance, Subject: Bombsight Requirements, dated April 16, 1929, National Archives.

[3] Letter from Bureau of Ordnance to Inspector of Ordnance Naval Proving Ground, Subject: Bombsight Mark XV, dated February 9, 1931, National Archives.

[4] The Navy's Mark XV (Norden) Bombsight. Its Development and Procurement, 1920-1945. Page 85, National Archives.

[5] M Series Bombsight Maintenance and Calibration, Bombsight Maintenance Division, Lowry Field, Denver, CO, dated September 1944, page 107.

[6] Ibid, page 108.

[7] Letter from Adm. Leahy Bureau of Ordnance to the Naval proving Ground, Subject: Bombsight Mark XV, dated February 9, 1931, National Archives.

[8] United States Department of Commerce, Patent Office.

[9] Letter from Adm. Leahy Bureau of Ordnance to Naval Proving Ground, Dahlgren, Virginia, Subject: Bombsight Mark XV, Low Speed of Servo Motor, dated February 26, 1931, National Archives.

[10] Letter from Naval Proving Ground to Bureau of Ordnance, Subject: Mark XV Bombsight Best Spotting Methods for Deliberate Bombing of Fixed Targets, dated September 4, 1931, National Archives.

[11] Ibid.

[12] Ibid.

[13] Letter from Naval Proving Ground to Inspector of Bureau of Ordnance, Subject: Tests of Mark XV Bombsight, dated September 8, 1931, National Archives.

[14] The Navy's Mark XV (Norden) Bombsight - Its Development and Procurement 1920-1945. Page 92, 93, National Archives.

[15] Letter from Naval Proving Ground to Bureau of Ordnance, Subject: Life test of Mark XV Bombsight Exp-1 and Mark XI Stabilizer #79, dated November 9, 1931, National Archives.

[16] Ibid.

[17] Letter from Naval Proving Ground to Bureau of Ordnance, Subject: Further Analytical Studies of the Mark XV Bombsight, dated November 9, 1931, National Archives.

[18] Letter from Naval Proving Ground to Bureau of Ordnance, Subject: Mark XV Bombsight, dated January 21, 1931, National Archives.

[19] Letter from Bureau of Ordnance to Inspector of Naval Aircraft, Central District, Wright Field, Dayton, Ohio, dated September 13, 1932, National Archives.
[20] Letter from Naval Proving Ground to Bureau of Ordnance, Subject: Production model of Mark XV Bombsight, dated October 27, 1932, National Archives.
[21] Letter from Bureau of Ordnance to Chief of Bureau of Ordnance, Subject: General Review of Various Equipment Inspected at Wright Field, Army Air Corps, dated February 16, 1932, National Archives.
[22] Letter from War Department, Air Corps to chief Material Division, Air Corps Wright, Subject: Bombsights, dated March 3, 1933, National Archives, MS.
[23] Memorandum from Chief of Air Corps to Deputy Chief of Staff, No Subject, dated November 23, 1934, National Archives.
[24] Letter from Bureau of Ordnance to Chief of Bureau of Ordnance, Memo: General Review of Various Equipment Inspected at Wright Field, dated February 16, 1933, National Archives.
[25] United States Department of Commerce, Patent Office.

Plate #1 Norden Bombsight Computer Assembly
Plate #2 Mark XV, Solution of Bombing Problem by Rate End
Plate #3 Mark XV, Simplified Cross-Trail Mechanism and Cross-Trail Problem
Plate #4 Mark XV, Exp-2 Bombsight, Bombsight Vibrated, Naval Proving Ground
Plate #5 Speed Controlling and Speed Measuring Centrifugal Mechanism
Plate #6 Sighting Apparatus
Plate #7 Synchronizing Bomb Sight
Plate #8 Synchronizing Bomb Sight

VIII

The AAF and the Navy Mark XV Bombsight

The complete story of the Norden Bombsight (NBS) cannot be written without showing the part the U.S. Army Air Forces (War Department) played in using, suggesting modifications, engineering and design changes, etc. The two services were so closely intertwined in the modification, production, and use of this instrument that it cannot be overlooked. Although the Navy developed the sight, the Army Air Corps were the greatest users of the NBS, rather than the Navy, who had spent many years in its development. By 1933, the Army Air Forces (AAF) was struggling to develop a bombardment force. The Engineering Division, Wright Field, Air Corps wanted to procure 35 Mark XV sights. This was later reduced to 23 because of a lack of funds. The contract for this quantity was awarded June 29, 1934.[1] The first model to be furnished to the Army Air Corps for testing was to arrive in April 1933, and if any modifications were required, they would be incorporated in the new production model being produced.[2] In late March 1933, the Bureau of ordnance invited representatives from the Materiel Division, Wright Field (AAF) to attend the acceptance tests being conducted at the Naval Proving Ground, Dahlgren, Virginia. The Mark XV bombsight #105 was tested in the shop and in the air. Twelve bombs were dropped on April 5, 1933, from an altitude of 5,000 feet under disturbed air conditions. Results were excellent and fully up to the level to be expected of the sight under perfect air conditions.[3] The sight was hand delivered to the representative of the Material Division, Wright Field, Army Air Corps on April 6, 1933.

The results of the tests performed by the AAF were so favorable that on April 10, 1933, the Navy Department was requested by the Air Corps to procure 78 additional Mark XV, Mod 1 bombsights. However, this was later reduced to 77 due to lack of funds. Since the preponderance of correspondence, visits, and contacts were with the Bureau of Ordnance and Bureau of Aeronautics AAF were to use these contacts instead of the Navy Department, very little contact was made at Command and Cabinet levels. Henceforth, all transactions were identified as to the office of origination. It was the Bureau of Ordnance's responsibility to obtain ordnance material for all naval activities. It was not their purview to obtain and provide for the Army or the Army Air Forces. The supplying of this material was being accomplished by cross production (i.e. Army and Navy using the same weapon). Between the two contracts, the AAF would now have 100 sights of the Mark XV, M-1 configuration (23 previously delivered plus 77 on order). The necessity for the strictest secrecy, agreed upon by the Bureau of Ordnance, Air Corps, and the manufacturer, which was in order to prevent disclosure of any design features of the sight, made the issuance of very definite yet workable sets of instructions regarding shipment, storage, and use imperative, and also required a special course of instructions to be carried out by specially selected officers in the care, maintenance, and operation of the sight. This preliminary course of instruction was carried out in July 1933. These promulgations continued to force until the Spring of 1944. As a consequence of this course, it was found that additional ballistic data for bombs was necessary for proper calibration of the sights not yet delivered. This data was immediately secured and furnished to the manufacturer so that it could be incorporated into bombsights on the production line.[4]

The third sight (#106) for the Air Corps was tested on October 15, 1933. Up to April 1934, eight sights had been furnished the Air Corps. Then twelve sights were delivered to the Air Corps in April 1934 and three in May 1934 for a total of 23 sights. Security on the new sights was so tight that the officers assigned were responsible for further instructions to personnel at their respective home stations. General Benjamin Foulois stated that this sight was the most important military secret project under development by the Air Corps. At the time there was only one manufacturer in the United States capable of producing this sight. General Foulois, Chief of the Air Corps, was cognizant of the problems in obtaining bombsights. He advised his subordinates, who were very impatient, on the slowness of the delivery of sights, the reason being the intricate and precise workmanship required and the unusual amount of hard work. Production was much slower than if these sights could be turned out by machine. This feature was to haunt the bombsight program not only in the years before WWII, but through the war until late 1944. He forcefully pointed out that the Norden Company manufacturing facilities were restricted in order to reduce to a minimum the possibility of disclosure of any feature of the sight to unauthorized individuals or foreign governments. In addition, the Bureau of Ordnance also had concurrent contracts for the same right, causing serious slippages in the scheduled delivery. Air Corps personnel on their own initiative requested and visited the Norden plant, which was in direct violation of the agreement between the U.S. Army and the Navy department. This created consternation for both services, resulting in new strict and mandatory guidelines in visiting the Norden Company. One of those pushing for accelerated delivery of new bombsights was Lt. Col. Henry "Hap" Arnold, who would later become the Chief of the Army Air Forces during WWII. The Air Corps wanted the most up-to-date sight available and all personnel involved with the weapon to be thoroughly indoctrinated in the care, maintenance, operation, and security of the sight. The Air Corps directed that all stations and unit commanders would be thoroughly indoctrinated with this new policy, and under no conditions would there be any deviation.

Since the workload for the C.L. Norden Company was increasing rapidly, and to protect the Navy, in July 1933 a performance bond was issued in behalf of the C.L. Norden Company to manufacture 53 bombsights, spares, tools, etc., in favor of the U.S.A. (Navy). The bond was issued by the Hartford Accident and Indemnity Company in the amount of $20,000.00.[5] In addition, the contract was classified "secret," and specific items such as work performance, amount paid to the contractor, percent completed, date of completion, number completed, etc., could not be furnished to the Company, except to state that the contract was either satisfactory or unsatisfactory. As of that date everything was more than satisfactory.[6]

In keeping with the wishes of the Navy Department concerning security of the sight, and to keep to an absolute minimum the number of personnel who would be permitted the sight in a Secret status concerning the manufacturer, personnel assigned to the program, and its progress to prevent the disclosure to a foreign government, no visitors were allowed in the plant, except by the authorization of the Bureau of Ordnance. This provision was rigidly enforced, even with commissioned officers of both services.[7]

Over the years, contracts for the bombsight program had been awarded to the Norden Company without advertising for bids. This was in violation of existing regulations concerning the purchase of ordnance material. The Bureau of Ordnance was well aware of this policy, but had extremely good working relations with the manufacturer and did not see the need to obtain another source at this time. Also, the manufacturer had developed the Mark XI and was the only facility that was mass producing precision bombsights, as all other competing bombsight manufacturers were not able to meet the precision qualities of the Mark XV. However, in May 1934, the Comptroller General of the U.S. contacted the Secretary of the Navy for awarding of contracts to the Norden Company for not advertising for bids, and showing facts that this transaction represented fair and reasonable compensation to the U.S. for the purchase of bombsights. The Bureau of Ordnance responded to the Comptroller General detailing the reasoning behind this problem. The contracts on the Mark XV sight were classified "Secret," and since the experience of the Bureau of Ordnance extending over 15 years and confirmed by the U.S. Army Air Corps showed conclusively that no other firm in the U.S. was capable of developing such a system. At the time the sight was the most advanced in the world, and the accuracy surpassed any other bombsights being used or in the process of being developed by other manufacturers. The unit cost was comparable to, if not lower than, other types of sights. Under the terms of the contract, the Navy was to acquire complete patent rights.[8]

The disclosure by advertising of the Navy's procurement of such a system would be prejudicial to National Defense, and the very existence of the contract should be held secret. If these contracts were submitted to unrestricted bidding, it would disclose valuable military secrets, plus the fact that this company had a proven record and was the only in the country capable of producing such an instrument. Further, Section 569, Title 34, U.S. Code provides that, "The Provisions which require that supplies shall be purchased by the Secretary of the Navy from the lowest bidder, after advertisement, shall not apply to Ordnance...." This ruling was to stand until the end of WWII.[9]

Since the ruling by the Comptroller General and as prescribed by law, an audit was required on all government contracts. To satisfy this requirement, the services of a Certified Public Accountant were obtained by the Norden Company. Since much correspondence on this subject was not available for research, one cost study was found which was conducted in November 1934 showing that labor, overhead expenses, and additional expenses to be incurred resulted in the operating cost per sight being $3,683, and after adding 10% for profit the per unit cost was $4,051. Based on an estimated increase of 25% in labor and materials, the selling price would be $4,970 per unit. The net profit for Norden Company for the Mark XI and the Mark XV for six and one half years, from 1928 to June 30, 1934, averaged about 10%. It must be noted that neither Mr. Norden nor Mr. Barth received a salary. The effect of an increase of hourly wages and contemplating on going to a 40 hour work week as soon as work on contracts awarded under the National Industrial Recovery Act auspices was commenced caused an increase per unit of the Mark XV-1 (Mod 1). Because of the delivery of 81 sights to the Air Corps and 72, for the Navy for a total of 153 sights, the unit price dropped from $4,725 to $4,555.[10]

By this time several hundred Mark XV sights were now being used by both the Navy units and the Army Air Corps. The usual type of casualties were being reported now on a large scale. In January 1935, the Norden Company received a letter from the Bureau of Ordnance concerning a large number of "sticking brush" casualties in the Mark XV sight that had been submitted by the using activities (Navy and Air Corps). This problem was of extreme concern to the manufacturer, as well as to the Bureau of Ordnance and the Air Corps. Norden engineers immediately began working on this problem to determine the cause of these excessive failures, which had not been apparent in the past. These casualties had a critical effect on the gyro, causing the bombsight to not function properly and thereby destroying its precision capability. A sticking brush would cause the gyro commutator to foul up. There are many reasons that can cause these casualties, such as defective soldering, rough bars, excessive vibration, oil splashes, etc. All gyros were given a severe test at the factory and then a rigid inspection and acceptance tests at the Naval Proving Ground. At the factory all gyros averaged approximately 30 hours of operation at normal operating speeds of 7,500 R.P.M., plus or minus 250 R.P.M. The brush carbons used throughout the entire testing and rotor balancing operation at the factory remained in the gyro when it was shipped to the Naval Proving Ground.[11]

Inspection and research revealed that the "stuck brush" invariably turned out to be a stuck brush spring. The result of this investigation and findings was an 11 page letter from the Norden Company to the Bureau of Ordnance explaining in detail the reason for this high rate of casualties. Invariably, most of the reported defects were due to improper maintenance by personnel assigned to repair and maintain the sights in operating condition. The clearance between the spring coils and the brush holder tube wall was approximately .005 (five thousands of an inch). The spring had to seat properly on the carbon brush and on the brush holder screw plug. Brush carbons were not interchangeable in their brush holder tubes. The brush carbon, when properly worn in, conforms to the radius of the commutator. The brushes used in the Mark XV were of hard grade for better performance. Improper spring pressure on a brush caused excessive friction, arcing, wear, and abnormal deposits of carbon dust, causing the rotor to slow down to more than 1,000 R.P.M. The clearance on a properly fitted brush carbon and the wall of its brush holder tube should never exceed .002 inch (two thousands of an inch). Fitting a new brush carbon, although a fairly simple operation, required careful attention to established maintenance procedures. It was also found that too frequent use of crocus cloth (a very, very fine abrasive) for cleaning the commutator surfaces produced rough edges on the bars, and much of the problem originated from this practice.[12] It was recommended that this cloth be used about once in six months, which would not be harmful.

Another reported casualty was the appearance of an "oil smudge" resulting from an oil splash. It was found that excessive lubrication of the gyro rotor bearings was the cause. One drop of oil is sufficient to lubricate a bearing for more than 100 operating hours. A large drop, or two drops of oil instead of one, caused the excess oil to be centrifuged from the bearing and sprayed on to the commutator.[13] *Note*: the author, as supervisor of the Bombsight machine shop, assigned one machinist to turn the gyro commutators on a precision lathe. The machinist was instructed to produce the smoothest cut possible. Each bar was honed with an Arkansas white very fine stone, removing all of the burrs or rough edges. This was extremely tiring work, as each segment on the commutator had to be perfectly honed. At times additional personnel were required to do this work, as so many commutators were being received that the assembly line personnel could not assemble the sight until the reworked commutators were received. This part of maintenance was taken very seriously by maintenance and repair personnel. The same problem was encountered in 1941 as it was in 1935. Maintenance line mechanics (field) were "sloppy" when doing this repair work by not following established maintenance standards.[14] Explicit instructions in the form of Technical Orders for corrective action were immediately forwarded to the using activities to comply with the manuals when installing or replacing brushes in the gyro. With the correction of this problem, tests continued on the Mark XV sight at the Naval Proving Ground. These tests were very successful, and proved beyond a doubt that this instrument was an extremely accurate sighting device when used against fixed targets.[15] It now appeared that the Navy was within grasp of having an accurate sight that could be used against moving targets! The Army Air Corps, on the other hand, was more interested in fixed land targets. It was thought that the Mark XV, Mod 1 could not be more accurate. However, in the following years many modifications were made which vastly improved the accuracy of the sight.

Although the Mark XV was extremely accurate, much thought was given to the development of an automatic pilot. The Army Air Corps had done some experimentation with an automatic pilot and had some success with it. Mr. Norden, and for that matter, the Bureau of Ordnance were not satisfied with the existing Pilot Director. In the early years of the Mark XI many types of directors were developed, redesigned, re-engineered, and modified, but they were not the best. It had always been Mr. Norden's opinion that there was nothing wrong with the bombsight and that it was correctly engineered, but that the trouble was in the Pilot Director as a stabilized bombing platform, which was possible with this instrument only under ideal circumstances. The use of an automatic pilot was not something new, as they were being developed and in use by other manufacturers, such as the Sperry Gyroscope Company. The Navy was giving a lot of thought to developing a system of their own. Under this climate the Bureau of Ordnance awarded a contract to Carl L. Norden, Inc., in December 1933 for an experimental model for automatic flight control equipment to be used in conjunction with the Mark XV sight. Not much thought had been given to this revolutionary concept of replacing a human with an automatic pilot during a bomb approach. This equipment was designed by the Navy and called the "Stabilized Bombing Approach Equipment" (SBAE). The SBAE procured under contract #34103 from C.L. Norden, Inc., was received at the Naval Proving Ground on February 14, 1935. The installation of this equipment in a XT3D-2 airplane was completed on March 28, 1935, and flight and bombing tests were conducted as expeditiously as possible. As with all new systems, preliminary reports from the Naval Proving Ground exhibited the good and the bad points of this new instrument. The strong points were: the directional flight gyro was superior to the Sperry gyro (it was compared to a Sperry gyro-pilot installed in another plane); it produced sufficient stability about the lateral and longitudinal axis for bombing approaches; power was at least twice of what was needed; and the directional and flight gyro were markedly superior to the Sperry gyro pilot. Some of the deficien-

Chapter VIII: The AAF and the Navy Mark XV Bombsight

cies were: it was not quite as simple to operate as the Sperry and a proportionate banking feature had not been successfully demonstrated. Ruggedness and dependability had not been determined for this new equipment, but it was compared with many sensitive elements that were similar to the Mark XV sight, which had withstood over 1,000 hours of testing very satisfactorily. However, the adjustment of the rudder control to give automatic flight sufficient steadiness for bombing approach was not satisfactory. Acceptable bombing accuracy was done from 1300 and 1600 foot altitudes at 70 knots airspeed using the low altitude bombing device. No attempt was made to bomb at higher altitudes. The combined weight of this automatic flight control system was 126 lbs., and with the Mark XV-2 bombsight totaled 178 lbs. The required base for this equipment was 30 x 30 inches. The starting current was 12.5 amperes, and maximum current with all clutches slipping was 17.0 amperes. Although this was not an auspicious start, considering it was the prototype the results were somewhat encouraging. It would be several years before an operational unit was developed. As in the past, Mr. Norden and Mr. Barth were not unduly alarmed with this performance, as they had gone through this with the Mark XI and XV bombsight. As usual, they immediately started working to perfect this new system.[16]

On April 7, 1936, the Bureau of Ordnance was advised by the Norden Company that they had prepared engineering drawings, data, and charts for use in studying the installation of the SBAE. These drawings showed the approximate dimensions of the mounting surfaces required for the SBAE units. The dimensions were liberally estimated and were in excess of the actual mounting surfaces the units would actually occupy. It was suggested that the SBAE units as well as the flight gyro could be mounted on the chart when these units were combined under one housing or grouped close together. It was also pointed out that it might be advantageous to mount each servo unit in a different location, for instance, close to the control cables on either side of the fuselage. Any available space in the fuselage could be used for the location of the vertical flight gyro.[17] On April 18, 1936, the Norden Company forwarded drawings showing the exact dimensions for the major sub-assemblies. the drawings showed two possible installations. However, the units could be mounted independently or rearranged differently to suit conditions in the airplane. It also was pointed out that it would be desirable to have the units located or arranged so that the follow-up cables could be kept within a reasonable length, thereby avoiding excessive cable elasticity and friction loads, which the use of additional sheaves would introduce.[18] Testing and modifications would continue on the new system.

At about this time the Bureau of Ordnance once again wanted a low altitude bombsight. At the present time the only sight available was the Mark XI and Mark XV, both of which were of the high altitude type. In April 1929, the Bureau of Ordnance had formulated their requirements for bombsights. One of the items was the development of a low altitude bombsight. The Mark XV was their high altitude precision type. The services were still using the Mark III, which was a WWI wartime development by the British and subsequently underwent many modifications. In order to determine the practicality of using the Mark III versus the Mark XV, in April 1936 a Low Altitude Bombing Comparative Test of the Mark III and the Mark XV was undertaken by the U.S. Fleet, Aircraft Base Force, VP Squadron One-F, Fleet Air Base, Pearl Harbor, Hawaii. The operation of the Mark III was very simple in contrast to the Mark XV. Through extensive tests, it was found that the Mark III's only advantage was that it was more readily adapted for use by inexperienced personnel in time of war. Although the Mark XV was not designed for low altitudes, it could easily be adapted to low altitude bombing by the use of techniques developed by the above squadron. Most service bombs were more or less similar enough in their ballistic coefficients to permit using tables and curves already in use, and could be readily usable for all altitudes. Also, the Mark XV could be utilized with the stabilizer gyro locked in during the bumpiest of weather without sacrificing accuracy. With the Mark XV calibrated to permit use down to 1400 feet, the superiority of the sight precluded the necessity to carry both the Mark III and Mark XV in patrol planes in time of war. During these tests a total of 10 bombs were dropped at altitudes of 1000, 1500, and 2000 feet. The results of these drops conducted by VP Squadron 6-F showed the average radial errors in feet for the Mark III at different altitudes was 97.9 and for the Mark XV was 62.2. The average mean radial errors in mils was was 47.6 for the Mark III as opposed to 39.5 for the Mark XV. In all instances the Mark XV outperformed the Mark III. The tests were made by the same bomber experienced in using the Mark XV, but had no previous experience with the Mark III. However, the simplicity of the Mark III made its use comparatively easy. Plate #1 and #2 graphically show the pattern of bombing. These tests proved that there would be no advantage in using the Mark III and the Mark XV in the same plane, and it was abandoned.[19] In previous years no thought had been given to a contingency plan should war become a reality. In 1936, the resurgence of Germany under Adolf Hitler as a military power was looming in Europe and was of grave concern to the United States and its Allies. With this situation, the Bureau of Ordnance asked the Norden Company to formulate a "wartime" emergency condition and what their capabilities would be. In January 1936, the Norden Company forwarded a letter to the Bureau of Ordnance stating that within a year's time the production of the Mark XV would be about 75 sights beginning the first month and rising to about 400 sights per month by the 12th month. Of course, the gearing up for production would require the expansion of facilities, hiring new personnel, training, acquiring new production machines, sub-contractors, etc. These figures were based on the allocation of the International Projector Corporation's plant being allocated to Carl L. Norden, Inc., to assist in the manufacture of bombsight parts.[20]

Chapter VIII: The AAF and the Navy Mark XV Bombsight

The Air Corps was now beginning to acquire the Mark XV in quantities, but the shortage of the sight to the Air Corps was still critical. The Air Corps had the Sperry C-4 and the Estoppey D-4. The D-4, when new and in perfect condition, and providing all other conditions were also perfect, attained very good results at an altitude of 8,000 feet, but above 8,000 feet the errors became excessive. As for the C-4 sight, it was very unsatisfactory. In April 1935, the Commanding General, Hawaiian Department, contacted the Adjutant General, Washington, D.C., explaining his position regarding the defense of the Hawaiian Islands. After observing some very accurate bombing by the Navy at medium altitudes, arrangements were made locally with the Navy to borrow several Mark XV sights for the purpose of executing bombing missions at service ceilings of about 12,000 feet. Ten Air Corps officers were given short but intensive training. It readily became apparent that the Navy sights were vastly superior to what the Army was using. They found the Mark XV easy to operate, and bombing teams with relatively little practice secured excellent results. Due to the geographical location of the Hawaiian Islands and their isolation, it was of necessity that this department be equipped with the very best and latest bombsights and accessories, including spares, without delay. Reacting to this problem, the War Department, Air Corps Materiel Division, Wright Field wrote to the Chief of the Air Corps advising that 12 type M-1 Mark XV bombsights should be allocated immediately to the Hawaiian Department during the calendar year 1935.[21] Then sufficient bombsights to equip each bombing airplane would be shipped during calendar year 1936 as soon as deliveries from the contractor would permit. In addition to the Air Corps contracts, the Navy Department also had contracts for the same sight, which could effect deliveries, and which the Air Corps had to take into consideration.

The fact that the Air Corps was dependent on the Navy for the Mark XV was creating a problem. The Materiel Division, Wright Field, War Department forwarded a letter to the Chief of the Air Corps outlining the problems encountered in obtaining the Mark XV sight.

In 1932, the Chief of the Air Corps ordered the purchase of 25 sights, which were to be delivered as soon as possible. By January 1936, the Air Corps had purchased a total of 206 M-1 Mark XV Norden bombsights from the Navy and, by the end of 1936, only 100 had been delivered. The Bureau of Ordnance could not give the Air Corps any definite delivery date. At this time the Bureau of Ordnance was not able to provide sufficient bombsights for both services without expanding the Norden manufacturing plant. It was becoming a critical issue, as the Martin B-10 Bombardment airplanes were currently being received at a rate of three a week. The Air Corps was promised a delivery of 10 sights per month, and if such a delivery had been carried out, it would have been barely adequate to meet the Air Corps' requirement. Much of the problem laid in the secrecy surrounding the bombsight. The secrecy complications surrounding the sight were so involved that any tactical unit did not report of what was accomplished. Another pitfall was that any changes by the Air Corps could be accomplished only by Navy agreement. The handling of the transfer of the sights from the Bureau of Ordnance to the Materiel Command (Air Corps) was conducted in extreme secrecy. No notice of this action was forwarded to the proper supply department. When making an inventory, the bombsights were listed as "found on post." All distribution of the bombsight was personally handled by the Chief of the Air Corps. [22]

It was readily becoming apparent by January 1936 that a single source of supply was becoming highly unsatisfactory, since the Navy was trying to furnish the sight for themselves and the Air Corps. Further, it was the opinion of the HQS, GHQ Air Force that bombardment aircraft should be designed around the bombsight rather than have the bombsight installed in an airplane as a removable adjunct. Past practice was to have the sight installed after the bomber was built—almost like an afterthought. The construction of the sight lended itself to a built-in installation, capable of complete shock proof mountings, as well as constant temperature and moisture control regardless of altitude. The issue of the comfort for the bomber/bombardier was brought up again. It was the belief of the Chief of the AC that the bombardier should sit in a natural, comfortable position with no movement of his body other than his hands, so that he would be capable of precision equal to that of the sight. Otherwise, physical exhaustion would induce errors which no sight could eliminate. The same problem existed with the Mark XI. The nose of the airplane was modified to install the sight after the plane was built, and contractors were required to make this modification. The Commanding General of the Army Air Corps felt that the present practice of removing all of the bombsight system was wrong. It was felt that damage would result with this constant handling, especially in the sub-assemblies. It was directed that all future manufacturers be actively encouraged in this development and that all future procurement of bombardment airplanes, beyond present contracts, provide for the permanent installation of sights as part of the airplane's structure, with the nose designed for maximum comfort for the bombardier and unlimited approach vision while seated at the sight table. With the advent of the B-17 "Flying Fortress," the bombardier nose was completely re-engineered to provide the best possible access to the sight and sub-assemblies.[23]

Chapter VIII: The AAF and the Navy Mark XV Bombsight

```
GUNNERY, 19
SHEET No. 35, 36                    BOMBING DATA
(ONE COPY REQUIRED FOR THE DEPARTMENT)
                         INDIVIDUAL      LOW ALTITUDE
              PART 1  {~~SQUADRON~~}    {HORIZONTAL}  BOMBING PRACTICE No. 1
                      {~~GROUP~~}       {~~DIVE~~  }               
                         (Scratch two)    (Scratch one)            RUN No. 1-10
~~SQUADRON~~
~~DIVISION~~
PLANE No. 6-P-9                 Type of plane PM-1      Date 1 April, 1936.
       (Scratch two)
```

Flight Comdr. or Pilot **Sears, AMM1c (NAP)**
Bomber **Lt-(JG) G.C. BRIANT.**
Air speed (at release) **80**   Wind velocity { Surface **12** / Bombing altitude **14** }
Bombs dropped **10**
Number of live bomb duds (total) ....
Bombing altitude (at release) **1500**
Mark of bomb and fuze **MK VII 100 lb.**
Type of bomb sight **MK III**

SKETCH OF FORMATION OF PLANES FOR FORMATION PRACTICE
ELEVATION VIEW | PLAN VIEW
DIRECTION OF FLIGHT →

Hits		Average deflection error (perpendicular to (5)). (6) **62** ft.
Number of planes bombing		Radial error, $\sqrt{(5)^2+(6)^2}$ .... (7) .... ft.
Hits per plane (score) .... (1)		Control factor, $\frac{200-(7)}{200}$ .... (8) ....
Standard score .... (2)		Pattern { Range axis .... (9) .... ft.
Hit factor, (1) ÷ (2) .... (3)		{ Deflection axis .... (10) .... ft.
Merit (for hit factor only), (3) × 100 .. (4)		Pattern factor, $\frac{400}{(9)} \times \frac{100}{(10)}$ .... (11) ....
Average range error (along line of approach) (5) **79** ft.		(neither term shall exceed unity)

DIRECTION OF TARGET COURSE FOR TOWED TARGET—
DIRECTION OF APPROACH (OR DIVE) FOR BOMBING FIXED TARGET——→
(See instructions on reverse side)

Sight setting data for horizontal bombing: Scale ...., Trail ....
Cross trail ...., Dropping angle ...., Drift angle ....

Submitted .... Checked ....
              (Date)   (Date)
.................., U.S. Navy,   **W.M. McDADE** .... U.S. Navy,
               Chief Observer.   Commanding Officer

Plate #1 - Bombing Data, Mark III

GUNNERY, 1935-36		BOMBING DATA	
SHEET NO. 16			

PART 1 { INDIVIDUAL / ~~FORMATION~~ } { LOW ALTITUDE / HORIZONTAL / ~~DIVE~~ } BOMBING PRACTICE NO. **1**
(Scratch two)   (Scratch one)

RUN NO. **1-10**

PLANE No. **6-P-9**
(Scratch two)

Type of plane **PM-1**   Date **4 May, 1936.**

Flight Comdr. or Pilot **SEARS, AMM1c (NAP)**
Bomber **Lt. (jg) BRIANT, G.C.**
Air speed (at release) **80**   Wind velocity { Surface **15** / Bombing altitude **17** }
Bombs dropped **10**
Number of live bomb duds (total) ....
Bombing altitude (at release) **1500**
Mark of bomb and fuze **MK VII 100 lb.**
Type of bomb sight **MK XV**

SKETCH OF FORMATION OF PLANES FOR FORMATION PRACTICE

ELEVATION VIEW	PLAN VIEW
DIRECTION OF FLIGHT ——→	

Hits ....
Number of planes bombing ....
Hits per plane (score) .... (1)
Standard score .... (2)
Hit factor, (1) ÷ (2) .... (3)
Merit (for hit factor only), (3) × 100 .... (4)
Average range error (along line of approach) (5) **31.5** ft.

Average deflection error (perpendicular to (5)). (6) **17** ft.
Radial error, $\sqrt{(5)^2 + (6)^2}$ .... (7) .... ft.
Control factor, $\frac{200-(7)}{200}$ .... (8)
Pattern { Range axis .... (9) .... ft. / Deflection axis .... (10) ft. }
Pattern factor, $\frac{400}{(9)} \times \frac{100}{(10)}$ .... (11)
(neither term shall exceed unity)

DIRECTION OF TARGET COURSE FOR TOWED TARGET——→
DIRECTION OF APPROACH (OR DIVE) FOR BOMBING FIXED TARGET——→
(See instructions on reverse side)

Sight setting data for horizontal bombing: Scale ...., Trail ....
Cross trail ...., Dropping angle ...., Drift angle ....

Submitted .... (Date) ...., U.S. Navy, Chief Observer.

Checked .... (Date) ...., U.S. Navy, Commanding Officer

W.M. McDADE.

Plate #2 - Bombing Data, Mark XV

Chapter VIII: The AAF and the Navy Mark XV Bombsight 99

1   Letter from War Department, Air Corps, Material Division, U.S. Army, Procurement Division, Subject: Bombsights, dated March 6, 1933, MS, National Archives.
2   Memo for Deputy Chief of Staff from Gen. Foulois, Chief of Air Corps, dated November 23, 1934, National Archives, MS.
3   Letter, from Naval Proving Ground to Chief of the Bureau of Ordnance, subject: Bombsight #105 Acceptance, April 6, 1933, National Archives.
4   Memo for Deputy Chief of Staff from Gen. Foulois, Chief Air Corps, dated November 23, 1934, National Archives, MS.
5   Ibid, page 3.
6   Letter from Hartford Accident and Indemnity Company, New York, N.Y., Bond Number 1098695, No Subject, dated June 27, 1933, National Archives.
7   Letter, from Naval Proving Ground to Bureau of Ordnance, Washington, D.C., Subject: Letter of Hartford Accident and Indemnity, July 1, 1933.
8   Letter, from Acting Chief of the Air Corps to Chief of the Bureau of Ordnance, Subject: Maintenance of Secrecy of Bombsights, May 11, 1935, National Archives.
9   Letter from Comptroller General of the U.S. to the Secretary of the Navy, No Subject, May 10, 1934, National Archives.
10 3rd Endorsement from Bureau of Ordnance to Supplies and Accounts, Subject: Contract #34167 and #35336 with Carl L. Norden, Inc. dated May 23, 1934, National Archives.
11 Letter from I.J. Dolgenis, Public Account and Auditor to C.L. Norden, Inc., No Subject, dated November 3, 1934, Memo of telephone conversation with T.H. Barth of C.L. Norden, Inc. dated November 8, 1934. Letter from Chief of Bureau of Ordnance to Air Corps, U.S. Army, Subject: Bombsight Procurement, dated November 17, 1934.
12 Letter from C.L. Norden, Inc. to Bureau of Ordnance, No Subject, concerning Brush Failures, dated January 10, 1935, National Archives.
13 ibid, page 5.
14 Ibid, page 8.
15 Ibid, page 10.
16 The Navy's Mark XV Norden Bombsight, Its Development and Procurement, 1920 - 1945, page 107, National Archives.
17 Letter from Naval Proving Ground to Bureau of Ordnance, Subject: Contract No. 39353, dated May 28, 1935, National Archives.
18
    Letter from C.L. Norden, Inc. to Bureau of Ordnance, No Subject, RE: Contract No. 39353, dated April 7, 1936, National Archives.
19 Ibid, April 8, 1936.
20 U.S. Fleet Aircraft Base Force to Chief Bureau of Ordnance, Subject: Bombsights Mark III and Mark XV Comparative Tests, VP Squadron 1-F and 6-F, Fleet Air Base Pearl Harbor, Hawaii, dated 14, 18, 26, and 30 May and 24 June 1936, Same Subject, National Archives.
21 Letter from C.L. Norden, Inc., No Subject, dated January 22, 1936, National Archives.
22 Letter from Hq Hawaiian Department, Office of the Department Commander, Fort Shafter, T.H., Subject: The Adjutant General Washington, D.C., dated April 29, 1935, and 1st Endorsement from War Department to the Chief of the Air Corps, dated June 12, 1935, National Archives, MS.
23 Letter from General Arnold Air Corps, Subject: Bombsights, Permanent Installation, to the Chief of the Air Corps, Washington, D.C., dated January 18, 1936, National Archives, MS.

Plate #1 Bombing Data, Low Altitude, mark III, April 1, 1936.
Plate #2 Bombing Data, Low Altitude, mark XV, May 4, 1936.

IX

Production Increases and Problems Escalate

In the allocation of bombsights and SBAE/AFCE, the most pressing issue facing the Navy and Army Air Forces was the critical shortage of fire control equipment. In June 1940, the Bureau of Ordnance requested what the requirements were for the AAF concerning President Roosevelt's request to Congress for 50,000 airplanes, and the number of bombsights that would be needed. The Norden Company stated that production could be increased to 120 sights a month on a three month notice. However, the AAF felt that, due to conditions in Europe, they would require 316 sights in 1941, 3,645 in 1942, 5,000 in 1943, and 5,115 in 1944. In view of the inflated requirements, the Bureau of Ordnance expanded the Norden facilities to produce 400 sights per month as soon as possible to provide the AAF with the 1942 schedule. In order to do this, a new plant was needed. A new site was selected in Indianapolis, Indiana.

By the summer of 1941, the situation was growing desperate because the AAF requirements would be at least double of the amount planned in the summer of 1940. To meet these new requirements, the Bureau of Ordnance determined that at least $5 to $6 million in additional tools and equipment would be required by the bombsight manufacturers. By December 1941, the airplane production schedules were rapidly reaching the point where they surpassed the bombsight production rate. The allocation of this equipment was not working. By December 1942, the problem had increased in intensity to a very serious critical stage. The AAF could not properly conduct its mission in bombing enemy targets in the European and Pacific theaters of war. The Deputy Chief of Staff, AAF, was of the opinion that further correspondence with Admiral King would only invite the same remarks that were forwarded to General Marshal in November 1942. The next step was to present this situation to the Joint Chiefs of Staff. However, the Chief of the Air Staff felt

Chapter IX: Production Increases and Problems Escalate

that, since, on a number of occasions, the Joint Aircraft Committee (JAC) handled the allocation of bombsights and, in view of the successful compromise decisions reached by the JAC, it was of sufficient importance to be forwarded to them for adjudication.[1] Admiral King's idea concerning the AAF production of the Norden sight for its use and the production of precision bombsights, other than by the Norden Company, was inaccurately descriptive. The AAF did have facilities—the Victor Company and the Cardanic Corporation—however, these factories had not yet produced a single complete unit and were experiencing considerable difficulties, and delays were encountered before they would be in a position to produce satisfactory bombsights in quantity. The Sperry Gyroscope Company was producing precision bombsights that were coming off the production line at the rate of 200 per month. The Sperry sight had not yet proven satisfactory. The AAF strongly urged the Navy that additional allocations be made available, particularly during the first six months of 1942.[2] (See footnote No. 1)

Since no action was taken by the Navy by December 30, 1942, the AAF requested that this again be made a case for the JAC. The request was based on the fact that the present allocation was not in accordance with the needs of the respective services.[3] An agreement was reached with the Bureau of Ordnance to increase deliveries to the AAF. The revised allocation was in accordance with the agreement reached as a result of the withdrawal of the proposed action by the AAF to the JAC.[4] This would allow the AAF to plan their bombsight installation and determine in which airplane the sight would be installed. The new schedule was based on the War Production Board Report 8-L Schedule dated November 30, 1942 (See footnote No. 2). The Bureau of Ordnance provided a statement as of the 25th of the month which showed the estimated production for the succeeding month. This increase in sights would not begin to satisfy AAF requirements.[5] As the months rolled by, the increases in deliveries of the NBS did not materialize. The undelivered balance of bombsights as of July 1 from the Navy was 12,382, and the AAF 10,432. The estimated total sights to be produced in 1943 for the Navy was 6,902, and for the AAF it was 7,536. The 1944 production was estimated for the Navy at 5,305, and the AAF only 2,896, or 35%. It was expected that the Non-Norden facilities would fill this huge gap. Again, this allocation looked promising, but it would fall very short for the anticipated requirements of the AAF.

In keeping with every attempt to produce sufficient Mark XV sights, the Navy Bureau of Ordnance advised the Commanding General of the Army Air Forces that they were extremely cognizant of the AAF's additional requirements, but due to the uncertainty of financing, no commitment was made at the time of the November 1942 meeting. Orders at the Norden Company were estimated to be completed by February 1944. Due to the time cycle, approximately 10 months planning was involved in the production of sights (acquisition of material, parts, training of personnel, etc.), therefore it was necessary to plan months in advance to keep production lines going. Further orders would have to be placed no later than March 1943 to prevent interruption in the production of bombsights. In view of this, the AAF was requested to furnish accurate schedules, giving M-Series bombsight requirements on a monthly basis in order to help the Bureau of Ordnance plan the utmost use of their facilities. The AAF immediately honored this request. In addition, the Navy also requested the AAF to furnish schedules showing their M-Series requirements, prospective monthly production rate from AAF facilities, and monthly quantities of sights received from Navy controlled facilities. Also, it was requested that the number of any bombsights delivered to the British by the AAF or contemplated deliveries to them be included. Aircraft with bombsights scheduled for delivery to the British under Lend-Lease were to be included in these totals. The records of allocation of the total bombsight production from Navy facilities showed that the Army Air Forces had not received less than 65% of all production in the preceding six month period.[6] In February 1943 the Bureau of Ordnance forwarded to the Commanding General, Army Air Forces, the total number of bombsights under contract, manufacturers' estimated production schedules, and an estimated allocation of the total production between the services. The quantities shown included 20% for spares. The Bureau of Ordnance also informed the AAF that the listed quantities would vary depending on the actual Navy monthly aircraft production, bombsight production from Navy facilities, and the diversions of aircraft to the British.[7]

With the acquisition of several contractors and sub-contractors for the production of Norden sights and AFCE equipment, the AAF and the Navy issued a policy governing the classification of Navy, M-Series, and SBAE/AFCE. Due to difficulty in handling this equipment under secret status, the Navy lowered the classification to CONFIDENTIAL.[8] This action made it easier for service personnel to work with the equipment, thereby releasing very much needed personnel for other jobs. A detailed account of security for the bombsight is discussed in another chapter.

From the very beginning, the Norden sight was subjected to continual changes in order to improve its bombing accuracy. Since the inception of the Mark XV in 1933, much attention was devoted to the bubble problem. This referred to the two bubbles located on top of the bombsight gyro. It was immensely important, because precision bombing depended on the sight having a level platform for fore, aft, and athwartship. The bombsight gyro kept the sight as accurately level as the bombardier set it. First, it was necessary to set it accurately level, and this depended upon the leveling bubbles, and the bombardier's ability to interpret them. The Navy and the Norden Company were very much aware of the "bubble problem." Both the Proving Ground and Norden Company conducted many experiments in an effort to solve this problem. For a time it was felt that it was a mathematical item. If this were the case, it would require such close control of the levels as to the accuracy of the bore, diameter, and radius of curvature of the vial that the cost of pro-

Chapter IX: Production Increases and Problems Escalate 103

duction would be prohibitive, and the rate of displacement was greatly influenced by the type of fluid and size of the bubble. The bubbles shrink in size as the temperature inside the bombsight housing rises. With all the investigations conducted on this item, even with outside commercial specialists, research agencies, and universities, no answer was ever found for this erratic fluctuation. All of this research led to the development of the Automatic Erection System (AES), hoping that this would solve the problem. The use of AES would automatically erect the gyro without use of the leveling bubbles. It was felt that this "bubble chasing" could be corrected by the development of the AES. This system was desired because experience in both the AAF and Navy had amply demonstrated that only the most experienced bombardiers could avoid the temptation to "chase bubbles," and it was satisfactory to de-clutch the self-erecting features for a bomb run if the need arose. The AES consisted of a multitude of gears, terminal blocks, pendulum, bale, limit switches, solenoids, mercury switches, etc. It was hoped that this would solve the problem. The use of the AES would automatically erect the bombsight gyro without use of the leveling bubbles. Initial testing indicated the system would reduce errors for both experienced and beginning bombardiers. Investigation and experimentation continued until 1939 when it was installed in the Mark XV, Mod 4.[9] In July 1940, the Mark XV, Mod 4, equipped with the AES, was designated the Mark XV, Mod 5. This system was devised by Minneapolis-Honeywell, and it seemed to work in the laboratory. It was decided to give the AES tests under actual conditions at Midland, Texas, using a M-7 bombsight.[10] The initial tests seemed encouraging, but further tests were required. The AAF recommended that 30 AES be purchased from the contractor and be given a thorough test at the AAF Proving Ground, Eglin Field. In December 1942, the AES was eliminated from the Mark XV, Mod 5 sight, and the 24-26 volt sight produced without the AES was designated Mark XV, Mod 7 type, which was to be produced commencing in December 1942. The modification was not a major change, but nevertheless it caused considerable trouble. Maintenance was a factor, as servicing personnel were already being burdened with maintenance requirements with all parts of the sight and AFCE. Further tests revealed that it was not successful, and a decision was made to eliminate it due to tactical inefficiency and additional maintenance required. In January 1944, it was to be either removed or made inoperative in accordance with Technical Order 11-30-27, dated January 22, 1944.[11] Complete removal would be accomplished by the repair depots when sent in for a major overhaul. At the same time, the Minneapolis-Honeywell type SBAE/AFCE with electrical instead of mechanical follow-up system was designated, with the 12 volt type as Mark 2 and the 24 volt type as Mark 2, Mod 1. The Mark 2, Mod 1 was adopted as standard equipment by the AAF, and the Navy was using it in selected airplanes.

    The perennial problem on patents arose again. Patents, patent liability, royalties, and license to manufacture came up on an electronic attachment for the SBAE/

AFCE. In October 1942, an objection was raised by Minneapolis-Honeywell as to patent provisions in the purchase orders from the Norden Company. The electronic attachments to the servo motors was part of the automatic pilot that was invented by them. This was the first time that electronic equipment was installed in airplanes. It was a revolutionary approach to control the flight of an airplane. This would have tremendous impact on future use of this equipment in commercial application, with an expected explosion in air travel after the end of the war. The Norden Company agreed to waive this provision and they, in turn, requested modification of their prime contract through the Bureau of Ordnance. No separate license from the Minneapolis-Honeywell impliedly licensed the Norden Company, to the extent of any patent or other right they had, to install the first 3,000 attachments furnished to them.[12] This problem, of necessity, had to be resolved immediately, as the installation of these units were held in abeyance pending resolution. This was not the only war contractor that was faced with the same momentous problem. It would be necessary to modify the prime contract to relieve the Norden Company from patent infringement and liability on the first 3,000 units to be modified. In this manner, the patent liability clause was waived. It was not only for the Norden Company, at the present time, but also in the future, including other subcontractors who would be installing this equipment. By November 4, 1943, this problem was resolved by the Navy (Bureau of Ordnance), Army Air Forces, Wright Field, and the Minneapolis-Honeywell Company, in which they granted reproduction rights to the government on a royalty free basis for the duration of the war. This equipment was to be used solely for the benefit of the government and its Allies. There would be no commercial application.[13]

As was to be expected, many reports of bombsight failures were coming in from combat areas. Extensive investigation was conducted by the Naval Proving Ground with various Dash Pot fluids. This was an important part of the automatic pilot. The Directional Panel is mounted on the side of the Directional Stabilizer. Its primary function is to control the airplanes direction of flight. The Dash Pot is part of this panel and contains oil in the cylinder. This was not a new problem and had existed from the very first use of this equipment. Over the years, several types of fluids had been used and tested under severe conditions. Strained Prestone had been used, but it was not too satisfactory. In July 1942, the Bureau of Ordnance indicated that dibutylphathalate was superior to Prestone and other liquids used under combat conditions that had not proven successful. This new chemical did not corrode the Dash Pot metals as did the water based Prestone mixtures, and had better viscosity-temperature characteristics than oil (i.e., it did not thin excessively at high temperature nor thin out at low temperatures). It was tested extensively and found to operate satisfactorily between +70° and -40° C. Precautions were to be taken to keep the Dash Pot free of water, as water and dibutylphathalate did not mix and, at temperatures below freezing, ice would form and prevent Dash Pot opera-

Chapter IX: Production Increases and Problems Escalate

tion, possibly putting the airplane in jeopardy. The Bureau of Ordnance issued Circular Letter V-3-43 authorizing this new liquid for use in all Dash Pot SBAE/AFCE, and that it was available from Naval Supply. These instructions were binding on the AAF equipment, as well as the Navy.[14]

Another item that was of concern for the services was corrosion control, especially in hot and humid areas. The prevention of rust in any equipment must be controlled or it will destroy the use of the item. Many types of rust compounds were tried on the Mark XV bombsight and stabilizer, and other parts, as well. Previously, the Bureau of Ordnance had advised the manufacturers to spray bright metal parts to prevent corrosion with a light coating of Tectyl #511, which is a rust preventative or inhibitor, with a thin film (Polar Type), Grade III. Tectyl #511 was designed primarily for application to metal parts which were flooded with water, and it was only a temporary preventative. Continued testing at the Naval Proving Ground found that rust preventative compound, thin film (Polar Type) Grade II, Tectyl #502 proved more satisfactory than the other types that were in use. This type of rust preventative compound gave a more lasting protection and was easily removed with any common solvent or lubricating oil. Accordingly, this new compound was authorized for use on all equipment.[15] Another corrosion problem was caused by salt water immersion, especially for Navy units. Active corrosion starts immediately, on most parts, when exposed to salt water. It was extremely important that immediate steps be taken as soon as possible. The bombsights and SBAE/AFCE were to be immersed in fresh water in order to keep them from contact with the air. The instrument was disassembled and all traces of salt removed from all working surfaces, followed by dipping in boiling water. While still *hot*, the part, or parts, were dipped in carbon tetrachloride or alcohol and then coated with a preservative such as Vaseline (white petroleum) or similar compound. The Navy issued Circular Letter No. V-33, dated March 3, 1939, which was binding to all who had the Mark XV bombsight and SBAE/AFCE, and was in effect until December 1944 when the above preservatives were authorized for use.[16]

A more vexing problem on the Mark XV bombsight was developing a satisfactory lubricant or oils for the various bearings in the bombsight and stabilizer. This problem was more serious than the corrosion of metal, and it was an ongoing experimentation since the development of the Mark XI in the late 1920s. Even with the new Mark XV in 1933, this deficiency existed. Many different types of oil were used and subjected to severe testing by the Naval Proving Ground. What looked promising at the time turned out as a disappointment. Lubricants that were available were being used as no other oil was obtainable. The Norden Company, the Bureau of Ordnance, and the AAF were continually investigating and testing new oils. Due to faulty operation of bombsights during cold weather, even with the advent of heating blankets, they recommended a new "light" oil. This type was specifically manufactured by the Socony-Vacuum Oil Company of New York City, and was called RD-7-100. It was on the order of transformer oil, to which rust and oxidation and inhibitors had been added. This new oil was a considerable improvement over other previously used oils, which had no rust inhibitor included. The flash point of this oil was 305° to 320° F, indicating that the oil would evaporate quite easily, necessitating oiling the sight bearings at regular intervals, especially in warm weather. The pour point was -55° F, and the oil could be used effectively at temperatures below -40° F, preventing freezing up conditions previously encountered with the old oil. The new oil was to be used in all places, except the gyro bearings, which were oiled with the heavy type oil presently furnished. It was strongly stressed that oiling the bombsight be in accordance with service conditions and instructions, as far more damage could result in these instruments through excessive oiling. To make sure that the application of this new product for bombsights was correctly carried out, the Bureau of Ordnance issued Circular Letter No. V-32-42, which was binding on all activities, including the AAF. The Socony-Vacuum Oil was substituted for the product that was in use, and the old oil would be used until stocks ran out. All bombsight overhaul shops were supplied by the Supply Annex, Naval Ordnance, Indianapolis, Indiana. Field activities and squadrons were to replenish their requirements from the nearest overhaul shop.[17]

Note #1: So many issues were developing on a monthly, weekly, and even on a daily basis, that some form of organization was needed to solve them. The solution was the formation, in September 1940, of the Army-Navy-British Purchasing Commission. Later, it was reduced to the Joint Aircraft Committee (JAC). This organization included very high level officials from all facets of government. The AAF and Navy were also members. This agency was established high enough in the chain of command to speak with authority. Its decisions were binding, and there was no appeal. Their scope of operation was not limited to aircraft alone. They could and did oversee all phases of war material being produced. The issues were sent to them when they could not be resolved by other means.

Note #2: The following are excerpts from the law governing PATENTS. "The contractor shall hold and save the Government, its officers and agents harmless from patent liability of any nature and kind, including costs and expenses for or on account of any patented or unpatented invention made or used in the performance of a contract, including the use or disposal thereof by or on behalf of the government: Provided that the foregoing shall not apply to inventions covered by applications for the United States Letters Patent which, on the date of execution of this contract, are being maintained in secrecy under the provisions of Title 35, U.S. Code (1940 ed), section 42, as amended."[18]

Chapter IX: Production Increases and Problems Escalate

Notes:

1 Headquarters AAF Routing and Record Sheet, Subject: Allocation of Bombsights, dated December 22, 1942, National Archives, MS.
2 Memorandum for Admiral E.J. King, Subject: Allocation of bombsights, dated 30 Bombsights, dated December 30, 1942, National Archives, MS.
3 Memorandum for Recorder of Joint Aircraft Committee, Subject: Allocation of Bombsights, dated December 30, 1942, National Archives, MS.
4 Letter to Chief of Bureau of Ordnance from Hqs Army Air Forces, Subject: Bombsight Production and Allocation, dated January 27, 1943, National Archives, MS.
5 Letter from Bureau of Ordnance to Commanding General, AAF, Subject: Production and Allocation, dated January 29, 1943, National Archives, MS.
6 Memorandum from Admiral Blandy, Bureau of Ordnance to Commanding General AAF, Subject: Bombsight Procurement, dated February 6, 1943, National Archives, MS.
7 Letter Bureau of Ordnance to Commanding General, AAF, Subject: Bombsight Procurement, dated February 6, 1943, National Archives, MS.
8 Policy Governing M-Series and S-Series Bombsights and Automatic Flight Control Equipment (Stabilized Bombing Approach Equipment) dated July 1, 1942, National Archives.
9 The Navy's Mark 15 (Norden) Bombsight. Its Development and Procurement 1920 - 1945, page 174.
10 War Department, Air Corps, Memorandum Report, Test of Automatic Erection System Installed on M-7, bombsight at Midland, Texas, dated February 2, 1943, National Archives.
11 Technical Order No. 11-30-27, Bombsights-Discontinuance of Automatic Erection System - M-Series, dated January 22, 1944, Library of Congress.
12 Letter to Minneapolis-Honeywell Regulator Company to Navy Bureau of Ordnance, Subject: License to Manufacture Electronics for SBAE, dated October 29, 1942, National Archives.
13 Letter from Minneapolis-Honeywell Regulator Company to Bureau of Ordnance, Subject: Electronic Attachment for SBAE, License to Manufacture, dated November 4, 1942, National Archives.
14 Letter from Bureau of Ordnance, Circular Letter No. V-3-42, Subject: New SBAE Dash Pot Liquid, dated July 9, 1942, National Archives.
15 Naval Proving Ground to Chief of the Bureau of Ordnance, Subject: Proposed Application of Rust Preventative Compound to Bombsight and Stabilizer, dated December 9, 1944, National Archives.
16 Navy Department, Bureau of Ordnance Circular Letter No. V-33, Subject: Care Following Salt Water Immersion, dated March 3, 1939, National Archives.
17 Navy Bureau of Ordnance Circular Letter No. V-32-42, Subject: Bombsight Mark 15, Mod 5, Lubricant for, dated November 5, 1942, National Archives.
18 Memorandum, Chief of Bureau of Ordnance to Commanding General, AAF, Subject: Minneapolis-Honeywell Electronic Attachment for SBAE/AFCE, dated November 23, 1942, National Archives.

X

Pearl Harbor: The Navy's Mark XV Goes to War

That fateful day of December 7, 1941, forever changed the conduct of waging war. Air power was suddenly pushed to the forefront. The advocates of air power, such as Douhet, Mitchell, et al, would finally be vindicated, although they were not alive to see it happen. The Japanese air attack on Pearl Harbor showed what air power could do to battleships, cruisers, airfields, hangars, etc. Air power could no longer be ignored. The German Air Force vividly displayed the destruction of cities by aerial bombardment. This horrendous destruction was clearly shown during the Battles of France and Britain. Fortunately, the United States was on its way to increasing the Army Air forces as a bombardment power, but it would be quite sometime before an efficient strategic and tactical air force was developed to compete with enemies that had been in the war business for several years.

All of the plans previously made by the Navy and the Army Air forces (AAF) in regards to the production of Norden bombsights and SBAE suddenly became obsolete. Immediately, new plans for the sight and auto pilot were developed. The quantities agreed on in early 1941 were out of date within months, and sometimes in days. The production of sights would no longer be in the range of tens or hundreds, but in tens of thousands. These quantities were staggering with no manufacturing facilities were available to immediately convert to war work. The expansion of bombsight production had increased slowly during peace time, as there was no urgent need at the time. These plans were now woefully inadequate for a global war. In early January 1942 high level meetings were held between the Army Air Forces and the Navy to reach an agreement with the Air Corps to find and develop a source or sources for the production of bombsights and SBAE under license from C. L. Norden, Inc. Representatives of the Air Corps visited the Norden Company for the purpose of becoming familiar with production techniques required in the

Chapter X: Pearl Harbor: The Navy's Mark XV Goes to War

manufacture of this extremely precise equipment. After the selection of prospective manufacturer or manufacturers, representatives were to visit the Norden factory for familiarization and techniques required in their manufacture. The sources selected would be licensed by the Norden Company, and they would provide technical services, such as complete engineering data, drawings, parts lists, process sheets, tooling data, sample parts, and complete sight and engineering advice.[1] In addition, the manufacturing field was being canvassed as to who might be suitably equipped to produce this type of equipment. Among those that were solicited were the A-C Spark Plug Company, Ford Instrument Company, Remington Rand Company, etc., who indicated that they might be interested. A letter from the Secretary of the Navy outlined to the Chief of the Army Air forces the situation in regards to the production rates for the Navy Norden Mark XV bombsight, which indicated a serious shortage against their requirements for the remainder of the calendar year reaching a deficiency of about 1,200 units by July 1942. A letter from the Secretary of the Navy to the Secretary of War outlined the following: "In early 1941, the Bureau of Ordnance took extraordinary steps to increase bombsight production to an expected rate of 800 units per month with an expectation in arriving at this quantity by February 1942. It was realized that the new Ordnance plant at Indianapolis would not begin full production until September 1942".[2] Actual production was not until April 1, 1943. Training of personnel in producing and assembling these parts could not be done overnight. Locating and training skilled machinists and mechanics in performing this work was difficult and time consuming, making it necessary to train personnel on the job. This in turn reduced the manufacturers' capability to produce acceptable items. The author is very cognizant in obtaining qualified personnel. It was necessary to retrain these people to work to closer tolerances than ever before attempted on a production line basis—a lengthy learning process. There was such a critical shortage of bombsights that it was necessary to make some compromise, and emergency allocations between heavy, medium, light bombardment airplanes, bombardier schools and tactical units, and to foreign countries.[3] In January 1942 the Norden bombsight was released to all Allied nations to which AAF aircraft were being diverted to, except Russia. The authority for this release was the Assistant Secretary of War for Air and the Under Secretary of the Navy. Bombsights would be delivered to the respective services from the contractor plant. In order to proper evaluate any decisions to be made by the Chief of Staff, Under Secretary for War (Air) and the Chief of the Army Air Forces were to be fully advised by the completion of a comprehensive summary of bombsight procurement. The summary of requirements was based on the latest production and allocation figures.

HQ Wright Field, AAF established that production facilities for "M Series" sights required a total capacity of 2,000 sights complete with AFCE per month, that the existing facilities would be unable to meet these requirements, and that new and extensive factories be erected and production implemented. Deliveries from existing facilities amounted to about 350-400 units per month, or about 4,000 per year.[4] The Bureau of Ordnance made every effort to supply the AAF with 1,000 units per month by September 1943. In February 1942 Admiral Blandy agreed that the AAF should be authorized to procure sights from any source that they desired, and that they would deal with the Norden Company.

In 1942, to alleviate the shortage of sights it was decided that the Norden Company would revive one of their unused subsidiaries, known as the Cardanic Corporation. The AAF would place its prime contracts with Cardanic. The Navy selected the Burroughs Adding Machine Company as a source for bombsights, and the Air Force selected the Victor Adding Machine Company. The AAF also selected the Minneapolis-Honeywell Regulator Company for the automatic pilot as prime contractor for the AFCE.[5] How this selection was made is not shown in the examined files. In order to arrive at maximum production the Bureau of Ordnance and the Air Corps in February 1942 considered the simplification and standardization of the bombsight. Prior to the war the development and improvement of auxiliary equipment occupied an important place, however, the main objective was now to eliminate "non-essentials" in the interest of increased production. Both services were in full agreement on this. The automatic erection system was not working as was expected. Even with all these efforts it was quite clear that there would not be enough Norden bomb sights to go around. The Navy required additional sources, and after reviewing many activities and manufacturers, selected Remington Rand Company of Elmira, New York, as a sub-contractor to Norden Company. Top priority was given to the expansion of present factories, building new plants, and acquiring factories that could be converted to making bombsights. There was no change in plans for 1,000 sights a month from Navy facilities which were: C. L. Norden Company, Lukas Harold Corporation, and the Remington Rand Corporation as a sub-contractor. To alleviate the critical shortage of bearings, the Bureau of Ordnance consulted the S. K. F. Bearing Company to provide necessary steps for expansion of this company to furnish precision bearings for the AAF and for Navy facilities. Another serious bottleneck was optics. The Air Corps and the Navy made every attempt to find suitable suppliers for this highly specialized item. The Office of Naval Inspector of Ordnance (NIO) directed that the NIO at the Bausch and Lomb Optical Company provide the Eastman Kodak Company with all information pertaining to the production of Mark 15 bombsight optics necessary to fulfill Bureau of Ordnance contracts. Eastman Kodak was already producing optical material for the Navy.[6] This was the situation in early 1942 concerning the status of the Norden bombsight. Quantities that appeared sufficient would be out of date within months. Revision of requirements increased by leaps and bounds.

The situation regarding the manufacture of the automatic pilot was so confused that by April 1942 action was taken by the AAF and the Honeywell Company to formulate plans for one standard set of modified SBAE/AFCE for all bombardment airplanes in which the equipment was used. This was necessary because installation of this equipment resulted in the use of six different types of auto pilot's control boxes and four different types of junction boxes. In order to obtain maximum production it was mandatory that only one standard set of SBAE/AFCE be manufactured at one time. Specifications were drawn up, and only one type of pilot's control box, Type A, would be produced by Minneapolis-Honeywell. Installation Specification Number XC-24755 was developed to be used in all airplane installations for the standard modified AFCE. This directive applied to all airplanes being installed with this new equipment, as well as airplanes under procurement. This new equipment was as compact as possible, with the pilot having easy access to all controls, such as the master switches, turn control, tell-tale lights, and sensitivity and ratio knobs.[7] Minneapolis-Honeywell also furnished terminal blocks in the pilot's control box, whereby the airplane manufacturer would connect wiring direct to this unit with the use of an Army-Navy connector. Further, they conferred with the Victor Adding Machine Company at the Chicago plant to discuss production design for the M-Series stabilizers (used with the bombsight), which both companies were in the process of preparing for production. The adoption on one standard AFCE produced by Minneapolis-Honeywell would greatly facilitate their production.[8]

Since there was only a limited supply of M-Series sights it was necessary to replace the M-Series sights with other types. No Norden Mark XV or Sperry S-1 was to be given to the Russians, but they would receive the Estoppey D-8 sight. Aircraft allocated to the Chinese would receive the D-8 sight installed prior to transfer. The Estoppey D-8 bombsight was made by the International Business Machine Company (IBM) and was to be considered as a stop gap measure, which would be replaced as soon as other sights were available from production. Hqs Army Air Forces devised the following priority system for the scarce M-series sights as follows: First priority, for American operated heavy bombardment airplanes on a basis of one per plane; Second priority, for Air Corps Advance Flying Schools (Bombardier and Maintenance), in accordance with schedule of requirements showing minimum number required for these schools; third priority to heavy bombardment units in the Zone of Interior (U.S.) not later than the beginning of the last six weeks of their intensive training in preparation for combat duty to the extent of one sight per airplane; fourth priority the M-Series bombsights were provided to all other heavy bombardment airplanes designed for this type of equipment; and M-Series bombsights were allocated on the fifth priority to medium bombardment airplanes, B-26 airplanes of this type being the first medium bombardment airplanes to be considered. Airplanes assigned to the British were to have the SBAE/AFCE installed, but were not to have the "M" Series bombsight installed.[9]

Due to an increasing number of Unsatisfactory Reports (UR) being received concerning the operation of the Norden type AFCE, the AAF took action through the Directors of Bombardment, Military Requirements, Flying Training Command,

and other activities to resolve this problem. General Arnold directed the following actions be implemented:

1) Through Extensive testing it was found that the Minneapolis-Honeywell modified AFCE was very suitable for AAF airplanes, and that it be made standard equipment.

2) In equipping airplanes with the modified Minneapolis-Honeywell AFCE and allocating the units, the following priority system governed their allocation:

a. First priority—production of heavy and medium bombardment type airplanes;

b. Second priority—production of bombing training type airplanes;

c. Third priority—50% of the bombing training airplanes at the bombing training schools;

d. Fourth priority—two pilot airplanes (one principal and one co-pilot) with range requiring a time period of six hours or more;

e. Fifth priority—One pilot airplane (no co-pilot) operating time of four hours or more.[10]

The attendees at this conference were unanimous that the modified AFCE was an extremely excellent automatic pilot. This opinion was based on a considerable amount of experience obtained with equipment in the AT-11 airplanes at the Bombardier Training Schools. However, it was felt that the unmodified AFCE was better than no auto pilot at all, since there was little hope that the Sperry A-5 auto pilot would be available for many months, if at all. To further reduce confusion, the Directors also issued a directive that the electronically modified AFCE be given a type number and name of its own to keep service personnel from associating it with the unmodified AFCE.[11]

The production of Norden sights and AFCE was of grave concern to the AAF. The U.S. Army requirements were outstripping total production, due to the increased airplane deliveries. Considerable confusion existed between the various contractors and sub-contractors for bombsights, AFCE, and stabilizers manufactured by Robbins and Myers, and bombsight contractors such as Minneapolis-Honeywell and Norden Company. Unable to resolve the problem, and due to the growing importance of this equipment, in May 1942 it was forwarded to the Joint Army-Navy Munitions Board for resolution. The air arm of the U.S. Army was not prepared when war came. The plans for mobilization were faulty and inadequate.

Note: The Secretaries of War and Navy by administrative action created the Joint Army and Navy Munitions Board (ANMB), which became the authoritative source for Joint mobilization plans. The general procedures were worked out by the ANMB and the services for handling priorities to deal with individual cases as they arose. To plead for priorities before the ANMB, the Chief of the Air Corp established a Priorities and Allocation Section in the Office of the Chief of the Air Corp, to serve as a special mediator and advocate for all air arms requests. It was on the order of a civilian super agency for the purpose of mobilization planning. One July

Chapter X: Pearl Harbor: The Navy's Mark XV Goes to War

1, 1939, President Roosevelt issued a military order in his capacity as Commander in Chief and placed the Joint ANMB and other Boards be placed directly under the leadership of the Executive Office.[12]

The expansion of the Robbins and Myers Company was to take care of both SBAE (Navy) and AFCE (Army requirements which were predicated on the production of 1,500 units per month) by Fall of 1942. At the present time they were producing 400 stabilizer units per month and felt that a capability could be developed of over 2,000 units per month by November 1942. The Navy took the total production of 400 units per month, giving the AAF 200 of these. The Navy requirements were 600 units per month when full production was achieved. The AAF could count on 1,400 units a month when this production was underway. In view of the expansion of Robbins and Myers and the Minneapolis-Honeywell project, the whole situation was stopped by the Army and Navy Munitions Board until it could be thoroughly reviewed.[13]

Again, in May 1942 the AAF bombsight requirements were revised upward, consisting of 10 contracts for 14,394 bombsight units on order from the Norden Company and the Lukas-Harold Corporation. Actual deliveries from September 1941 to April 1942 was 1,572. Requirements for AFCE/SBAE consisted of 10 contracts for 10,515 units. Actual deliveries were 1,830 sets for the same period. This serious shortage was primarily due to the acceleration of airplane deliveries from the manufacturer. Although facilities were expanded, a considerable time was required to realize a pronounced increase in production rates due to the degree of precision required. Airplanes without any fire control equipment would be of no value in a "shooting war." It was mandatory that airplanes be fully equipped before leaving for combat.[14] Within six months into WWII many critical shortages began to appear in the production of bombsights and automatic pilots. By June 1942 an acute shortage developed for highly precision anti-friction ball bearings. The S. K. F. Industries bearing factory was able to supply only a limited quantity of good bearings required to assemble bombsights. The Norden plant was assembling more bombsights per month than S. K. F. could supply bearings for, and the AAF could not obtain any urgently needed bearings on order with the Navy Department. With no bearing supply there would be no bombsight production. It was that simple. The precision bearing situation had not been solved. In April 1943 the AAF and the Bureau of Ordnance agreed to form a *Steering Bearing Control Committee*. Personnel selected were high level officers and would be located at the Barden Company, Danbury, Connecticut. This committee advised the Government Presiding Officer of the Bearing Industry Committee as to allocation of orders and distribution of products from bombsight bearing manufacturers.[15] This super bearing committee covered all aspects of bearing production, such as total quantity of bombsight bearings by company, inventory, position, and disposition. Procedures for distribution were formulated and forwarded to the manufacturers, including criteria for inspection, spares, and allocation. Immediate action was taken to contact all bearing manufacturers—Norman-Hoffman, S. K. F., New Departure, Federal, Fafnir, etc., as well as sub-contractors and other manufacturers to determine if any of these companies could take immediate action to manufacture additional precision bearings. New sources of supply were urgently needed in order to surmount this critical bottleneck in the manufacture of the Mark XV, Mod 7, and later the M-9 series. Much planning would be required to surmount this program. In keeping with the urgent requirement for precision ball bearings, the Navy and the Norden Company were scouring various locations for a new plant. The supply of precision high-speed bearing supply was limited to about 10% (one in ten) that was suitable for critical application. Because the bombsight needed these extremely high-precisioned bearings, a new industry was born in the United States. After extensive investigation and research, the Navy (Bureau of Ordnance/Aeronautics) and the Norden Company selected a site in Danbury, Connecticut, for two reasons: First, it was available, and second, it was close to New York City where the Norden plant was located. The Tweedy Silk Mills was selected for the new plant. It was to be an affiliate of C. L. Norden, Inc. Approximately 55,000 square feet of space was available for renovation. The estimated cost for refurbishing the plant was about $200,000. Plant protection, fencing, reconstruction work, painting, etc., was about $42,000. The acquisition cost of special machine tools was over $207,000. This was the only company that had an accurate cost of conversion available. The company is now a recognized leader in producing highly specialized precision bearings on an international basis. The company was recently acquired by a foreign concern. The name was made up of the first three letters of Theodore H. Barth and the last three letters of Carl L. Norden, hence, BAR+DEN= BARDEN. The task of complete reconversion of a building, obtaining machinery, tools, engineering fixtures, and training personnel and inspectors almost appeared to be an insurmountable task. These problems were accomplished in nine months. The renovation of the building started in mid August 1942, and by December 11, 1942, the first precision bearings were delivered. They were the three platform bearings the sight rotated and pivoted on. This was a monumental task, as no precision equipment for bearing production of this type existed at the time. A total of 61 precision bearings were used in the sight. The original plans called for 50,000 complete sets of good bearings a month with inspection of perhaps three times that quantity, and in order to achieve this figure for critical bearings would require production of 90,000 to 100,000 bearings per month.[16] The company went on to win a Navy "E."

    The Navy, in a moment of weakness, authorized the Army Air Forces to obtain other sources for manufacture of the Navy Norden bombsight and SBAE/AFCE. The AAF selected the Victor Adding Machine Company of Chicago, Illinois. Victor started to convert to war production immediately, and it was expected that deliveries of the sight would start in November 1942 to provide for a peak production of

100 units per month. Minneapolis-Honeywell Regulator Company was the facility selected for SBAE/AFCE equipment located at the Chicago, Illinois, plant, with an expected production and delivery of 1,000 sets per month. Every effort was being made to arrive at the above commitments. By the end of June 1942, the situation had deteriorated to a point that the Director of Military Requirements, Hqs AAF, initiated a study encompassing all types of bombsights. This study revealed that the period of June to December 1942 inclusive, installation of Mark XV sights in heavy bombers was 1,469, training program 455, and spares 147, for a total of 2,071. M-Series type production for the same period totaled 1,924, or a deficit of 147. The Sperry S-1 bombsight and A-5 pilot requirements for this same period for installation in heavy bombers was 227, training program 252, test models 20, and spares 23, for a total of 522 units. Production of Sperry equipment for the same period was 766, or a surplus of 244. The Minneapolis-Honeywell AFCE production met the needs for heavy bombers and training during this period, and a limited number of AFCE units were available for installation in medium bombers. The company was encouraged to expand its production capabilities at the same ratio as the Norden sight to provide units for medium bombers. The D-8 Estoppey sight was available for installation for the above airplanes. This sight was not a precision instrument and was at best capable of a fair degree of accuracy at low altitudes, however, it was not gyroscopically controlled. Unfortunately, M-Series bombsights, Minneapolis-Honeywell AFCE/SBAE, Sperry S-1 sights, and A-5 pilot production facilities, present and future, were not expected to meet the requirements of heavy and medium bombers and training programs prior to June or July 1943. Extreme pressure was directed on the manufacturers to accelerate the production of the above instruments from existing facilities and those under construction.[17]

With several contractors producing bombsights, the question of interchangeability arose. Each new contractor was furnished with a complete set of engineering drawings of the bombsight, including a physical article taken from the production line. Also included was a maintenance requirements and a spare parts list. The Victor Adding Machine Company adhered to every detail in the specifications (blue prints). The author remembers quite well when a contingent of personnel led by the son of the owner (A. C. Buehler) were assigned to the AAF Base where he was working to obtain all types of information on machining, teardown, disassembly, assembly, and final testing. He was offered a top position, but it was politely declined, as he expected to be drafted in the military.

In searching for additional factories to produce bombsights, the Commanding General AAF was informed that the Cardanic Corporation, in addition to previously named AAF manufacturers, was to be used as an AAF facility for the manufacture of Norden bombsights and SBAE. However, it was necessary for the AAF to go through the Cardanic Company to contact the Norden Company. This was time consuming and unwieldy, causing untold delays. Production on this contract was to start in February 1943 and reach a rate of 1,000 units per month. The Bureau of Ordnance in July 1942 initiated a contract for the procurement of 9,000 sets of AFCE type C-1 equipped with Minneapolis-Honeywell (MH) electronics, with production to start in September 1942. The AAF was to furnish engineering drawings, tooling data, and information relative to the change in the design from the standard SBAE to the MH type of AFCE for this modification to the Navy at the earliest possible date for transmittal to the Norden Company.[18] The AAF made every effort to provide the Navy with a complete set of drawings covering the manufacture of the MH electronic equipment in order to fabricate these items and turn over to the AAF complete SBAE agreeing with the AAF configuration.[19] A considerable number of meetings and correspondence was generated in determining what and who would be responsible for these modified AFCE units. The problem was that the SBAE produced by the Navy (C. L. Norden Company) needed to be modified with the MH electronics. The Navy did not use the modified sets. In July 1942, the manufacturing control of the MH conversion of the Norden AFCE passed from the AAF and returned to the Navy. The Navy lacked sufficient facilities to support AAF requirements and were woefully inadequate in this area. The MH Company did a remarkable job of engineering and production in producing a successful version of an entirely unsatisfactory system within a total period of *six months*. This action by the Navy was difficult to understand in supplying the critically needed fire control equipment during the very desperate conditions that existed after Pearl Harbor. To avoid further confusion in the types of SBAE/AFCE, the designation for 13 volt AFCE would be Type B-1, and 26 volt would be Type C-1.[20] Reports were being received from the South Pacific theater of operation indicating that the AFCE (unmodified) was not being used due to losses during combat of men experienced in maintaining this equipment, lack of competent personnel to perform this function, and frequent malfunctions rendering this equipment inoperative. It was known that the Navy SBAE/Army AFCE had inherent unsatisfactory characteristics, but it was the only one available at the time. The advent of electronic modification had proven extremely satisfactory to date, both in flight and lack of maintenance required—especially during flight—and made the AFCE a very excellent automatic pilot.[21] In the beginning, to make the SBAE operate properly the AAF selected the MH to do the modification. Later, the AAF requested and received skeletonized units.[22] Navy SBAE with MH installing the electronic equipment. By October 1942 the MH Company was able to modify approximately 1,500 sets of stripped AFCE per month, and this could be increased to 3,000 per month if stripped units were available. In addition, there were 11,000 sets of AFCE to be delivered to the MH Company factory from the Bureau of Ordnance upon current contracts being furnished in stripped condition. However, delivery of stripped AFCE was totally dependent on the availability of the Bureau of Ordnance to provide these

Chapter X: Pearl Harbor: The Navy's Mark XV Goes to War 117

sets.[23] Commanding General AAF directed the Materiel Command, Wright Field to purchase an additional 9,000 sets of AFCE, electronically modified from the Bureau of Ordnance. This new contract was to take precedent over the 11,000 already on order. By late October 1942 it was found that with deliveries of 1,000 AFCE sets per month starting in November 1942, there would be no substantial increase from previous deliveries of these sets and that these quantities could not be met. To further compound the problem the AAF facilities—Victor Company, Cardanic Corporation, and MH Company—had not been placed under a licensing agreement with the Norden Company.[24] Because of the severe shortage, the allocation of bombsights and SBAE/AFCE was placed in the Office of the Commanding General H. H. Arnold, Hq AAF. Letters had been written to the Secretary of the Navy for an increase in sights and AFCE. It appeared that the Navy requirements would be taken care of before the AAF was allocated any equipment. In November 1942 the problem became so acute that General Marshall, Chief of Staff, wrote to Admiral King—Commander-In-Chief, U.S. Fleet, Navy Department—stating: "Dear King, see if you can do anything for us—," requesting his help to bring this subject before the U.S. Joint Chiefs of Staff.[25] No answer was found for this request. Also included was a letter from the Secretary of War to the Secretary of the Navy. The letter from Admiral King pointed out that the AAF had other sources for bombsights, that the Navy used only the Norden sight, and that all sources needed to be included in the grand total. He also felt that the policy of the Secretary of the Navy regarding allocation of Norden bombsights should be continued in effect. In other words, the Navy was to receive about 20% of the production and the AAF about 80%. The availability of Norden sights and AFCE was so severe that M-series sights and C-1 automatic pilot would be installed only in all B-17 type airplanes, and the S-1 Sperry sight and A-5 auto pilot would be installed in all B-24E airplanes, pending if the equipment proved satisfactory.

Since the shortage was so acute, Hq AAF directed that 25% of the AT-11 airplane trainers would be engineered for the installation of the Sperry sight. There would be no automatic pilot in these planes. The remaining 75% would be engineered to receive the M-Series bombsight and automatic pilot as soon as available from production. All B-25 and B-26 airplanes were designed for installation of the M-Series sight and would be allocated as rapidly as they became available, that is, after requirements for heavy bombardment, training airplanes, and training programs were met. The AAF directed the Air Service Command, Wright Field, to provide a sufficient number of S-1 bombsights for all B-18 airplanes assigned to the bombardier training schools to receive this equipment. No M-Series sights were to be allocated or given to any light or medium bombardment airplanes for any reason whatsoever until all requirements for the heavy bombardment and training programs were met.[26]

Notes:

[1] Memorandum, Minutes of Conference, Procurement Division, Bureau of Ordnance USN, dated January 8, 1942, National Archives.

[2] Letter from Secretary of the Navy to Secretary of War Concerning Bombsight Allocation, dated January 26, 1942, National Archives, MS

[3] Letter, Subject Bombsights, Hqs AAF, War Department to Chief of the Air Corps dated January 26, 1942 National Archives, MS

[4] Ibid, 1st Indorsement, dated February 3, 1942.

[5] The Navy's Mark 15 (Norden Bombsight) Development and Procurement 1920-1945, Page 273, National Archives.

[6] Office of Naval Inspector of Ordnance at C. L. Norden Plant to Naval Inspector of Ordnance at Baush and Lomb, Subject: Production of Bombsight Optics and Illuminated Mirrors by Eastman Kodak Company, dated January 21, 1943, National Archives.

[7] Memorandum, War Department, Air Corps, Materiel Division, Conference with Minneapolis-Honeywell Regulator Company personnel on Modified AFCE April 17, 1943, National Archives, MS.

[8] Ibid, page 3

[9] Technical Instructions Hq AAF, Materiel Command, Bombsight Allocation, April 27, 1942, National Archives, MS.

[10] Letter, War Department Hg AAF, Washington, D. C., Requirements for AFCE Equipment, dated May 6, 1942, National Archives, MS.

[11] Letter, War Department Hq AAF, Materiel Division, Memorandum, Conference on AFCE, dated April, 30, 1942, National Archives, MS.

[12] U. S. Army In World War II, Buying Aircraft: Materiel, Procurement for the Army Air Forces, Center of Military History, pages 151 and 259.

[13] Memorandum, Army and Navy Munitions Board, Subject: Robbins and Myers, Inc. Springfield, Ohio and Minneapolis Regulator Company, Minneapolis, Minnesota, dated May 14, 1942, National Archives, MS.

[14] Letter, Bureau of Ordnance to ANMB, Subject: Acceleration of Production of Ordnance Items Procured by the Navy for the AAF, Smoke Tanks and Bombsight Deliveries, dated May 20, 1942, National Archives, MS.

[15] Letter from Wright Field, to Bureau of Ordnance, Subject: Bearings, Norden Bombsight, dated March 1943, National Archives.

Chapter X: Pearl Harbor: The Navy's Mark XV Goes to War

[16] Letter from Barden Corp. Additional Facilities for Production of Ball Bearings for Bombsights and Related Equipment July 7, 1942, National Archives.
[17] Letter, War Department, Hq AAF, Study of Bombsights Requirements, dated June 29, 1942, National Archives, MS.
[18] Letter, Navy Department, Bureau of Ordnance, Memorandum, Minneapolis-Honeywell, Electronics for AFCE Request for Drawings and Pertinent Data, July 8, 1942, National Archives, MS.
[19] Inter-Office Memorandum, Hq AAF, The Materiel Command to Materiel Center, Wright Field, Dayton, Ohio, dated July 14, 1942, National Archives, MS.
[20] Inter-Office Memorandum, Hq Materiel Command, Washington, Change in Type Designation of SBAE, dated August 5, 1942, National Archives, MS.
[21] Inter-Office memorandum, War Department, Air Corps, Norden AFCE in Operational Airplanes, dated August 20, 1942, National Archives, MS.
[22] Chassis manufactured by Norden Company but without the new electronic device. Stripped meant the same thing.
[23] Letter, Production Engineering, Wright field, to Commanding General AAF, Materiel Center Washington, D. C., dated October 12, 1942 and 1st Indorsement October 21, 1942, National Archives, MS.
[24] Ibid
[25] Memorandum for Admiral King from General George Marshall, Chief of Staff, dated November 20, 1942. Letter from Admiral King, U. S. Fleet Hqs, Commander in Chief, Navy Department, Bureau of Ordnance, Table of Mark XV Bombsights and allocations, November 29, 1942, National Archives, MS.
[26] AAF Technical Instructions, Hq Materiel Command, M-Series and S-1 Bombsights Conservation and Allocation, December 4, 1942, National Archives, MS.

XI

AAF Expansion and the Norden Mark XV Bombsight

In 1936, the Air Corps was beginning to receive the new Martin B-10B twin engine bombardment aircraft at the rate of three per week. They wanted the new Mark XV bombsight for each plane, but since the Navy was not able to produce sufficient instruments, the quantity they did receive was barely adequate to meet the requirements for six bombsights per squadron, and was not only causing tactical, but also training problems.[1] The Army Air Corps wanted their tactical units to have this new bombsight so that their bombardment units could obtain bombing experience in order to perform their wartime mission or for any emergency should it arise. Since the AAF could not deal directly with the Norden Company, it was necessary to go through the Bureau of Ordnance, which was very time consuming and not satisfactory. In addition, the orders from the Chief of the Air Corps regarding secrecy, compelled the procurement of these bombsights to be made through the Navy Department. The sights were purchased by the Navy Department and shipped to the Naval Proving Ground, Dahlgren, Virginia, for testing. On completing the air tests, an airplane would secretly be dispatched to that station from the AAF at Wright Field and then locked in a secret vault. These rigid instructions were modified slightly during the summer of 1935 so that shipments could be made more easily to the stations for which allocations had been set up. The Procurement Section was still forced to operate on a Confidential basis with the Navy Department. All inspections and acceptances of the new and contractor-repaired sights were accomplished by the Bureau of Ordnance, with delivery schedules subject to Navy jurisdiction. The result was that the Bureau of Ordnance took Air Corps money and delivered a sight which would or would not be satisfactory to the AC, and delivery was accomplished at the convenience of the Navy. The Air Corps could not distribute written reports or Technical Orders as all information was handled through the Navy on a "Confidential" basis, which was comparable to the Army Air Forces' "Secret" instructions. Even unsatisfactory reports could not be handled through normal channels. This prohibited the Air Corps from learning whether or not the sight was suitable for their needs. The purchase of replacement parts could not be made without prior approval of the Bureau of Ordnance.[2]

The use of bombardment airplanes was of primary concern for the Air Corps, as the General Headquarters (GHQ) Air Force was stressing bombing training at altitudes above 18,000 feet during 1936. The Air Corps wanted the authority to permit them to deal directly with the manufacturer on the M-1 bombsight instead of going through the Bureau of Ordnance. To remedy this situation, a liaison officer was assigned to work with the Bureau of Ordnance. Although there was an agreement between the Air Corps and the Bureau of Ordnance on the procurement of bombsights, it could only be accomplished through the Navy. The M-1 sight was an excellent instrument. However, there were several objectionable technical features that were determined through extensive testing by the 1st Wing of the GHQ Air Force. However, even with these deficiencies, the M-1 sight was vastly superior to any other sight then available to the services.[3] Regardless of the problems in obtaining bombsights, the Army Air Corps still wanted them very badly and were willing to wait because of its proven accuracy. The Air Corps also had the Sperry bombsight, which was not proving satisfactory. The Army Air Forces had been experimenting with an automatic pilot, but had not yet developed a workable model. The Materiel Division, Wright Patterson AFB, was cognizant that the Bureau of Ordnance had designed an automatic pilot and was extremely interested in obtaining a complete unit. In September 1936, the Bureau of Ordnance awarded the C.L. Norden Company a contract for 72 sets of the Stabilized Bombing Approach Equipment (SBAE) to be completed by May 1937.[4] A decision was reached by the Chief of the Air Corps to assign a B-18 airplane for the purpose of installing and testing of the Navy SBAE. Information regarding the control of the airplane forces and flying characteristics was furnished to the Bureau of Ordnance and the manufacturer.[5]

In 1937, it became evident that the requirements for bombardment aircraft for both the Navy and the Air Corps needed to be addressed. Accordingly, a joint conference was convened for the purpose of consolidating the performance requirements of the Air Corps with those of the Navy for bombsights to be used in bombardment airplanes for the next several years. With the advent of new and higher speeds of airplanes, it was agreed that the altitudes would be from 1,500 to 30,000 feet; air speeds would increase to 400 miles per hour above 5,000 to 15,000 feet, and 250 miles per hour from 1,500 to 5,000 feet. A new range computing and optical system was to be installed in the present bombsights, extending the operating limits by increasing the forward vision to 80 degrees, the release angle to 64 degrees, and cross trail allowance with single setting for trail and cross trail. Another item that might apply was a stationary eye piece, providing greater magnifying power of the optics, trail angles up to 150 mils, and the possible inclusion of a glide approach in a single path angle.[6]

The Air Corps initiated a contract in 1936 to the Navy for 82 Norden Mark XV sights, which also contained an option to purchase SBAE equipment at a cost of $2,400 per set. This option was not exercised because there had been no opportunity for testing the suitability of this equipment in Air Corps airplanes. Since the expiration of this option, the cost rose to $3,000, with indications that further increases were in the offing. By April of 1937, the Air Corps had 156 Mark XV sights on hand and 130 on order. This did not include the Sperry or the Estoppey sights. Also, in 1937, an expected 10 YB-17 test airplanes were to be received. Although the Air Corps had some Sperry and Estoppey sights, they were not considered satisfactory. It was expected that procurement of bombardment airplanes in fiscal year 1937 on hand and on order would be 279 modern airplanes. This did not include the 63 Keystone airplanes on hand leaving an excess of six Norden M-1 bombsights for all airplanes, plus the 20 Sperry sights for a total of 26. This brought the grand total to 342 airplanes on hand or on order. The Keystones were procured for the Army Air Corps from 1930 to 1936. The latest model was the B-6A. Top speed was 121 miles per hour, gross weight of 13,000 lbs., and a crew of five.[7] It was a twin engine airplane and was considered obsolete. The Air Corps was pushing very hard to obtain an SBAE set so that they could start their own testing. In addition, during this period $805,000 was allotted for bombsights, of which $330,000 was for Sperry, and the rest for Norden newer M-Series bombsights.[8] The Bureau of Ordnance expected that delivery of the SBAE set would be by August 15, 1937. The results obtained by the Bureau of Ordnance in reducing bombing errors during extensive testing observed by the AC justified standardization for both services. The use of SBAE was already justified by the Army Air Corps for use at the earliest opportunity. Their plan was to obtain sufficient quantities as a service test by equipping all B-17s and one squadron of B-18 airplanes to be procured as soon as possible.[9]

Again, in 1936 the Chief of the GHQ Air Force brought up that modern aircraft should be *designed around the bombsight*, rather than having the bombsight installed in the airplane as a removable adjunct. The bombsight had been designed for a built-in installation, with complete shock proof mounting, constant temperature, and moisture control, regardless of altitude. It was also stressed that the new aircraft be designed to give the bombardier a comfortable position with full forward vision above and to the sides of the sight. Although the bombardier was supplied with oxygen, it was an absolutely unnecessary operating handicap to continue the cramped position now given him, with no movement of his body other than his hands. The bombardier, in a comfortable and natural position, would be capable of precision equal to that of the sight, otherwise physical exhaustion would induce errors which no sight could eliminate. It would be several years before this advice would be taken seriously.[10]

Chapter XI: AAF Expansion and the Norden Mark XV Bombsight

The Air Corps immediately made available sufficient funds to provide for the purchase of 102 SBAE sets as follows: 14 sets for B-17 airplanes; 18 sets for the B-18 airplanes; 26 sets for installation in the B-17B airplanes; and 44 sets to be procured as Government furnished equipment and installed in B-18 airplanes. The Navy had already adopted this equipment as standard in the PBY flying boats, it had been in service many months before delivery, and any defects that existed would be corrected by the Bureau of Ordnance before the Air Corps equipment was delivered.[11] Mr. Barth, of the C.L. Norden Company, advised the Bureau of Ordnance, in October 1937, that the new optical system for the Navy Department was completed. This new system would give extreme forward vision of 87°, extreme forward position of athwartship wire of 76°, a field of view of 22°, a magnification of 3 plus, maximum dropping angle of 63°, and maximum trail setting of 150 mils. The new optical system could be installed on the present Mark XV sights by personnel in the field. For purposes of simplification, the present rate end was designated as the Mod 3, and the two modified rate ends were identified as Mod 4 and Mod 5. The Mod 5 rate end had provisions for a glide bombing feature, and was a redesign of the present sight. The estimated cost of the Mark XV, Mod 5 sight with low altitude attachments, integral tachometer, and provisions for SBAE was $5,240.00. Costs for the Mark XV, Mod 4 sight was an additional $390.00 for new optics and cradle, and $380.00 for Mod 4 rate end, for a total of $6,010.00.[12]

In November 1937, the Navy announced that a squadron of PBY-2 airplanes were to fly from San Diego, California, to Coco Solo, Panama, non-stop using three sets of SBAE. Another group was scheduled to fly non-stop from San Diego, California, to Pearl Harbor, Hawaii, also with three sets, about January 15, 1938. This would provide an opportunity to demonstrate the automatic flight control features under a long and arduous flight. The Norden Company had subjected the auto pilot to severe tests, which were equivalent to about 1,500 hours of continuous functioning. The flight to Panama on December 8 through 9, 1937, was a success, and all of the equipment performed satisfactorily during the 22 hours of flight. The same results were reported on the flight to Pearl Harbor in January 1938. All of the pilots were satisfied with its performance. It now appeared that the use of the SBAE was a reality, and was very promising as a navigation instrument. The Bureau of Ordnance felt that the minor items that had developed could readily be dealt with immediately. About a year later, in February 1938, the equipment was at a point where the SBAE was now becoming standard in both the Navy and Army Air Forces' airplanes.[13]

With the success of the Navy flights, the Air Corps was very anxious to try it on their B-18 airplanes. The equipment was installed in the airplane by AAF Materiel Division, Wright Field personnel on May 27, 1938. Testing consisted of navigation flights and bombing. These tests were not long flights, but limited from Wright Field to Langley Field, to Dahlgren, Virginia, to Baltimore, Maryland, and return to

Wright Field. The flights were conducted in very bumpy air, and much of the flights were in overcast weather. In spite of the very rough weather, the average precession was about 1° for each 30 minutes of flight. On the four bombing missions conducted, only one, the 12,000 foot mission, was performed under excellent conditions. In spite of clouds interfering with the other three missions, the average results of the four missions were equal to the best bombing done by the bombardier, in spite of the fact that he had dropped only 30 bombs the previous six months. There were some minor recommendations suggested to the Bureau of Ordnance, and these could be rectified on newer models. The Army Air Corps was so impressed with this system that the procurement of SBAE units and installation for all bombardment airplanes and reconnaissance airplanes was recommended to be expedited. They further recommended that B-17 airplanes be given first priority for procurement and installation of the SBAE. At the same time, a school was to be established for thorough instruction of enlisted aerial armorers prior to installation of SBAE units in tactical airplanes. Without trained personnel, the equipment was completely useless. A handbook on the equipment was to be supplemented by a short and concise set of operating and adjustment instructions for the pilot and bombardier.[14] In accordance with the request to install SBAE in B-17 airplanes, in September 1938, the Materiel Division, Wright Field personnel prepared drawings for installing this equipment in Y1B-17 (Air Corps No. 36-149). However, in order to use this equipment, it was necessary to remove the Sperry automatic pilot which had been previously installed. Upon completion of this installation, all necessary data was to be forwarded to Field Service Section in order that they take necessary steps to have this equipment installed in all B-17 airplanes.[15]

At about the same time, personnel from the Materiel Division, Wright Field visited the Norden Company concerning the production of SBAE sets. The manufacturer stated that between 50-60 sets suitable for installation in the B-18 airplane would be ready for shipment by January 15, 1939, and that all of the SBAE now on order for the Air Corps would be ready for shipment by April 1939. Trouble was being encountered with oil in all units presently in use. A sample of new oil to be used in the dash pots of the sector panel was obtained from the C.L. Norden Company. This new oil was to be analyzed by the AAC Materiel Division personnel so that specifications could be drawn up as to a source of supply that could be established and so that it could be purchased for replacing the oil in existing sets presently in use. This special oil was identified as "porpoise oil." It had no gumming qualities, was free from acid and impurities, and had a constant viscosity between -20° C and +35° C. Although not the best, it was usable since there was nothing else. The oil was of limited quantities until such time as another source was established. However, continuing research was being conducted in hopes of finding a better product. Three fourths of an ounce was required for each part. At this same meeting, the Air Corps expansion program was discussed in detail with Mr. Barth of the

Chapter XI: AAF Expansion and the Norden Mark XV Bombsight

C.L. Norden Company and the Bureau of Ordnance concerning the possible expansion of their production line capability to meet this added requirement. Mr. Barth stated that, within two months after award of contracts, if SBAE in large quantities were desired, present production could be tripled and between 120-150 units a month could be produced at no extra expense or expansion of facilities, purchase of new machinery, or obtaining new tools. At an additional cost of approximately $200.00 per instrument, the delivery of sets could be doubled. These quantities were in addition to the Navy requirements. The bombsight production could be increased to 100 sights per month within three months after an order was placed, and in six months this could be increased to 150 instruments a month. It was estimated that 1,500 units could be produced in a year at no additional cost for expansion. The Norden Company had obtained outside assistance in the form of sub-contractors for SBAE units, which provided a means of increasing production without expanding the Norden factory.[16]

Following the recommendations of the Conference on Bombsight Requirements, the Bureau of Ordnance experimented with two sets of new optics. The experimental optics were received by the Naval Proving Ground about April 1938, and the results of the new optics showed that the mechanical operation of the telescope in the bombsight was satisfactory and the light transmission through the telescope was excellent. There was only a slight amount of optical parallax, and the apparent magnitude of the movement of the cross lines were half the width of the lines. The magnifying power of the optical system was 2X. At the suggestion of the Services who were clamoring for better optics in the telescope for the Mark XV bombsight, thought was given to increasing the power of the telescope. Over the years, the Bureau of Ordnance and Aeronautics were also concerned. The telescopic power had been increased through the years. The Bureau of Ordnance and Bureau of Aeronautics requested the Naval Gun Factory to prepare a 3.2 power telescope and have it installed in the bombsight. The new design was reviewed by the Bureau of Ordnance personnel, and then it was released to the NPG for extensive testing on December 7, 1943. In the meantime, studies were underway to develop a telescope of about 3.4 power. Considerable studies were conducted by a number of expert military and combat personnel. The studies showed that the improvement in quality with a 3.4 power over the 3.2 power was so small that it was considered so slight as to scarcely be worthwhile in developing the new telescope.[17] The Bureau of Ordnance and Aeronautics were advised of the finding with a view of increasing the power of the new telescope. It was brought out that further improvement could not be made without extensive redesigning of the entire bombsight. Continued research showed that the 3.2 power scope was performing satisfactorily even at high altitudes, the scope in use in the bombsight was of sufficient power, and that no further work was to be done in this area. The Eastman Kodak Company was producing bombsight optics and aluminized first surface mirrors. Due to the criticality of optics, the Naval Inspector of Ordnance directed the Bausch & Lomb Company to furnish Eastman Kodak all information pertaining to the Mark XV bombsight optics in order to fulfill outstanding Bureau of Ordnance contracts.[18]

The maximum forward vision obtainable using the forward view device was 81.5°, which was adequate and satisfactory with the present telescope. The field of vision was 15.4°, as against 16° for the present Mark XV telescope. This very slight reduction of field of vision did not detract from the effectiveness of the optics. With the very satisfactory results obtained with these new optics, the Proving Ground recommended to the Bureau of Ordnance that the contractor be directed to proceed with the manufacture of these optics for the Mark XV, Mod 4 sights, which were equal to or better than those now presently installed. From the initial procurement of bombsights from 1932 to 1938 by the Navy and the Army Air Forces, during the past six years the Bureau of Ordnance had developed many improvement features through the manufacturer that would be applicable to the available sights now in use, which would enhance their value by extending the operating limits with respect to both altitudes and speeds, and also provide means for use of the SBAE equipment. A conference between representatives of the Army Air Forces, the Bureau of Ordnance, and the manufacturer provided an informal quotation of $3,500 per sight to cover the complete overhaul, necessary changes, and modifications to all existing instruments, as well as to incorporate the latest change and make them conform to the present Mod 4 sights, which were currently being produced by the Norden Company. In addition to providing increased range in improved operations, this would permit interchangability of parts between existing instruments and those on procurement. The Navy service units, and especially the AAC, wanted the Bureau of Ordnance to provide them with the exact changes to be made and the quantity that could be returned to the manufacturer per month. The Bureau of Ordnance had already taken action to upgrade Navy sights to the M-4 configuration. The Army Air Corps requested that action be taken to retrofit 124 sights.[19]

With the advent of new modifications being incorporated in the sight, a troublesome radio interference developed when using the Mark XV, Mod 4 sight.[20] The Norden Company was contacted, and their engineers immediately started to work on this problem. After considerable research, the manufacturer was able to develop a "fix" to eliminate this. The Norden engineers by April 1938 found that radio interference could be reduced to a point where it would no longer be a problem. By the use of shunting resistors of 500 ohm across the inductance coils, contacts, and the telescope motor brushes, most of the trouble was eliminated. Through continuing research, it was found that the trouble emanated from the above mentioned sources. In the future, all Mark XV, Mod 4 would be equipped with a 500 ohms resistor. The Norden Company would equip all existing Mod 4 sights with this resistor at no cost to the government. At about the same time another vexing problem was encountered. This problem had existed, to some extent, for the past several years, but it was not considered a serious problem. With newer airplanes being received, the problem was exacerbated. Excessive vibration was destroying the precision capabilities of the sight. The Bureau of Ordnance called on the manufacturer and the Bureau of Aeronautics for assistance. Many different mounts were tested by the Naval Proving Ground, but no satisfactory solution had been found. The mount that was presently being used was supported by trunions, and considerable difficulty was encountered in mounting it in certain types of airplanes. The Bureau of Ordnance considered a single point suspension bombsight mount used in the TBD-1 (B-25) and PBY 1, 2, 3, and 4 with only slight modification to the adjustable plate, which was supporting the "Lord" (Lord Manufacturing Company design) type of shock absorbers, and would occupy about the same amount of space. In order to use this bombsight mount it was necessary to change the airplane structure. However, the Bureau of Ordnance and Aeronautics decided to have this mount tested by the U.S. Fleet, Aircraft Scouting Force, Patrol Squadron Eight, Pearl Harbor, Hawaii. Test runs were conducted at 10,000 feet, and the behavior of the sight was still unsatisfactory due to excessive vibrations of large amplitude. Subsequent tests carried out showed that the vertical vibration was also excessive. Although a one point suspension mount reduced the number of vibrations per second about 60%, it increased the effect caused by acceleration and centrifugal forces. This resulted in a telescope "sweep."[21] The amplitude of vibration was not reduced appreciably to warrant a change in the present installation. From the results of testing of the sight mounts, it was recommended that this type of mount not be installed for service activities. The "sweep" of the telescope was very disconcerting to the bombardier, and it was necessary to eliminate this troublesome item.[22] Continued testing and development of new types of mounts began to produce acceptable mounts. The results of all these tests by the Naval Proving Ground by June 1940 engineering personnel had developed a new mount using modified Lord mounts provided by the airplane manufacturers. The Douglas modified mounts were now being used by all Service activities (Army and Navy).[23]

With the ever increasing troubled conditions in Europe, and with the Japanese in the Pacific, war clouds were beginning to rapidly form. The Chief of the Army Air Forces was extremely concerned about their mobilization plan should war break out in the very near future, and wanted all bombardment airplanes to be equipped with the latest Norden bombsights. The Bureau of Ordnance was also cognizant of this war threat. As a result of these concerns, a high level meeting was held at the Norden Company plant to discuss costs of bombsights and SBAE in connection with the proposed Army Air Corps airplane expansion program. The AAF wanted cost figures for bombsights and SBAE/AFCE equipment in various lots of unheard quantities of up to 700 units of each type. The current price for both units was $10,000. The AAF order amounted to an unprecedented $7,000,000. This was in addition to the 319 sights the AAC had on order with the Navy. The aggregate AAF order was nearly equal to the grand total production of the Norden Company during the time that the Bureau of Ordnance conducted business with this single manufacturer for the past fifteen years. Although this cost was somewhat high to the Navy, it was needed to have a reasonable margin of safety against increased labor costs. The Navy also had about three million dollars worth of contracts with other suppliers in addition to the AAF. In the past, the contracts with the Norden Company averaged several hundred thousand dollars a year. A decision had to be made whether to continue on a proprietary basis or adopt some other arrangement. The Norden Company stated that large orders involved larger risks, but their main interest was not a large profit, but insurance against unusually heavy losses. Since they were the only ones manufacturing the Norden bombsight, which was the most accurate in the world, the company had to forego any opportunities for commercial application or export business. The life of the company depended on the Navy and AAF for contracts. One solution was to apply the Vinson Act (named after Congressman Carl Vinson), which would allow Navy contracts to charge a "safety price" that would insure the company against losses.

In September 1939, Poland was devastated by Nazi Germany within a few weeks. Then, a short time later, the Low Countries and France were occupied. With these conditions in Europe, the Norden Company was now faced with a tremendous quantity of sights to be manufactured without sufficient facilities. The Army Air Corps by October 1939 had placed a contract for 319 bombsights and SBAE with the Navy which had not yet been completed. With all these astounding orders for sights that were on order, the Navy (Bureau of Ordnance) and the Norden Company were scrambling to find sufficient facilities to produce the bombsights on order, and for those in the future. To meet this challenge, the Norden Company on its own took extraordinary action to purchase raw materials and located suitable sub-contractors in advance of any new firm contracts. To satisfy the expansion of orders for bombsights, the C.L. Norden Company committed $500,000 of their own funds far in advance of firm contract awards in their continued efforts to meet contract delivery dates for assembly and manufacturing space. Also, another $250,000 was allotted for tooling up for production quantities, as well as in the production of many of the actual parts of the Mark XV, Mod 4 sight without official knowledge that the new production model would be accepted by the Bureau of Ordnance or the AAF. The Norden Company, realizing the seriousness of the situation that existed in Europe, had taken immediate steps to meet the Mark XV, Mod 4 sights' delivery dates. These actions, since July 1939, included the doubling of the number of trained workers in the bombsight assembly plants. In addition, to support this tremendous explosion of all types of fire control equipment, the C.L. Norden Company sent approximately 125 trained workers and keymen for fabrication of parts and the assembly of the Mark XV, Mod 4 sight sub-units to the Robbins and Myers Company in Springfield, Ohio. This factory also produced practically

Chapter XI: AAF Expansion and the Norden Mark XV Bombsight 129

all of the principal units of the SBAE. Agreement between the management of Robbins and Myers Company, the Navy, and the Norden Company, this company was reserved and groomed for large scale "emergency" production of bombsights and SBAE. The facilities of this plant were well suited for bombsight and SBAE production. The production of the Mark XV, Mod 4 bombsight stabilizer unit was being accomplished by this plant. In addition, for the past two years, the International Projector Corporation, New York City, was producing, under contract, a number of bombsight and SBAE educational materials. Approximately 125 Norden "trained workers" sent to the International Projector Company under supervision of Norden factory keymen were engaged in the production and assembling of the complete SBAE Sector panels, Mark XV Low Altitude Devices, and many other bombsight and SBAE accessories. This company had complete modern machine shop equipment consisting of about 90,000 square feet of manufacturing space, employed 125 skilled workers on Norden contracts, and had facilities to expand and employ approximately 800 mechanics. Between the Norden Company (50,000 square feet and 140 employees), Robbins & Myers, Inc. (500,000 square feet and 125 employees), International Projector Corporation (90,000 square feet and 125 employees), and Manufacturers Machine & Tool Company (25,000 square feet and 200 employees), a total of 665,000 square feet of factory space was available and 590 employees were working on bombsights. With these facilities the total number of direct workers on bombsights could be expanded to 2,400 on short notice.[24]

With the war raging in Europe and with more airplanes coming off the production lines, additional bombsights were needed to meet this demand. As previously mentioned, a new ordnance plant was to be built in the interior of the country (Indianapolis, Indiana). In late 1940, it was decided to increase the capacity of the present Norden Company. Accordingly, four new floors were rented for installation of machinery and machine tools at the Norden Company's New York plant, and one new floor was rented for assembly, giving the Norden Company eight floors where formerly they had three. Two hundred seventy-five thousand dollars worth of machine tools were installed and in operation. Another $325,000 worth of machine tools were ordered from the manufacturer but not yet received. There was also $1,600,000 worth of machine tools in which the Norden Company purchased tools on their own responsibility. With all this expansion, the Norden Company, by December 1940, had bombsight production up to 120 sights per month.[25]

With the increased production, different means of transporting the sights to the Naval Proving Ground for testing needed to be addressed. In the past, all shipments of sights from the Norden plant were via White Plains, Maryland. The sights were delivered to White Plains by rail and then transported to the NPG by government transportation. This procedure involved undesirable delays in shipment due to waiting for scheduled barge trips. The NPG designed and procured a delivery truck capable of transporting sights without damage. Beginning in December 1939, all bombsights from the Norden Company would be delivered to the NPG via Fredericksburg, Virginia, by government truck. All existing contracts were modified so that this provision was carried out. This was agreeable to the Norden Company, and there would be no additional cost to the government.[26]

Service activities using SBAE equipment had been using "porpoise oil" because there was no other agent available. Research had been conducted on a continuing basis on the use of this oil for the dash-pots and found that it was not the best and was causing maintenance problems. Extensive investigations by the NPG and the manufacturers of various substances resulted in the adoption of strained "Prestone." Before using the Prestone, it was strained through a chamois skin to remove gummy substances that were present in the commercial product. It had the least change of viscosity with change in temperature of any substance tested, which was an advantage in making dash-pot adjustments. It was suitable at all temperatures between +43.3° C and -37.2° C. At -42.2° C strained Prestone crystallized and became a solid. Bureau of Ordnance Circular Letter No. V-49 directed all activities that approximately every three months the dash-pots were to be drained and cleaned, the Prestone re-strained, and the unit refilled with Prestone.

Notes:

[1] Letter, Army Air Corps, Materiel Division, Wright Field, Subject: Bombsights, dated January 18, 1936, National Archives, M.S.
[2] Ibid.
[3] Headquarters, GHQ Air Force, to Chief of the Army Air Corps, Washington, D.C., Subject: Bombsights, Permanent Installation, January 21, 1936, National Archives, M.S.
[4] Letter, Carl L. Norden, Inc. to Bureau of Ordnance, Subject: Stabilized Bombing Approach Equipment, dated September 18, 1936, National Archives.
[5] Letter from Chief Materiel Liaison to Contracting Officer, Subject: Stabilized Bombing Approach Equipment, dated September 18, 1936, M.S. National Archives.
[6] Letter from Army Air Corps Materiel Division, Wright Field, Dayton, Ohio, Subject: Conference on Bombsight Development, March 1, 1937, National Archives.
[7] Installation of SBAE equip. in YB-17 airplane, Sep 17, 1938 Nat Archives MS
[8] Memo to Chief of the Air Corps, from Materiel Division, Wright Field, Subject: Bombsight Procurement, April 8, 1937, M.S. National Archives.
[9] 2nd IND, Headquarters GHQ Air Force, Langley Field, to Chief of the Army Air Corps, Subject: Stabilized Bombing Approach Equipment, dated July 26, 1937 National Archives M.S.
[10] Letter, HQS GHQ Air Force, to Chief of the Air Corps, Subject: Bombsights, Permanent Installation, dated January 21, 1936, National Archives.
[11] 2nd Wrapper Ind from Hq GHQ Air Force, to Air Corps Materiel Division, Wright Field, Subject: Stabilized Bombing Approach Equipment, August 21, 1937, National Archives M.S.
[12] Conference of Mark XV (M-1) Bombsight composed of Representatives of the Army Air Corps, Navy and C.L. Norden Company, dated October 20, 1937, National Archives, M.S.
[13] The Navy's Mark XV (Norden Bombsight), Its Development and Procurement 1920 - 1945, page 131.

[14] Letter from Second Bombardment Group to GHQ AF to Commanding General 2nd Wing, Subject: Report of test of SBAE, dated July 20, 1938, National Archives, M.S.
[15] Letter from Army Air Corps, Materiel Division, to Chief Engineering Section, Subject: Installation of SBAE in Y1B-17 airplane, dated September 17, 1938, National Archives. M.S.
[16] Letter, Army Air Corps, Materiel Division visit to C.L. Norden, Inc. plant, Re: Stabilized Bombing Approach Equipment, dated December 14, 1938, National Archives, M.S.
[17] Report of telephone conversation between the Bureau of Ordnance (Lt. M.L. Sandall) and the Naval Gun Factory (Mr. F.J. Miller) dated March 23, 1944, Subject: Increase in Power of Telescope, Bombsight Mark XV, Mod 5, National Archives.
[18] Letter Naval Proving Ground to Chief of Bureau of Ordnance from Inspector of Ordnance in Charge, Subject: Report on Experimental Optics for Bombsight Mark XV, dated April 1, 1938, National Archives.
[19] Letter from Chief of the Army Air Corps to the Bureau of Ordnance, Navy Department, Subject: Modernization of M-1 Bombsights, October 19, 1938, National Archives.
[20] Letter from C.L. Norden, Inc. to Bureau of Ordnance, Subject: Radio Interference with the Mark XV, dated November 10, 1938, National Archives,
[21] Letter from Bureau of Ordnance to Bureau of Aeronautics, Subject: Single Point Suspension Bombsight Mounts, dated February 16, 1939, National Archives.
[22] Letter, U.S. Fleet, Patrol Squadron Eight, Fleet Air Base, Pearl Harbor, Hawaii, Subject: Test of Anti-Vibration Mounts, dated March 28, 1939, National Archives.
[23] Letter, U.S. Naval Air Station, San Diego, California to Chief of Bureau of Ordnance, Subject: Bombsight Shock Absorber Mounts, dated June 14, 1940, National Archives.
[24] Letter from C.L. Norden, Inc. to Bureau of Ordnance, Subject: Expansion of Manufacturing Facilities, October 3, 1939, National Archives.
[25] Letter, Navy Department, Subject: Expansion of Facilities for the Manufacture of Norden Bombsights, dated November 20, 1940, National Archives.
[26] Letter from Bureau of Ordnance to Bureau of Supplies and Accounts, Subject: Shipment of Bombsights from Norden Company to the Naval Proving Ground, Dahlgren, Virginia, dated December 18, 1939, National Archives.
[27] Navy Department, Bureau of Ordnance, Circular Letter No. V-49, dated April 1, 1940, Subject: SBAE Dash-Pot Liquids, National Archives.

XII

Preparing for Expansion and Production

The German "blitzkrieg" in Poland and later the invasion of the low countries and France caused great concern for U.S. military planners. It was now evident that eventually the U.S. would be drawn into the conflict; as it was, only Great Britain was left opposing Hitler. There was also a need for Hemispheric defense against the Axis powers. The Air Corps in 1939 had completed a study for requirements to combat this threat. They had contacted the Bureau of Ordnance and Norden Company with the view of dramatically increasing bombsight production. While this was going on, President Roosevelt, in his speech before Congress in May 1940, asked for at least 50,000 airplanes. This impassioned plea for such a huge number of airplanes fired the imagination of the nation. There is a great difference between asking for 50,000 planes and actually having that number operational in such a short time. It became clear that drastic action had to be taken since the disaster in Europe. The resulting furor of this announcement left the military seriously groping for answers. How many of this figure would be heavy or medium bombers? How many bombsights would be needed? A partial answer to this was that the Bureau of Ordnance, Army Air Corps, and C.L. Norden, Inc., felt that the production of sights as outlined by the Norden Company in a previous conference had stated they had sufficient resources to cope with any large scale orders. To meet this tremendous expansion for bombsights, a new Navy Ordnance factory was being built in Indianapolis, Indiana, and would be completed sometime in September 1941. The name of the company was Lukas-Harold Corporation. The name was obtained by using the middle name of Carl *Lukas* Norden and Theodore *Harold* Barth, hence the name Lukas-Harold. However, no information on this name was available for research, so it is assumed that it is correct. It would not be until Sep-

Chapter XII: Preparing for Expansion and Production 133

tember 1942 that the Lukas-Harold Corporation would commence production, and that the production rate of 300 per month would be reached in June 1943.[1]

Production of sights and SBAE was steadily climbing. For example, in April 1940 the Naval Proving Ground reported they had received 62 Mark XV, Mod 3 and 4 from the contractor for testing and shipped 90 sights to the using activities, as compared to having received 72 sights in March 1940 and shipped only 40 to the activities. *Note:* At this time all bombsights were tested at the Naval Proving Ground prior to being released for use. The build up of the Mark XV bombsight was picking up at an accelerated pace. Expansion possibilities were explored with the Sperry Company, but they could not give any figures on production of the sights in any quantities. This was the opposite to the SBAE. During 1940, considerable time and funds were expended on SBAE units on further refinements and design for better performance.

The Army Air Forces had named the Navy SBAE the Automatic Flight Control Equipment (AFCE). Severe testing was starting to reveal many deficiencies in the Air Corps AFCE, and by mid 1940 they were creating a problem. All of these deficiencies were analyzed by both the Navy and AAF. The analysis resulted in a recommendation for a more durable and rugged construction, provisions for selective adjustment of limit cams, synchronization of applications of controls to permit recovery from banks without skip or skid, and provisions for controls in the cockpit to permit adjustment of ailerons, rudder, etc. There were even discussions of abandoning this system if it could not be made to operate efficiently.

Because of the security restrictions, it was difficult to disseminate information on the system. Information concerning instructions, maintenance spares, and installation of new equipment from the factory was sorely lacking. Immediate steps were taken to improve these conditions.[2]

Even with these items addressed, Headquarters GHQ Air Force felt that redesign was necessary to obtain satisfactory results. In May 1941, action was initiated to change the classification of the SBAE, the Flight Gyro, Rudder, Elevator, and Aileron Servo Motors, Banking Motor, and Sector Panel Assembly from CONFIDENTIAL to RESTRICTED. It did not include the bombsight stabilizer, nor did it affect the classification of the bombsight whatsoever. All correspondence which discussed design characteristics, performance, and malfunctions of this equipment remained CONFIDENTIAL.[3] The principle purpose of this change in classification was to facilitate the installation of automatic flight equipment in aircraft at the aircraft manufacturer's plant. The descriptive pamphlet on SBAE O.P. No. 630 retained its CONFIDENTIAL classification since it contained information on the bombsight stabilizer.[4] The Secretary of the Navy informed the Secretary of War that the Navy Section of the Joint Aircraft Advisory Committee, Chief of Naval Operation issued a directive stating that the NBS would not be released to any foreign government.[5] The NBS and SBAE would be removed from Navy planes released to Great Britain. Request was made that a similar directive be promulgated for the War Department.[6] The standardized installation would be to install Sperry A-5 automatic pilots in all bombardment airplanes; install M series bombsights in all U.S. airplanes; and install Sperry type S-1 bombsights in all British airplanes. All bombardment airplanes were to have provisions for installation of both M series and S-1 Sperry bombsights.[7] Provisions were made to have mountings, such as brackets, wiring, etc., installed in the event that when SBAE equipment became available, it could readily be installed. To further the subject of SBAE/AFCE, a joint Army-Navy conference was held at the Naval Proving Ground on June 3, 1941. It was attended by high level representatives from the Chief of the Air Corps, Chief of Bureau of Ordnance, Chief of Materiel Division, Wright Field, Dayton, Ohio, and the Norden Company. The Navy (NPG) stated that they used the AFCE as a bombing aid and installed it in airplanes with speeds of 120 knots (135 m.p.h.), and did not use it as a navigation system. It was designed to provide a stabilized platform for bombing. The Bureau of Ordnance concurred with the Air Corps in improving this equipment. Immediate action was taken by the contractor to begin the necessary modifications.[8] It was agreed that, henceforth, all bombsights and AFCE be equipped with A and N standard electrical connectors for uniformity. The Bureau of Ordnance directed that a heating blanket be designed by either in-service or by commercial companies. The delay brought about by this policy created havoc in the area of engineering drawings, as parts lists were not being received for bombsight and AFCE equipment up to a year after receipt of the instruments. To compound the problem, many changes were made on the original drawings by the manufacturer and the AC were not notified, which retarded the procurement of badly needed spare parts. The solution was for the AC to obtain authority from the Navy Department to contact the manufacturer directly.[9]

Conditions presently existing required almost every action be coordinated through the Bureau of Ordnance or receive their approval for any Norden materiel, regardless of the equipment. Procurement was delayed, and many obstacles existed in the efficient handling of this equipment. To alleviate this bottleneck, an enactment of a law or change in the Navy policy was suggested. It was such an urgent matter that General Arnold recommended that a letter from the Secretary of War (Air) be forwarded to the Under Secretary of the Navy outlining the problems resulting from the Navy requirement concerning all negotiations between the AAF and the Norden Company. A letter was forwarded to Mr. James Forrestal, Under Secretary of the Navy, outlining the complicated procedure with respect to bombsights, up-to-date information on drawings, engineering changes, etc. It was suggested that an attempt be made to work out new procedures to expedite the handling of this critical equipment.[10] It was becoming a problem for the AAF, and a high level meeting was held to rectify or find a solution to this dilemma. Accordingly, a conference was convened, held at the Norden Company, New York City, on July 8, 1941, for the purpose of discussing changes, deficiencies, production, and allocation of the present bombsight problems. It was attended by representatives of the Bureau of Ordnance, Naval Proving Ground, Chief of the Air Corps, Chief of the Materiel Division, and Norden people. Both the Navy and Army Air Corps desired changes.[11]

These changes and modifications were to correct present undesirable features and modernize or redesign sights to meet present and future needs for increased airplane speeds and altitudes. The Navy was cognizant of and admitted that the AC requirements were more extensive then those of the Navy. As new bombers were being built, the AC would have, in the future, airplanes with speeds in excess of 300 miles per hour and service ceilings of 40,000 feet. It became clear that the design of an all purpose bombsight to cover a wide range of bombing requirements, such as altitudes, speeds, and bomb characteristics, would be almost impossible without sacrificing accuracy and simplicity of operations. The Norden Company was conducting a study of increasing the limits of the sight to about 60,000 feet. This change would be accomplished by an attachment changing the disc speeds and possibly increasing trail and cross trail to about 180 to 200 mils without redesigning the sight. Due to freezing temperatures at high altitudes, the bombsight could not perform efficiently. The General Electric Company was asked to develop a heating blanket for all bombing planes, and some models were being tested by the Navy and AC.[12] The sensitive question of the AC dealing directly with the contractor or obtaining a manufacturer under AAF control was not discussed and referred to higher headquarters. The Bureau of Ordnance could not see any reason for limiting the Norden factory to production of identical bombsights for both Air Corps and Navy. This was a big concession, as the Air Corps had different requirements as to altitudes, speeds, bomb characteristics, and targets. The Instruction Manual was a confidential publication and vitally needed for operation of the bombsight by all service personnel for maintenance, testing, and calibration. The manual was revised and brought completely up-to-date; drawings, sketches, schematics, and engineering changes were received from the factory and the Army Air Corps to be advised immediately when the publication was completed. They also requested the Army to make early and positive commitments on this equipment. The letter also brought out that on occasions the Bureau of Ordnance let the Army handle the new construction of an explosive plant even though it was intended solely for the Navy and the British, so it was felt that this cross production would not be a problem. They felt that the Norden bombsight was of the greatest usefulness against moving targets, and it was of vital importance that it remain with the Navy. For this reason, as well as the Navy developing the sight, the Norden Company financed and supervised the expansion. The Department of the Navy preferred to continue the present practice of handling all contracts concerning the bombsight and AFCE equipment.

Also, to further expedite the processing of equipment, the Bureau of Ordnance placed additional naval aviators qualified in ordnance duty to be ordered to the various ordnance plants. The Navy Department thought that these changes would solve the past difficulties and find little or no problems in the future concerning the procurement of sights and accompanying equipment.[13]

Despite all the assurances from Naval Headquarters, the solution to the problem was not to be and was not working. Much of 1941 was spent in trying to improve the AFCE and obtaining sights. The Navy SBAE had many shortcomings that were brought out during AAC testing. An unusual amount of unsatisfactory reports were being received from the operating activities. Some of the more serious were the high maintenance needs requiring highly skilled personnel to make flight adjustments necessary for every changing condition of flight for the best results. The human pilot did not have, under his control, means of trimming or adjusting the AFCE other than to make one preset turn and to level the airplane longitudinally. The auto pilot was satisfactory only for bombing and not for navigation. The limits of mechanical adjustment made it doubtful that a satisfactory flight could be obtained from the airplane in wide ranges in speed and altitudes. The present limitations prevented satisfactory use on the B-17 airplanes at low speeds or high altitudes. It was impossible to change the sensitivity of the AFCE, and consequently, in rough air, controls reacted violently. The location of the AFCE, in the forward part of the airplane, demanded long control cables. The tension of the control cables decreased with low temperatures at high altitudes and could not be overcome by the AFCE.[14] The most pressing problem confronting the services was the auto-pilot problem. The development of the auto-pilot was under the jurisdiction of the Bureau of Aeronautics and the Bureau of Ordnance, and they were aware that the Norden SBAE was not totally successful, which reflected opinions of some of the operating squadrons of both services. The Bureau of Ordnance and Aeronautics were cognizant of the Sperry auto-pilot and had purchased some for testing in Navy airplanes, as well as the Norden equipment. At the request of the Bureau of Aeronautics, Sperry equipment developed by Patrol Squadron FORTY THREE was immediately installed in an EPBY-1 airplane at the Naval Aircraft Factory and sent to the NPG for comparative tests with the present SBAE to determine its performance. The results of the tests showed that both systems had deficiencies, and in general the SBAE was not more complicated or less dependable than the Sperry.[15]

The Bureau of Ordnance and the Bureau of Aeronautics were aware that the Norden equipment was not combat worthy nor successful as a bombing and navigation aid, which reflected the opinions of the majority of the Navy and AAF operational squadrons of both services. By October 1941, the Navy Department ordered seven sets of AFCE to be modified by the addition of the Honeywell electronic system, including complete installation, flight test, and calibration for one airplane and six weeks instruction for Navy personnel. All of this brought the cost to $705.00 per unit. The installation, calibration, and flight checks raised the cost to $935.62 per unit.[16]

Included with the cost of the sets for the Bureau of Ordnance that were in the process of manufacturing, was a training course for personnel provided by Honeywell Company beginning November 1, 1941. It was expected that the course of instruction would last about six weeks, which included tying up airplanes (Navy and Air Corps). The Honeywell Company felt that service personnel could not obtain adequate installation and operational information necessary to perform the work unsupervised by company personnel due to the newness of the equipment. It was necessary to have one type of airplane available in which the equipment was installed.[17]

Since the Navy SBAE/AFCE was not operating satisfactorily under simulated combat conditions in early 1941, the AC had taken action to contact the Minneapolis-Honeywell Regulator Company (hereinafter called Honeywell) of Minnesota to demonstrate new various electronic devices. They suggested the replacement of the mechanical follow-up system with these new electronic devices. The addition of potentiometers or variable resistors to the gyro and servo motors would allow the system to maneuver the control surfaces smoothly. The rudder and elevator compensation resulted in perfectly banked turns, which was not previously available, and the sensitivity of all controls could be adjusted by the pilot to suit air conditions. Aileron compensation was also provided which permitted banking the airplane at any angle up to 30° consistent with desired operations or within limits of the bombsight. These new developments were installed in a B-17 airplane (S/N 38-269) by September 1941.[18] This revolutionary new instrument eliminated the biggest criticism of the AFCE, namely the use of highly skilled personnel to make the flight adjustment. In addition, there was no comparison for the ease with which these adjustments were made by the pilot. The Honeywell Company was conducting future engineering developments in the field of level flight control, second order control system, development of servo motors having much smaller steps, faster reaction action, and increased power.[19] For a short monograph, see chapter on development of AFCE.

With the war raging in Europe and Japan becoming increasingly defiant, the U.S. started a crash program in obtaining sufficient bombsights and updated automatic pilot equipment at the earliest possible time. But time had run out, as the U.S. would shortly enter WWII. To assist the hard-pressed British, U.S. bombers were being diverted to the British government without the Norden sight under the provisions of the Lend Lease Act. In the haste to produce all types of fire control equipment, the priority system that was established in allocating materials and machine tools was not working the way it should. All activities producing war material were clamoring for the highest priority they could obtain. Who would be assigned the much sought after A-1 or A-1-A? It came down to the point of which item was the most important—bombsights, or other types of armament. The result was that the War Production Board and the Joint Aircraft Committee would have to determine to whom the priorities would be assigned. By late 1941, airplane production schedules were reaching the point where they were surpassing bombsight production. All new bombing airplanes scheduled for delivery were now designed for operation on 24 volt current. Accordingly, the Mark XV, Mod 5 (12 volt) would no longer be manufactured. These sights would be used for training purposes.

Chapter XII: Preparing for Expansion and Production   139

Notes:

[1] Letter, Bureau of Ordnance, Chief of the Air Corps, War Department, Bombsight requirements for the Army Air Corps, December 23, 1941, National Archives, M.S.
[2] Memo, Assistant Chief, Material Division from Commander, Maintenance Command, Automatic Flight Control Equipment, April 2, 1941, National Archives.
[3] Technical Instruction, Air Corps, Material Division, Letter of Instruction No. LI-118, Reclassification of SBAE, May 2, 1941, National Archives.
[4] War Department, Chief of the Air Corps, Memorandum for General Brett, April 21, 1941. National Archives
[5] Letter of Instruction, Chief Material Division, Air Corps, LI-117, Provisions for Bombsights in Airplanes, April 29, 1941, National Archives.
[6] Department of the Navy, Office of the Secretary, Washington, to Joint Air Advisory Committee, from Frank Knox, May 12, 1941, National Archives.
[7] Letter of Instruction, from Chief, Material Division, Air Corps, LI-117, Addendum No. 1, Provisions for bombsights in Airplanes, June 3, 1941, National Archives.
[8] Ibid.
[9] Letter, Chief of the Air Corps, memo for Chief of the Army Air Corps, July 3, 1941, National Archives.
[10] Letter, from Assistant Secretary for War (Air) to Under Secretary of the Navy, July 24, 1941. National Archives
[11] Letter, War Department Air Corps, Materiel Division, Memorandum Report Conference of Norden Company, Air Corps, etc., August 6, 1941. National Archives
[12] Ibid, page 2.
[13] Letter, from Under Secretary of the Navy, to Assistant Secretary of War (Air) Addressed to Mr. Lovett, August 11, 1941, National Archives.
[14] Letter of Instruction, to Chief, Experimental Engineering Section, Wright Field, from Chief Armament Laboratory, September 15, 1941, National Archives, MS.
[15] The Navy's Mark 15 (Norden) Bombsight. Its Development and Production 1920 - 1945, Page 137 and 138.
[16] Letter of Instruction LI-118, Chief, Materiel Division, OCAC Modification of AFCE, November 3, 1941, National Archives, MS.
[17] Teletype from Armament Branch, Chief of Experimental Engineering Section, No. A 530 dated November 4, 1941 and Teletype No. T 909, November 6, 1941, National Archives, MS.
[18] Letter of Instruction, Chief of Experimental Engineering Section, Wright Field, from Chief Armament Laboratory, September 15, 1941, National Archives.
[19] Ibid, page 3.
[20] Letter, Bureau of Ordnance, Subject: Procurement of Bombsights Mark XV, Mod 5 (12 volt), dated December 5, 1941, National Archives.

XIII

The Navy's Norden Mark XV, M-9 Bombsight

Since the first contract awarded in 1922 to Carl L. Norden by the Navy Bureau of Ordnance, both activities continually strived to design and engineer a truly precision high altitude bombsight. With the advent of the Mark XV in 1933, they were nearing this goal. However, many shortcomings were evident over the intervening years from 1933 to 1942. It was thought that, by the time the Mark XV, Mod 7 was ready for production, it would be the ultimate in precision bombsight. In the early years of bombsight experimentation, airplane speeds were in the range of 60 to 120 knots per hour. In order to keep abreast of the rapid progress in the development of heavy (B-17, B-24, and later the B-29) and medium (B-25, B-26, etc.) bombardment airplanes, the Army Air Forces was struggling to develop a bombsight to match these higher speeds and altitudes. It was expected that bombardment planes would soon reach speeds up to 200-300 miles per hour and bombing altitudes from 1,500 feet to 20,000 feet. It was anticipated by 1945 that bombing altitudes would be 30,000 feet or more and speeds of up to 200-300 miles per hour. The AAF were vitally interested in these higher airplane speeds and a wider range of altitudes than the Navy. Much work had been done on the Mark XV, Mods 5 and 7 to achieve this advancement. The AAF wanted a sight that had the ability to bomb from higher altitudes in an attempt to get away from anti-aircraft fire. However, what the AAF wanted was still not available. It would be almost two and a half years into WWII before these new sights would be sent to the Naval Proving Ground for extensive bench and air testing before being put into production. When war was thrust on the U.S. in December 1941, the Norden bombsight had been perfected to a high degree of precision. Bombing tables and the ballistic coefficient of bombs had been developed and in use for many years. As was customary with the Bureau of Ordnance, Bureau of Aeronautics and Mr. Norden were not satisfied with the latest Mod 7,

Chapter XIII: The Navy's Norden Mark XV, M-9 Bombsight

Mark XV performance. In 1941 the Bureau of Ordnance requested the Norden Company to explore the possibility of bombing from a 60,000 foot altitude. However, at that time there were no airplanes in existence or on the drawing board that could reach this extreme altitude. This new project was given a high priority. It was thought that the present sight could be used with the addition of an attachment. Work was started on a modified version, and in early 1943 the new sight would be put on a production basis. At the time, the Mark XV, Mod 7 was the most sophisticated precision bombsight and far advanced of any other in existence. Due to the limitations of the Mark XV, Mod 7, it was necessary to design a new sight capable of reaching these objectives. In the past years, much thought had been given to a "1945 bombsight." It was readily evident that it was to drag on for several years. By this time newer and sophisticated airplanes would be produced by the airplane manufacturers. This super altitude sight could theoretically bomb from a 60,000 foot altitude.

The answer to this challenge was the development of the Mark XV, Mod 9 Series. The M-9 bombsight was designed so that, when properly operated by the bombardier, it would solve the bombing problem by determining the dropping angle at which the bombs would be released and the course on which the airplane should be flown to obtain a direct hit on the target. This was accomplished by mechanical means, which set up a "line of sight" between the airplane and the target, and measured the angle which this line of sight made with vertical. Gyroscopic stabilization was employed to permit instantaneous and continuous measurement of the angles which existed between the line of sight and the true vertical. To do this, the line of sight must be maintained on the target. The sights, having only the trail modification, were designated as type M-9A, and the sights having both trail and disc speed modification for high altitudes were designated M-9B.[1] The author has an M-9B sight and a C-1 Stabilizer on display at the local air museum.

The Navy and especially the AAF wanted to increase the trail setting to the maximum. At the insistence of the AAF in 1935, the Bureau of Ordnance approved the 25% change suggested by the Norden Company and the AAF. The trail setting was increased from 85 to 105 mils.[2] Mr. Norden stated that this was the maximum available without extensive alterations and would allow for bombing at true air speeds, which compared very favorably with those visualized as available bombing speeds in the near future. The consensus of opinion was that Naval bombsights would be in step with Naval airplanes of 1945 and should be capable of handling the bombing problem of bombing at about 1,500 feet altitude at about 250 miles per hour, and also be capable of bombing from 30,000 feet. It also was of the opinion that AAF bombers of 1945 that the bombsight would be capable of handling the problem at 3,000 feet at 300 miles per hour and going up to 30,000 feet.[3]

Extensive testing at the Naval Proving Ground proved that this increase in trail setting could be accomplished without redesigning the sight. As early as November

1942, the development of the M-9 was of great interest to the AAF and accordingly assigned a high priority to it for its development. The Mark XV, M-9 series was to become the standard Army-Navy production model, not only during WWII, but even many years beyond. However, the development and use of this sight is another story. Many of the deficiencies that were inherent to the M-7 series bombsight were deleted for facilitation in the production of this sight. All activities concerned carried out an experimental program to correct all unfavorable characteristics from the M-7 sight. New additions or changes that were developed were incorporated in the new M-9 series sights. One change of importance was the automatic trigger release for the bombardier as an easy way to arm the sight. During testing at the Naval Proving Ground, the automatic release was released from its "armed" position approximately 1,500 times. No variation in the interval, between closing and opening of the switch, was noted. The mechanism was satisfactorily operated at temperatures below -35° F. The escapement mechanism was designed to give approximately a 1/2 second interval between the closing and opening of the bombsight automatic release switch. The electro-magnetic relay was designed for 24-28 volt operation.[4] Production line sights were installed with this device. The author remembers installing this modification for those sights that did not have it. All service activities having bombsights were furnished retrofit kits for this installation. At the same time the Automatic Erection System (AES) was deleted from the present M-7 Series, and all service personnel were instructed to lock up the AES and render it out of commission on all units that were in service. A requirement also existed for continued production for low altitude attachment for the M-Series bombsights because the Flying Training Command was not instructing on this item due to its non availability, and the fact that only one unit of Combat Command was making use of this attachment. Since production was devoted to the primary high altitude bombsights, this attachment was of secondary priority.[5]

As a good will gesture, on June 15, 1942, the War Department, Army Air Corps forwarded to the Navy Department, Bureau of Ordnance, the sum of $1.00 to reimburse the Navy Department in accordance with Navy Bill No. 191-42 for the transfer of one type M-7 bombsight fabricated and assembled for presentation to the Army Air Corps by employees of the Carl L. Norden Company. The presentation was accepted by the Chief of the Army Air Forces, General Henry "Hap" Arnold.[6]

Since the M-9 had proven itself during testing at the Naval Proving Ground, the Bureau of Ordnance, in January 1943, requested the Norden Company for quotation for the production of 3,000 Mark XV, M-9 sights without the stabilizer with one Pilot Directional Indicator (PDI), plus a fixed fee at 6%, or $210, bringing the cost per unit to $3,710. The total cost of this contract was $11.1 million dollars. This did not include the manufactured parts, which were provided by the Manufacturers Machine Tool Company. The Norden plants did all of the assembly of the bombsight by their own personnel. An additional order of 900 Mark XV, M-9 sights

Chapter XIII: The Navy's Norden Mark XV, M-9 Bombsight

with one PDI, but without stabilizer, were to be produced by the Burroughs Adding Machine Company as sub-contractors at a cost of $7,102. The estimated unit cost of $7,102 by Burroughs represented the quoted unit cost of $6,700 plus a 6% fixed fee of $420. This contract amounted to $6.5 million dollars. This contract for $17.8 million dollars was one of the largest orders awarded by the Bureau of Ordnance to the Norden Company up to that time. The delivery was F. O. B. (Free On Board)[7] at the contractor's plant, concurrently with deliveries under existing contracts that were sub-contracted. To accomplish this huge order a priority of AA-1 was assigned the contract. The unit cost did not include Federal Excise taxes on the end product. This was based on cost-plus-fixed fee and contained the same Article[8] with respect to Federal, State, and Local taxes, which had been included in other cost-plus-fixed fee contracts that had been awarded to the Norden Company in the past. The Norden Company had been in business with the Navy many years and had two plants in New York City—one at 80 Lafayette Street, and the other at 30-40 Varick Street—and employed approximately 3,300 people. No additional facilities were needed to complete this contract.[9]

Due to the increasing number of factories manufacturing the Norden sight and related equipment, as well as the responsibility for adjusting defective material and workmanship, it became necessary to identify and serially number the bombsights made by the various facilities. In October 1942, the Office of Naval Inspector of Ordnance issued a directive to identify each bombsight and where it was fabricated. This was binding on all manufacturers. Beginning with the Mark XV, Mod 5, each was instructed to have a letter and number. The following facilities were assigned as follows: N - Carl L. Norden, Inc.; L- Lukas-Harold; R - Remington Rand; B - Burroughs Adding Machine Corporation; V - Victor Adding Machine Company; and C - Cardanic Corporation. The serial number assigned to the manufacturer would start with the number 1, such as L-1 for Lukas Harold Corporation. The name plate showed the name of the concern making the bombsight, and was not to refer to Carl L. Norden, unless the instrument was actually manufactured at 80 Lafayette Street, New York City. These instructions were to apply to the Army Air Forces, as well. This action would eliminate confusion as to where the sight was made and who was responsible for the instrument.

Fourteen months after Pearl Harbor the quantities of Norden bombsights had not materialized to sufficiently supply the AAF's needs. It was now estimated that orders for the AAF would not be completed before February 1944, and by that time the monthly production rate would drop off to one third. This would mean a reduction of bombsights to the AAF. The Bureau of Ordnance informed the AAF that in order to obtain sufficient bombsights their orders should be in the hands of C.L. Norden, Inc., by March 1943 so that production would continue uninterrupted. Admiral Blandy, Navy Department, Bureau of Ordnance, forwarded a letter to the Commanding General AAF explaining the bombsight procurement and allocation.

In order to arrive at a meaningful schedule the AAF was requested to submit their requirements for M-Series bombsights, including any quantities from AAF manufacturing facilities, and this included sights delivered to the British by the AAF. The allocations were based on the estimated monthly production on the War Production Board Report 8-L dated November 30, 1942. Included was about 20% spares. After filling Navy requirements, the AAF would be allocated the rest. Navy facilities showed that the AAF had received not less than 65% of the bombsight production in any six months during this period. The schedule of allocation as of June 30, 1943, showed an undelivered total of 12,382 for the Navy and 10,432 for the AAF for a total of 22,814. It would not be until late 1943 that sufficient quantities of bombsights would be available to the AAF. It was expected that bombsight production for 1943 from all sources would be 14,438, with 7,536 allotted to the AAF and the rest for the Navy. This amounted to a little over 52%, and in 1944 the Navy would take approximately 65% of the expected production, with the AAF receiving about 35% from all sources.

With the escalation of the bombing of Germany and Japan, this did not look favorable for the AAF. Unfortunately, the Navy could not predict more than one month in advance what increased allocations would be to the AAF. The situation on the Mark XV bombsight was fast becoming hand-to-mouth. At this time the stabilizer (part of the Auto Pilot) was as scarce as the bombsight. It was expected that by late May or June there would be substantial production from the Victor Adding Machine Company and the Cardanic Corporation. Again, should conditions become more critical, contingency plans were made to eliminate shipments to AT-11 Trainers. The lack of the critical stabilizers was fast becoming a problem for the Boeing Company. It was necessary to determine where the exact place in the production line the stabilizer must be installed to prevent stoppage.

With the introduction of the new M-9 bombsight, the Bureau of Ordnance instructed that all M-9 bombsights manufactured by the Norden Company must contain a serial number preceded by the letter "N." For the first Norden instruments for the Army Air Forces' bombsight M-9 Series, the serial number was to start with No. 4,124 preceded by the letter "N." The Navy bombsights Mod 7 were to start with the number 2,795. Note: the Navy Mod 7 sight was the same as the AAF M-9 sight. This letter was assigned to all Norden sub-contractors.[10] Explanation of these letters are explained and appear in a prior chapter. The design of the Army and Navy bombsight name plate did not allow for sufficient space for the letter "N." In order to forward the sights to the using activities (due to the critical shortage of bombsights) pending the redesign of the nameplate, only the serial number was painted on the bombsight rate end housing, with the letter "N" preceding the serial number. The Army Air Forces and Navy SBAE had sufficient space to allow for this letter.[11] The first M-9 bombsight shipped to the AAC that had the trail plate numerals marked with the new type fluorescent paint was Serial Number 4,433. It was sent to the

Army Air Corps Middletown Air Depot on February 6, 1943.[12] The chief of the Bureau of Ordnance on April 1, 1943, advised "the employees of the Lukas-Harold Company, Naval Plant, Indianapolis, Indiana, that the first bombsight produced by the Lukas-Harold Company passed all acceptance tests at the Naval Proving Ground and was installed in a Navy Torpedo bomber type TBF-1 (Air Force designation B-26), which was operating in the Pacific Area. The next test was in combat. Naval aviators would put the sight to good use—'sinking Jap ships.'" This was the beginning for this new ordnance plant. The plant had been gearing up almost over a year and a half before a single bombsight was assembled and accepted. They would go on to make many more thousands of sights before the end of WWII.[13]

Nineteen forty-three loomed as a year of critical bombsight shortages. The Navy was attempting in every way to provide the AAF with M-9 Series sights. They could not predict more than one month in advance what the allocation would be. The shortage of M-9 sights was being felt and was on a "bare-hand-to-mouth basis." Even the bombsight stabilizer was as scarce as the sight itself. There was no advantage to ship stabilizers to the airplane manufacturer in advance of the sight. In February 1943, M-9 sights for B-17 airplane requirements was 500 units for Army Air Corps from Navy production, plus additional quantities from the Victor Adding Machine Company within the next ninety days. The situation was accelerating to a critical situation that emergency measures were mandatory. Should this condition become more critical, shipments of sights to AT-11 Training bases would be stopped. Shipments would be made to airplane manufacturers from an emergency stock of stabilizers. This action would be taken as the last resort, because this procedure made it necessary to calibrate the sight in the field. This would create additional problems due to lack of skilled maintenance personnel. The sight and stabilizer shipped together eliminated this operation. To compound the problem, the Boeing Company reported on February 13, 1943, that they had 17 sights on hand and would be completely out the next day. An emergency allocation of 92 sights was made, which would keep airplane production moving for a few more days until more sights became available.[14] However, more sights were not forthcoming. To try to expedite the allocation of more sights, the AAF again made a formal appeal to the Joint Aircraft Committee (JAC) for a review of the problem to determine if the Navy Bureau of Ordnance was providing the AAF's fair share of the production of sights in line with the actual output of airplanes of the two services using the same equipment. Rather than appear before the JAC, AAF personnel met to discuss this situation with the Navy and the Norden Company. The AAF was planning to put the Norden Company on the Urgency List. Other bombsight contractors were already on it.

The critical shortage of bombsights continued to plague the AAF. For April 1943, the AAF shortage of 75 sights existed, in addition to the accumulated slippages of the previous months. There were also 53 airplanes at the various Modification Centers that were completed and ready for shipment to the various combat theaters for which no sights were available. Stabilizers were shipped in sufficient quantities so that all airplanes were equipped on the production lines. No shipments were made to the AT-11 Trainers or for any spare parts. Note: The author remembers the critical shortages of spare parts and bombsights. Every conceivable method was used to save every usable part as none were forthcoming from the factory. Special tools, jigs, and fixtures were made so that the repair personnel would not damage any part of the sight. Special bearing pullers were designed by the author to prevent damage in removing and replacing highly precision anti-friction bearings. There were none to be obtained from the manufacturer, as this was the most critical item in the production of bombsights. Every effort was made to save as many bearings as possible. Only the most experienced machinists with an outstanding background in precision machining were accepted. Age was not a factor. One of my best machinists was a retired person from a watch company. He was in his late sixties and outperformed many of his younger colleagues.

With such a critical shortage of bombsights, the question was "what caused the bombsight shortage." The Norden Company and all of the sub-contractors producing bombsights could not turn out maximum capacity because of a shortage of certain critical bearings. Every attempt was being made to increase this supply through the various manufacturers. The Army Air Force's sources, Victor Adding Machine Company and Cardanic-Burroughs, were late getting into production, the result of delays in obtaining specialized machine tools (lathes, milling machines, grinders, etc.), training the labor force, learning manufacturing techniques brought about by the extremely small tolerances (in the range of 0.0001" to 0.001", and in some areas 50 millionth of an inch), and the complicated nature of the instrument itself. Desperate conditions required drastic actions to relieve the shortages. To help alleviate this, a Bombsight Bearing Committee was set up with representatives of the AAF, Bureau of Ordnance, the War Production Board, and all bearing manufacturers to allocate bearings to the best advantage among the Navy, AAF, manufacturing plants, and all bombsight sub-contractors. The machine tool industry (lathes, mills, grinders, etc.) was finally catching up with the tremendous amount of orders placed on them due to the war. The demand was so great that it was almost impossible to keep up with orders pouring in. With more machines becoming available, the bottle neck for this equipment was finally broken and equipment started flowing to the various contractors in sufficient quantities that it was no longer a severe limiting factor. For all activities, such as the Norden Company, Navy sub-contractors, and AAF production sources, received all possible assistance in expediting instructions to their employees in the manufacture and assembly of parts.

Arrangements were also made for all manufacturers and sub-contractors to exchange information on their difficulties, and to sell to each other surplus parts and materials to obtain *absolute maximum production*. As rapidly as possible, 12 volt sights were pulled back from the theaters of operation, with some already at the Victor Company plant, being reworked into 24 volt configuration. Arrangements were also made to pull back sights from the production lines at Vega, Boeing, and Douglas so that all available sights could be used at the Modification Centers for airplanes going overseas. The Navy requirements were reviewed by Under-Secretary of the Navy Forrestal as to the Navy's needs for the M-9 bombsights. The situation was so serious that the AAF was considering the use of the Estoppey D-8 bombsight in place of the Norden M-9 on airplanes that could later be equipped with the Norden sight. Replacement of the D-8 sight with the M-9 when they became available would be an easy task. The D-8 sight was fairly accurate at low levels but not at high altitudes. Bombsights would be made available only to staging areas, so that all airplanes taking off for overseas duty were fully equipped. There would be a considerable time before Modification Centers could be completely covered, and an even longer period before bombsights could be furnished from the various production lines. Every effort was made to obtain an agreement from the Bureau of Aeronautics that lower group Navy (i.e. Group IV and V) airplanes would not be equipped with M-9 sights until sufficient units became available. In order to speed up to maximum production, each bombsight contractor made a direct appeal to the workers in the plants to do their utmost to obtain peak production.[15] The author remembers giving pep talks to workers who were asked to do their best in meeting or exceeding production goals. At that time, the war was not going well for us. It was indeed heartwarming to see both men and women responding to do their best regardless of working conditions, such as lack of parts, tools, equipment, long hours, etc. There was no sympathy for those who wasted time. Being late for work was frowned upon. We were able to keep abreast or beat work schedules, as we were the last U.S. staging base before bombers left for the Pacific combat zone. In January 1944, our Supply Division made a carload shipment of bombsights to overseas destinations. Carload shipments of bombsights of this magnitude were unusual.[16]

In spite of all the strenuous emergency actions taken by both the AAF and the Bureau of Ordnance, there was still a huge shortage of sights that would extend approximately to November 1943. The Victor Adding Machine Company, Lukas-Harold Corporation, and Cardanic-Burroughs were just starting to produce sights. Remington-Rand had not yet reached any sort of satisfactory production and was experiencing considerable trouble in getting their production lines started. Every effort was being made by the Navy and the prime contractor (C.L. Norden Company) to bring this plant to capacity production as soon as possible. The Bureau of Ordnance offered to transfer this company to the jurisdiction of the AAF. There was no advantage for them to accept this facility, as they were experiencing their own problems and there would be no advantages gained by taking it over, so it was turned down.[17] By May 8, 1943, Victor Company had produced 14 sights, and

Cardanic-Burroughs, 10. At the present rate, it would not be until the first part of 1944 that both contractors would reach their peak production of 700-750 per month as scheduled.[18]

The AAF was now receiving increased airplane production and the shortage of bombsights required for equipping heavy bombardment airplanes was daily becoming more acute. As a temporary expedient for alleviating this condition, the AAF requested the Bureau of Ordnance to authorize the elimination of the Naval Proving Ground at Dahlgren, Virginia, to shop and flight test all AAF sights, which at times delayed delivery up to four or five weeks. In the past, all bombsights were shipped to the NPG before they were released to the user. However, since there were so many thousands of sights produced, it was decided that one representative sight out of ten would be tested. For example, the production rate of 1,000 sights per month would mean that 100 would be awaiting testing. This quantity of sights could equip several squadrons of airplanes critically needed in combat areas.[19] The Bureau of Ordnance would not waive the Dahlgren testing, claiming that the inspection at the manufacturer's plant was not considered adequate for the final acceptance of the bombsights. Apparently, they did not trust their own inspectors at each of the plants. To eliminate long delays of AAF sights, the Bureau of Ordnance would continue to shop test all representative sights sent to the Proving Ground built by Lukas-Harold and Norden Company, promising that the shop test would be completed on the same day of receipt, and to flight test only such portion of the AAF representative sights that could be completed in the time to permit shipment to be made no later than 15 days after shop tests. AAF sights would not be delayed any longer than approximately 15 days. The Bureau of Ordnance felt that inspection at NPG was necessary to maintain the high standards required by the bombsight. Alleviating the shortage of bombsights for the Air Force did not materially increase production. Production was limited to the availability of acceptable bearings. Even with the establishment of the Bearing Committee, controlling all bearing production would eventually increase the monthly allotment of sights to the AAF.

Throughout 1942, the Bureau of Ordnance continued strenuous efforts to increase the output from Norden and to bring the Naval Ordnance Plant at Indianapolis into production. Further attention was given to the construction and equipment facilities at Remington-Rand so that it could start production of bombsights.

It became apparent that the total manufacturing capabilities of bombsight production by Navy controlled facilities by the fall of 1943 would be 1,800 bombsights per month, 1,600 SBAE, and 2,500 directional stabilizers, plus other related equipment. Production from AAF activities was to begin in February 1943 and reach maximum monthly production by September 1943, with 1,400 bombsights, 1,700 SBAE, and 2,400 stabilizers.[20] To further compound matters, the production of bombsights was seriously affected by the continuing shortage of extremely criti-

Chapter XIII: The Navy's Norden Mark XV, M-9 Bombsight 149

cal precision ball bearings. Bearings were in such short supply that a meeting was held jointly with the Army (Air Materiel Command), Navy (Bureau of Ordnance), the War Production Board, and all high level personnel of the ball bearing industries. The urgency of this meeting reflected the gravity of the situation of ball bearings from all sources, and to insure the best possible distribution of all available useable bearings.[21] In spite of all the shortages, the Navy was able to furnish approximately 70% to 80% of the bombsight production, but the full needs for AAF could not yet be met. The Navy (Chief of the Bureau of Ordnance) was very anxious to cooperate with the Commanding General, AAF, and to discuss the bombsight production situation in person or through representatives. The problem was non-availability of material, not production facilities. On June 5, 1943, the Bombsight Bearing Committee, with headquarters at the Barden Company in Danbury, Connecticut, received an urgent message sent by the Norden Company to the Bureau of Ordnance requiring 4,100 various types of bearings. If these were not forthcoming during the week starting immediately by June 6, 1943, they would not be able to keep production on schedule. One department was already completely out of bearings and was shut down.[22] At about the same time the Lukas-Harold Company also reported critical supply of bearing number 465369. From July 1 through September 14, 1943, they delivered to the Navy 1,015 sights. The rejection rate for this bearing was 22.5% and, by September 10, 1943, twenty persons were idle in the final assembly department for 16 hours because they had completely run out of this size bearing, roughly equivalent to 40 completed bombsights. It was important, from the standpoint of morale in the organization, that they did not lose the momentum that was gained in production. This, in turn, would cause serious problems in the production lines. The Bearing Committee was well cognizant of the problem, but there was nothing that they could do until sufficient bearings were available from the various bearing contractors.[23] As explained in a previous chapter, the rate of good precision bearings was one in ten and, as the Barden Company started production, it increased to one, to two and a half, to three. This committee was above any politics, and their purpose was to view the requirements and allocate bearings where they were needed the most. There is no question that the Bureau of Ordnance was most anxious to assist in this problem and cooperated with the committee in every way to fill the needs of both services for bombsights and related equipment. These shortages were to last for many months before it reached a point where bearings would become available in sufficient quantities to satisfy all of the bombsight contractors.[24]

In mid July 1943, the question of the failure in supplying maintenance spare parts came up. The production and delivery of maintenance spares for the M-Series bombsight was deficient to meet all demands, and there was no positive assurance that sufficient spares would be available to the operating units in the future. The Navy (Bureau of Ordnance) and the AAF (Air Materiel Command) were cognizant that insufficient consideration was given to producing quantities of spares concurrent with bombsight head deliveries. The AAF felt that the Bureau of Ordnance had not given sufficient consideration to producing adequate quantities of spares concurrent with bombsight deliveries. This resulted in an inadequate supply of maintenance spare parts to meet demands in the combat theaters, and there was no assurance that any would be forthcoming due to the critical shortage of bombsights. The Bureau of Ordnance and bombsight contractors had concentrated in producing complete bombsight systems. Before WWII, the spares program was not acute because the bombsights could be returned to the manufacturer for necessary repair or modification. Even during peace time, spares shortages were not as acute, but some maintenance spares were difficult to obtain, although the Navy repeatedly assured the AAF that every effort was being made to provide them with the necessary spares. The Navy always had priority over the Army when it came to allocation. Since the war began, the AAF had deployed bombardment airplanes all over the world, with some in very remote areas. It was not only infeasible, but impossible to return the sights to the manufacturer or repair depots since the distances and logistical problems were too great. Because of the shortage of bombsights, it came down to the point of what needed to be done first, "Did the services want bombsight production curtailed in order to get more spare parts, or did they want completed bombsights at the expense of spare parts?" The policy was to produce completed bombsights. It was felt that sights could be overhauled at a later date if the occasion should arise. To alleviate the shortage, the argument was advanced that bombsights could be cannibalized (this means stripping other sights for usable parts) to be used in repairing other sights. There are several arguments, pro and con, on this concept. This practice has been used by the military for eons of time on other weapons with good results. It was pointed out that certain vital parts of the bombsight, after they had been in operation, are "run in" and suitable only to that sight. This point was open to arguments by maintenance experts, and they did not believe that such a procedure was satisfactory. It had never before been tried, so there was no history for it. Because of the extremely close tolerances, these instruments did not lend themselves to widespread interchanging of parts. In theory, it was expected that all bombsight parts being manufactured by the various contractors would have interchangeable parts. Unfortunately, this was not the case. The author had first hand information of trying to use certain parts from different manufacturers. Although the Victor Company was not yet in full production, they were not able to produce maintenance parts for the AAF except in minor quantities. Production lines could be throttled at any time in favor of spare parts or vice versa, depending upon requirements. This method was to become binding on all bombsight contractors.[25]

In another attempt to obtain more bombsights to alleviate the shortage, bombsights manufactured by the Burroughs Adding Machine Company were inspected by company inspectors, followed by a preliminary inspection by the Resident Naval Inspector of Ordnance (RNIO). After they passed inspection they were shipped to the Norden Company for another final inspection by their RNIO before being shipped to the using activities. In addition, one in ten was forwarded to the Naval Proving Ground for shop and flight test.[26] This procedure was cumbersome, unwieldy, and time consuming, and the duplication of effort was causing storage problems at the plants. In the beginning of producing sights by other activities, this method was considered essential for all the manufacturers getting started due primarily to lack of skilled personnel at these plants. This procedure was set up only as a temporary expedient during the early production phase. Burroughs, as well as all the other bombsight manufacturers, were producing bombsights showing consistently good workmanship, sound manufacturing methods, and few fundamental faults in production and other errors, which were discovered and quickly corrected. In February 1944 their production was only subject to minor adjustments that could normally be performed in the field. Also, their defects were approximately the same percentages that occurred in sights produced by the Norden Company. In view of quality produced by Burroughs, and to help alleviate the shortage, this practice was discontinued, and final acceptance instituted at the plant. In addition, the BurOrd directed the NIO at the Burroughs plant be beefed up with additional inspectors so that the sights could be shipped directly to the services from the plant. The Norden Company provided full assistance to Burroughs in any area they needed help. In order to accomplish this new system, the Bureau of Ordnance ordered that inspection by Naval Inspectors at the Burroughs Company, of necessity, be of the highest type possible, since the present company inspectors were relatively inexperienced in comparison with those of the Norden Company. Immediate steps were taken to obtain these specialized personnel.[27] The practice of sending one sight in ten to the Naval Proving Ground was still in effect for shop and flight test.[28] The Navy operated Remington Rand plant was undergoing several problems in the production of sights. The training of workmen under Norden management would progress to the extent that bombsights produced at that plant would be entirely satisfactory. The administration of the plant was taken over by the Navy by Executive Order signed by President Roosevelt on November 25, 1944, and placed under Norden management to continue to produce bombsights.[29]

The lack of bombsights was not only hampering the Navy, but also the AAF. It would be 1944 before sufficient sights would become available for both services. As dark as the picture was, there was light at the end of the tunnel. Within the next few months, Victor, Lukas-Harold, and Cardanic-Burroughs would reach maximum production. In view of the seriousness of the situation, the AAF made a direct appeal to the Deputy Chief of Naval Operations (Air) seeking their assistance in obtaining relief of the bombsight shortages from Navy sources. The AAF's request for monthly projected estimate of sights available for periods greater than 30 days in advance could not be provided, which precluded definite constructive planning for theater commitments in order to coincide with planned combat operations, not only for the AAF's heavy bomber program, but for the medium and light bombers, as well. To make matters worse, the WPB canceled one bombsight contractor, Cardanic Company, in August 1943. The appeal was from General Arnold, Chief of the Army Air Forces to the Chief of Naval Operations (Air), and was forwarded to the Bureau of Aeronautics for action and in turn to the Bureau of Ordnance. Since no action was taken, it was felt that this problem was of sufficient stature to make it an item for the JAC for proper scheduling.[30] The Chief of Naval Operations (Air) did not feel that this problem should go to the JAC and could be resolved by working with representatives between the AAF (Air Materiel Command), Deputy Chief of Naval Operations, the Bureau of Ordnance, and with the War Production Board since production would be influenced by any new requirements which the board would decide. The AAF expected to get some relief from these high level meetings.[31]

As a result of the meetings held in October 1943, the Navy increased the peak production of the Norden and Lukas-Harold Corporations from 600 to 650 sights per month each, which would give the AAF an additional 100 sights. In December 1943, the AAC received 911 bombsights from Navy facilities. By January 1944, this would increase to over 1,000 sights per month and continue to increase up to over 1,500 sights per month. In January 1944, total production was 1,700 sights, the Navy took about 35%, and the rest went to the AAF.[32] It is not clearly understood why the Navy needed so many sights, since they were not in the business of conducting strategic or radical aerial bombing on such a vast global scale as the AAF.

The AAF goal was to equip all very heavy, heavy, medium, and light bombardment airplanes, including the A-20, that had bombardier equipped noses. Although the action to equip A-20 airplanes could adversely effect the M-9 bombsight's levels, these planes were being furnished with D-8 sights, and it was deemed advisable to supply them from current stocks rather than attempt to create a surplus of M-9s in order to meet a future requirement that may or may not develop.[33] Action was taken to install M-Series bombsights in A-20 airplanes with bombardier noses scheduled for AAF overseas assignment. This could not be accomplished on airplanes in the continental U.S. until there was a surplus of sights for all heavy bombardment airplanes. After the above requirements were met, the surplus sights would be installed at the rate of 25% of all medium and light airplanes with bombardier noses. In no event would more than 25% of these planes receive Norden sights until maximum production had been achieved. When the M-Series sights became available in excess of the heavy bomber airplane requirements, 30 M-9 sights were to be immediately shipped to the 341st Bomb Group in India and 5 to the 11th Bomb Group in China by the most expeditious means possible, with absolutely no loss of time.[34] In order to accomplish this objective as soon as possible, the Chief of the AAF was not

Chapter XIII: The Navy's Norden Mark XV, M-9 Bombsight

satisfied with the quantity of sights being received from the Navy, and contacted the Bureau of Ordnance. The Bureau of Ordnance advised the AAF of their best production estimates and that the required quantities would be reached perhaps by October 1943 from Navy controlled facilities, including the Burroughs Company, who were later awarded a Navy "E." The Navy, in reaching 1,000 sights per month, the quantity allocated to the AAF would only be about 700 units after Navy requirements. However, deliveries of SBAE/AFCE units to the AAF surpassed the AAF requirements by delivering a total of 1,378 units in July 1943.[35] After this requirement was satisfied, all surplus sights were to be stocked for installation in medium bombers requiring bombardier operated bombsights when airplanes departed for combat theaters. Spare parts and maintenance equipment were to be sent along with these sights. No Norden bombsights were to be installed in medium bombardment aircraft in excess of the 25% requirement.[36]

The war had been going on for almost two years, and full production had not yet accelerated to a point where there were sufficient sights for all the services. Manufacturers who had started from "scratch" or converted from peace time to war work would, very shortly, be getting into full production. New requirements were continually being generated by the changing conditions in the combat theaters. By January 1944, the Army Air Forces' requirements for sights and stabilizers for the rest of the year was 20,960 units. To support this quantity, the Navy estimated that production from October 1943 to January 1945 for both bombsights and stabilizers from all Navy and AAF sources (Norden Company, Remington-Rand, Burroughs, Lukas-Harold, and the Victor Company) was 32,024 sight heads, with the Air Force receiving 20,960 units, and that all the sights ordered on contract by the AAF would be delivered by November 1944. The stabilizer production for this same period was 31,024 units. However, using finished stocks on hand at the manufacturers' plants, this shortfall of 1,271 units could be met. At the same time, provisions were also made for 33,072 spare parts kits for the bombsight. The spare parts program, which had suffered in favor of sight heads, was now back on track, but it had been reduced from 20% to 10%.[37]

By December 1943, the flow of sights was increasing to a point where the Army Air Forces (Materiel Command) felt that no additional new facilities would be required, nor any expansion of present facilities beyond those that were in operation. It was felt that, if shortages occurred, the spares would be allocated to make up these shortages until production again became adequate to maintain the proper balance of spare parts. In the event that further quantities would be needed, any expansion would have to be borne by AAF facilities, as the Navy had squeezed their manufacturers to the utmost to provide the AAF the most production that they could possibly give. With this brightening picture of Norden sights, the AAF could begin installing the sights in all very heavy, heavy, medium, and light bombers with bombardier provided noses.

Beginning in 1944, the industrial planning for mobilization of the bombsight for wartime conditions was finally beginning to bear fruit. The contractor plants that were in operation before Pearl Harbor and those that were converted to war work seemingly took a lengthy amount of time to achieve peak production. Admittedly, it was almost three years before this event took place due to the fact that the highly technical nature of the instrument caused untold problems never before encountered that required resolution before trying to mass produce such sensitive parts. It now appeared that the AAF would receive the necessary quantities of sights and auto pilots needed to carry out the unprecedented aerial bombing of Germany, Italy, and Japan on a scale never before attempted. Much of the cause was that workers were not familiar nor had any experience with precision work. Every precaution was being exerted by inspectors so that no part or assembly which did not conform to specifications was released for use. There was no substitute for experience in working on such watch-like measurements. Some watches did not require such extreme and critical tolerances. It was a natural consequence that, in bearing down on the manufacturing force to produce nothing but quality work, and rejecting numerous parts that were borderline cases, an element of extreme caution crept into the entire organization, resulting in a slowdown in production, and it would not accelerate rapidly. However, production would improve as mechanics gained experience and self confidence.[38] A case in point, the author hired a highly competent machinist with an extensive background in movie camera repair from a Hollywood studio, Metro Goldwyn Mayer. After two weeks of working to extremely close tolerances, he gave up and left, stating that he would take his chances in the Air Force. His choice was either be drafted or work in a defense plant. He chose the first rather than work on the bombsight.

Increasing production also created more problems. In February 1944, the using activities complained of receiving inferior sights as compared to the previous two years. In the rush to obtain more production, the strict tolerances that were demanded were, in some instances, relaxed. On some parts it did not matter, but, in others, it caused serious problems. During testing by the 2nd Air Force, it was found that there was an apparent erratic precession tendency of the bombsight optics stabilizing gyro. In order for a gyro to work properly, it must stay within a limit of 3.5 mils ± 1 mil in one minute of time based on latitude of 40° and conducted on a vibration stand provided by the Navy, which had an amplitude of .005 of an inch and a frequency of 2,000 per minute. The Norden Company had relaxed the tolerance on the bombsight cardan tilting bearings (locked end) from an axial play of 0.0003-0.0005 inch (three to five one ten thousands of an inch) to 0.0005-0.0015 (five one ten thousands of an inch to one and a half thousands of an inch), which is almost 0.001 of an inch difference. This seemingly small increase in tolerance (approximately one third size of human size) was sufficient to cause much trouble. However, this relaxed tolerance simplified the acute bearing situation by using pre-

Chapter XIII: The Navy's Norden Mark XV, M-9 Bombsight 155

viously unacceptable bearings. In the haste to produce bombsights, production requirements permitted the acceptance of bearings that were not permitted to be used in the past when only a very few of the best bearings out of each group received were to be used by all bombsight manufacturers. The Barden Company was starting to get into full production of producing highly precision anti-friction ball bearings. Since they were the primary contractor for this type of work, the Office of Army-Navy Bombsight Bearing Committee was located at the Barden Company plant located at Danbury, Connecticut. Also, at the same time the use of improved oil was explored for use in these bearings to take up the difference in axial play.[39] It was also found that any foreign matter in the oil or large changes in viscosity by temperature caused sticking of the gyro bearings and, as a result, an erratic precession. In addition, a vibrator stand to permit calibration under simulated operating conditions went a long way in helping to solve this problem. The relaxation of the critical tolerances and specifications in manufacturing precision bearings, which the sight depended on, was not being obtained. It was definitely proven that inadequate inspection procedures caused inherent errors far greater than the engineering specification permitted to be passed by the inspectors. A survey showed that a greater number of sights were being accepted (in the neighborhood of 75 to 80%) that did not comply with the stated specifications. In one example, the specifications permitted a 2.5 mil error, whereas the sights accepted varied from 2.5 mils up to as high as 10 and possibly 12-14 mils. The solution to overcome this difficulty was in tightening up inspection of the production lines. The Navy immediately introduced a new inspection procedure, and the AAF also had the Victor Company immediately perform additional stringent inspection with more inspectors. Inspectors were required to closely check more parts and sub-assemblies to insure that quality parts were being accepted. Procedures were initiated by strengthening the inspection process that would be common to all bombsight manufacturers, and which was quickly taken by both the Bureau of Ordnance and AAF for implementation as soon as possible. The Norden Company's out of tolerance did not vary more than 4-5 mils. The next best company was Burroughs, followed by Remington-Rand, Lukas-Harold, and the Victor Company. Correction of these errors in the field were solved by the next major overhaul. It was estimated that it would take almost one to two years to rectify these discrepancies. In addition, the services felt that this problem was of sufficient importance that it could be susceptible to public criticism and also might be a morale factor on the bombardier using the famed "pickle barrel" bombsight.[40] Combat forces and training organizations were already aware of these discrepancies. Because of this, Technical Orders were prepared for immediate action instructing Depot Repair activities as to the source of the problem and corrective action to be taken. Note: The author remembers very well the quality of sights going "downhill." We could almost identify the manufacturer by the precision measuring of various machined parts. As always, the Norden built units were the best.

For example, a 10 mil range synchronization error did not mean that in actual practice the bomb release point would be 10 mils in error. It would mean that the crosslines (crosshairs) would drift out of the target, and the bombardier would need to make additional refinements to depreciate the effects of the error. However, it would not affect the accuracy of the bomb hits to any real extent, but it was an annoyance to the bombardier. The Norden displayed better accuracy because Navy inspectors had established a system to take care of this problem many years previously. Unfortunately, this information had not been applied to Norden subsidiary plants, such as Lukas-Harold, Remington-Rand, and Burroughs. This deficiency was readily corrected.[41]

With the increase in speeds and altitude, the Norden Company was attempting to provide 300 mils trail and 150 mils of cross trail. To accomplish this, the drift and gyro freedom limits would be compromised, i.e. drift might be reduced from 30° to 25° and gyro freedom from 23° to 20°. The AAF Armament Division investigated the possibility of making this compromise. The incorporation of the Victor increased trail and spotting device, along with the high altitude disc speed modification, was not incorporated on bombsights under contract with the Navy. Engineering drawings were prepared and forwarded to all concerned. The question of dry stabilizer clutch surfaces versus lubricated clutch was developing into a serious problem. Manufacture of greased surfaces facilitated machining, but they were more likely to collect dust and dirt. The author spent considerable time working on this item. Dry clutch surfaces required considerably more time in manufacture, but gave a better contact of surface. However, such surfaces were subject to rust and corrosion. It was determined that, in the future, dry surfaces were to be incorporated on all bombsights. The AAC was all for developing the increased trail in the M-9 sights. The Norden Company, to increase trail and cross trail in the M-9, was given a very high priority to accomplish this new design and was closely monitored by the AAC Armament Laboratory (Wright Field).[42] In September 1944, the incorporation of the Victor increased trail and spotting device, along with the high altitude disc speed modification, had not yet been made a part of the bombsight under contracts with the Bureau of Ordnance, but this change was being expedited as fast as possible on the production lines so that high altitude bombing could be possible.

To permit bombing from high altitudes, new reticules (crosshairs) were required to have an effective opening of 2 mils and had undergone tests at the AAF Eglin Proving Ground and at the NPG. The new design had a 2 mil opening in the crosshairs, each line being split in the middle, 1 mil from exact center. These new optics were crucial and badly needed for extremely high altitudes. The crosshairs presently being used completely obliterated a target the size of a small building at altitudes of 25,000 feet and above. Optical manufacturers stated that such a change could be incorporated very quickly in their production lines.[43] The Navy did not require bombing from these high altitudes and offered no objection in manufacturing this type of crosshair.

By June 1944, increased trail provision to 230 mils was discussed. This was required by the AAF for 400 mils of trail with a corresponding amount of cross trail. At the present time there were no provisions for such an increase. The samples of this design had been thoroughly tested by the AAC and proved that it worked. The AAF Assistant Chief of Air Staff advised the Deputy Chief of Naval Operations that they would furnish these to the Bureau of Ordnance. Decreased disc speed provisions would permit use of the present Norden sight up to 45,000 feet. A design was completed and tested, and a sample supplied to the AAF. This change affected the present lower limits to which the bombsight could be synchronized and could not be synchronized under 2,900 feet. This change was necessary to permit bombing at high altitudes, where the target was so small that present solid lines obliterated the objective.[44]

Because of the huge investment involved in the production of bombsights and related equipment, the AAF and Navy Budget Offices requested information on the cost of M-9 Series sights purchased by the AAF from the Navy and Victor Company. The first contract for 1,000 sight heads from Victor showed a cost of $7,625.00 per unit. Four thousand thirteen (4,013) sights purchased from Norden cost $8,800.00 per unit, and 4,065 cost $6,051.00 per unit. The total of these two contracts, averaged and weighted, was $7,560.00 each, or only $65.00 under the price of Victor Company. Victor had not yet achieved full production and was having some problems on their production lines. However, as production and experience increased, their cost would decrease and would be competitive with Norden built units.[45] Although there were some deficiencies, they were not serious and were quickly corrected.

By the latter part of 1944, an embarrassing condition was created when maximum production was finally realized, resulting in an excess of bombsights for Army and Navy requirements. The production of the NBS was a story of "rags to riches." Due to this situation, it was determined by the AAF (Materiel Command), Navy (Bureau of Ordnance), Headquarters AAF (Air Staff), Supply Division, and all other production people, that requirements vs. production existed to the point that there would be a surplus of bombsights. In addition to this, the Twelfth and Fifteenth Air Force also cut back their bombsight head requirements, adding to a further surplus. This did not affect the stabilizers for the M-Series sights. As a result of this over supply in June 1944, partial contract termination notices were sent out to all of the bombsight contractors and sub-contractors. Action was taken for the cancellation and transfer to the Navy of all Army (AAF) facilities except the Victor Company. This would leave the Norden Company and Lukas-Harold Company operating. Which one of these two companies was to continue to produce sights was to be determined at a later date pending the outcome of the war. The contractors that were retained would continue to produce sights on an orderly basis. However, the AAF would continue to need bombsights even after the fall of Germany. The air war then would be transferred to the Far East, and particularly the Japanese home islands. In December 1944, modification of M-9 sights to M-9B configuration to incorporate a trail spotting device and high altitude features was required. Two thousand four hundred (2,400) sight heads in stock required this modification at a rate of 300 per month starting in March 1945. The work was to be performed by the Norden Company.[46] Minneapolis Honeywell Regulator Company was eliminated in the production of stabilizers, while the Burroughs Company was transferred to the Navy and would be closed in December 1944. The Elmira, New York, plant of the Remington Rand Company was scheduled to be closed in October 1944. The availability of bombsights would not be affected because the planned peak production was not realized until the late part of 1944.[47]

The Strategic Air Force in Europe advised Headquarters AAF that the next 1,000 replacement B-17 and B-24 airplanes were to be shipped less bombsight heads, but bombsight stabilizers were to be included (part of the autopilot). After this requirement was filled, one bombsight head per every two heavy bombardment airplanes was to be provided. This did not apply to medium or light bombardment airplanes. The savings of bombsight head definitely took care of the sight shortage.[48] At the same time, requests were received from various Allied governments for allocation of Norden bombsights. The requests came from China, Netherlands, Australia, France, and Brazil. The Chinese already had B-24s and B-25s, and it was assumed that the Japanese had captured some of the Norden units. There was no reason for not allowing the Chinese to have the Norden sight. The transfer of Norden sights with Lend-Lease aircraft was authorized for the above nations, providing sufficient sights were available for the AAF's needs before transfer to these countries. The French, in the first part of 1944, had received 57 B-26 airplanes and would have Norden equipment installed prior to shipment.[49] Since there was a surplus, this did not present a problem. The United Kingdom and Canada did not want Norden sights, as they had their own Mark 14 (or T-1). The Russians did

Chapter XIII: The Navy's Norden Mark XV, M-9 Bombsight

not receive the Norden, but since there was a surplus of Sperry S-1 and A-5 auto pilot, they would receive the Sperry to be installed on planes allocated to the Russians, due to the conversion of B-24 airplanes from Sperry sights to Norden equipment.[50]

By October 1944, the AAF took action to cancel 2,365 sight heads. Attention was turned to the production of new equipment and attachments. Production for the C-1 auto pilot Formation Stick was increased to the later part of 1945. The Reflex Open Optic Sight for M-Series was also increased and was to end in November 1945. The Low Altitude Bombing Attachment for the M-Series sights was scheduled for production to November 1945. The Steering motor for the C-1 Auto Pilot production rates of 500 units per month to peak in January 1945 was inadequate and had to be increased to approximately 700 units per month until November 1945. The Bombardier's Control Knob was also being produced, but no schedules were found in the files. Remington Rand plant was closed the end of September 1944, and Burroughs was to stop its production of bombsights by the end of December 1944. The Norden Company was to stop producing bombsights after the December 1944 deliveries.

In the first part of December 1944, the Army Air Forces found that they would need more sights than they had planned on. The war in Europe was not going to end as early as predicted, and would continue into 1945. Additional bombers were being received from the manufacturer, and more bombsights were needed. The M-9B had the high altitude disc speed modification and were to be used mainly on B-29 airplanes. The earlier cancellation of 2,365 sights was reinstated. On December 4, 1944, the Bureau of Ordnance was informed the AAF would require approximately an additional 4,200 sights from Navy sources. The Bureau of Ordnance placed 2,620 sights with the Norden Company at a rate of 350 per month, and the rest went to Lukas-Harold for the same monthly output.[51] Production continued until May 1945. The AAF had awarded a contract for 6,507 M-9 sights and 4,000 stabilizers. With the end of the war in Europe, production was curtailed about 50%. The dropping of the Atomic bombs over Japan in August 1945 brought the end of WWII, and all bombsight producing facilities were shut down.

Notes:

[1] Handbook of Instructions for Bombsight Type M-9, Technical Order AN 11-30-69 dated June 5, 1945.
[2] The navy's Mark 15, (Norden) Bomb Sight, Its Development and Procurement 1920-1945, page 176.
[3] Ibid, page 178.
[4] Letter Naval Proving Ground to Chief of the Bureau of Ordnance, Subject: Automatic Release Mechanism for MK 15 Bombsights, Test of, dated January 23, 1942 Book 16, National Archives.
[5] Army Air Forces, Materiel Center, Memorandum Report on Bombing Equipment Conference, November 10, 1943, Minutes of Bombing Conference Held at Wright Field on October 13-14, 1942, page 6, National Archives.
[6] War Department, Air Corps, Purchase Order to Navy Department, Bureau of Ordnance, dated June 15, 1942, National Archives.
[7] Letter from C.L. Norden to Chief of Bureau Ordnance, Subject: Bombsight Type M-9 and Stabilizers-Request for Quotation, dated January 4, 1943. National Archives.
Free On Board. Item priced at the manufacturers plant. Cost of shipping and any other expenses to be borne by the purchaser. In other words, it is "cash and carry".
[8] Cost plus fixed fee reimburses a manufacturer for his expenses but limits his fee or "profit" to a figure rigidly fixed beforehand. This type of contract was used to assist small manufacturers.
[9] Letter from C.L. Norden to Chief of Bureau of Ordnance, Subject: Bombsight Type M-9 and stabilizers-request for quotation, dated January 4, 1943, National Archives.
[10] Letter from Naval Inspector of Ordnance, to the Chief of Bureau of Ordnance, Subject: Serial numbers of Bombsights and S.B.A.E., dated January 13, 1943, National Archives.
[11] Ibid.
[12] Memorandum to Chief of Bureau of Ordnance to Commanding General Army Air Forces, Materiel Command, Subject: M-9 Bombsight, Modification of, dated March 3, 1943, National Archives.
[13] From Chief of Bureau of Ordnance to Employees of the Lukas-Harold Corporation, Naval Ordnance Plant, dated April 1, 1943, message of congratulations, National Archives.
[14] Memorandum from Chief, Production Control Section, shortage of M-9 Bombsights on B-17 Airplanes, dated February 13, 1943, National Archives.
[15] Letter to Commanding General, Materiel Command from The Technical Executive, Subject: Shortage of M-9 Bombsights, dated April 16, 1943, National Archives (MS), page 3.
[16] Article, Subject: Bombsight Breaks Record On Huge Rush Order, no date, from the McClellan AFB Pacemaker. Author.
[17] Memorandum for General B.W. Chidlaw, from Armament Section AAC, Subject: Bombsights — M-Series (Norden) and S-Series (Sperry), dated May 20, 1943, National Archives, MS.
[18] Ibid.
[19] Navy Department, Bureau of Ordnance to Commanding General, Army Air Forces, Subject: Inspection of the "M" Series Bombsights at Dahlgren, Virginia, dated May 21, 1943, National Archives, MS.
[20] Navy Department, Bureau of Ordnance to the Commanding General, Army Air Forces, Subject: Bombsight Production for the Army Air Forces, dated May 18, 1943, page 3, National Archives, MS.
[21] Ibid.
[22] Western Union Telegram to Bureau of Ordnance, Navy Department, to Carl L. Norden, Inc. concerning bearings, dated June 5, 1943, National Archives.

Chapter XIII: The Navy's Norden Mark XV, M-9 Bombsight   161

[23] Letter from Lukas-Harold Company to Army-Navy Bearing Committee, the Barden Company, Danbury, Connecticut to Commanding Officer Naval Ordnance Plant, Indianapolis, Indiana, Subject: Critical Condition on Bearing 465369, dated September 15, 1943, National Archives.
[24] Navy Department, Bureau of Ordnance to the Commanding General, Army Air Forces, Subject: Bombsight Production for the Army Air Forces, dated May 18, 1943, National Archives, MS.
[25] Army Air Forces Headquarters to Assistant Chief of the Air Staff, Subject: Production Facilities and Delivery of M-Series Spare Parts, dated July 26, 1943, National Archives, MS.
[26] Navy Department, Bureau of Ordnance Memorandum for Bureau Files, Subject: Inspection of Bombsight Mark 15 Manufactured by the Burroughs Adding Machine Company and by the Navy Operated plant Elmira, dated February 16, 1944, National Archives.
[27] Letter from Bureau of Ordnance to Naval Inspector of Ordnance, Carl L. Norden Company, Subject: Inspection of Bombsights Manufactured by Burroughs Adding Machine Company at Manufacturers' Plant, dated March 10, 1944, National Archives.
[28] Office of Naval Inspector of Ordnance to Chief of the Bureau of Ordnance, Subject: Inspection of Bombsights Manufactured by Burroughs Adding Machine Company at Manufacturers' Plant, dated March 10, 1944, National Archives.
[29] Ibid, same as number 21, page 2.
[30] Letter to Vice Admiral John S. McCain, Deputy Chief of Naval Operations, Subject: Allocation of Norden Bombsight Production from Chief of the AAF, dated August 20, 1943, National Archives, MS.
[31] Routing and Record Sheet, Headquarters Army Air Forces, Subject: Allocation of Norden Bombsight Production, dated September 15, 1943, National Archives, MS.
[32] 1st Endorsement, from Vice Chief of Naval Operations to Commanding General, Army Air Forces, Subject: Bombsight Mark 15, Production and Allocation (BurOrd. Secret letter F-41-8 (PLc), 001446 dated October 1, 1943) dated October 2, 1943, National Archives, MS.
[33] Letter from Production Engineering, AAF to Chief of Bureau of Ordnance, Subject: Estimated Delivery Schedules M-Series Bombsights from Navy Facilities for AAF, dated October 19, 1943, National Archives, MS.
[34] Memorandum to Assistant Chief of the Air Staff from Requirements Division AAF, Subject: Norden Bombsights for A-20 Aircraft, dated February 19, 1944, National Archives, Frame No. 0484.
[35] Letter from Assistant Chief of Air Staff, Materiel Command to Commanding General, Wright Field, Subject: Allocation of M Series Bombsights - Re: CTI 586, dated June 30, 1943, National Archives.
[36] Bureau of Ordnance to the Commanding General, Army Air Forces, Subject: M-Series Bombsight and A.F.C.E. Deliveries to Army Air Forces, dated August 12, 1943, National Archives, MS.
[37] Hqs Army Air Forces, Routing and Record Sheet, from Assistant Chief of the Air Staff, Subject: M-Series Bombsight, dated June 21, 1943, National Archives.
[38] Bureau of Ordnance, Memorandum to Commanding General, Army Air Forces, Subject: Bombsight Production and Deliveries to the Army Air Forces, dated December 9, 1943, page 4, National Archives, MS.
[39] Letter from Remington-Rand Inc. to the Chief of the Bureau of Ordnance, no subject, dated September 22, 1943, National Archives, page 2.
[40] Letter from Chief Engineering Division to Headquarters Materiel Command, Wright Field, Subject: Bombsight — Data submitted by Headquarters 2nd Air Force, 1st Indorsement, dated February 22, 1944, National Archives, Frame No. 0477.
[41] Letter from Materiel Command to Chief Bombsight Manufacturing and Inspection Procedures, dated August 25, 1944, no subject, National Archives, Frame No. 0563.

[42] Army Air Force Headquarters, Engineering Division, Memorandum Report, Subject: Conference on M-9 Bombsights on August 16, 1944, dated September 8, 1944, National Archives, Frame No. 0570.
[43] Ibid, page 3.
[44] Memorandum Report from Armament Division, Subject: Bombsight development Conference — Materiel Command, dated June 5, 1944, National Archives, Frame 0521.
[45] Letter from Assistant Chief of Staff to Deputy Chief of Naval Operations (Air), Subject: Changes in M-Series (Mark 15) Bombsight, dated June 17, 1944, National Archives, Frame No. 0528.
[46] Memorandum from Budget Office, no subject, dated August 17, 1944, National Archives, Frame No. 0562.
[47] Letter from Procurement Division to Assistant Chief of Air Staff, no subject, dated December 7, 1944, National Archives, Frame No. 0603.
[48] Headquarters Army Air Forces, Memo For the Record, Subject: Present Bombsight Progress, dated October 7, 1944, National Archives, Frame No. 0583.
[49] Memo for General Chidlaw, Subject: Norden Bombsights, dated April 14, 1944, National Archives, Frame No. 0516.
[50] Routing and Record Sheet from Deputy Chief of Air Staff to Chief of Air Staff, Subject: Bombsights for French Lend-Lease Aircraft, dated February 16, 1944, National Archives, Frame No. 0501.
[51] Headquarters Army Air Forces Memorandum for the Chief of The Air Staff, Subject: Possibility of Putting the Sperry S-1 Bombsight in Russian B-25's, no date, National Archives, Frame No. 0509.
[52] The Navy's Mark 15 (Norden) Bombsight, Its Development and Procurement, 1920 - 1945, page 313.

XIV

Augmenting the Mark XV, M-9 Bombsight

A strategic and tactical air force that can put more and better equipped bombardment airplanes into the field holds a large advantage on destroying the industrial capacity of the enemy. Quantity, combined with quality, large numbers, and superior performance, gives a decided advantage. Continued superiority requires continual changes. Every new innovation introduced by the enemy must be countermatched, resulting in an endless procession of changes. Range, distance, higher altitudes, evasion of flak, etc., caused many changes, namely modifications and attachments. Nothing remains static in war. Fluidity of production lines is the life blood of mass production. Static lines have no place in modern warfare. Unfortunately, changes in design, engineering, and testing sometimes resulted in complete renovation of existing methods, and the retraining of personnel caused a loss of momentum in the production lines. To this end, rather than shut down production for retooling, retrofits or modifications after acceptance of the equipment affected by many engineering changes, new designs, ideas, etc., the use of many modifications was made on the bombsight proper. Some were extremely successful, while others did not work. Since the production of the sight had peaked, attention was turned to making the sight more accurate and easier to use during a bomb run. Many of these items had been thought of and were in the design and engineering stage, and some were in semi-production process in 1941. Previous to 1944, all efforts were directed to the production of complete bombsights and related equipment, with all the other equipment held in abeyance so as not to disrupt the production lines actively engaged in the production of the M-7 and M-9 Series bombsights. Due to the complex nature of some of these modifications or attachments, only a brief description will be given. The following are some of the major and important ones concerning the Norden bombsight:

GLIDE BOMBING ATTACHMENT

Commonly known as the GBA, this attachment was suitable only for flat dives of twelve to fifteen degrees, and its principal value was in reducing the effectiveness of anti-aircraft artillery by permitting engines to be silenced, and by upsetting calculations through changing altitudes. A sight for steeper glides is unnecessary for bomber types, since they would not withstand the air loads of steep dives and bombs would not clear the bomb racks. The GBA automatically computes and generates the disc speed for the bombsight throughout the range of altitudes from 5,000 to 26,000 feet, and a range of vertical velocity from 9,000 feet per minute glide to 2,400 feet per climb automatically corrects a bombing operation, which permits accurate bomb releases in climb, glide, or horizontal flights. It measures the altitude and vertical velocity of the aircraft with a self-contained altimeter and constant speed motor. This data is supplied to a computer, which positions a roller on a pair of constant speed discs. The roller, through a flexible shaft connection, drives the disc in the bombsight. It is powered from 26-28 volts Direct Current from the airplane's circuit. It is a very complex piece of equipment (Plate #1).[1] The instrument consists of two principle parts: The disc speed computer, which is contained in the base of the instrument, and the barometric element, with its computer for altitudes and vertical speed. It is almost like a bombsight in itself. A connection is made to the side of the barometric element changer to the pitot line of the airplane. Mr. Norden had plans for developing a GBA as far back as 1937 using the Mark XV, Mod 1. The GBA was not something new as an attachment for a bombsight. A patent was filed by F.W. Morganthaler and J. Garwood on January 17, 1942, and was issued on October 29, 1946, with assignors to the Sperry Gyroscope Company of New York City. This was predated by a co-pending application for a GBA Serial No. 269,838 filed April 15, 1939, in the names of C.A. Frische and G.N. Hansen. Mr. Norden, at a conference, stated in March 1937 that the modernization of the Mark XV, Mod 1 bombsight would permit horizontal and glide bombing. The Commanding General, Hq Army Air Forces was vitally interested in the development of a GBA. In March 1941 the Chief of the Air Force stated that a GBA had been tested and recommended for adoption.[2] However, production of this device interfered with the production of sights and was temporarily held up. The Bureau of Ordnance expected these units would be produced at the Lukas-Harold Corporation, the new Naval Ordnance Plant under construction in Indianapolis, Indiana, including special tools for this new device, and it would take several months before all engineering drawings would be completed and forwarded to the contractors for production of this new attachment. Extreme efforts were being made to obtain a limited quantity at an earlier date for testing and satisfying any needs that might arise during production.[3] Working models were obtained at a much later date, causing a long delay in production. It was also found that the sight, GBA, and certain units of the

Chapter XIV: Augmenting the Mark XV, M-9 Bombsight

SBAE/AFCE malfunctioned when temperatures were below -8° centigrade (18° Fahrenheit). It was essential that this equipment operate to a low of -40° centigrade and must be operational within 10 minutes of take off.

Heaters were not provided for the Bombardiers' compartment, and the compartment heating system in this area was, at best, a very insufficient way to heat this equipment. The Bureau of Ordnance provided integral or readily available electric heaters for the bombsight below temperatures of -8°. The contractors were to provide the Bureau of Ordnance data on weight, size, electric current requirement, and type of connector for each heater to be furnished as soon as it was established.[4] It was necessary to design heating equipment for the bombsight and, since SBAE was used as an automatic pilot, it was necessary that this equipment be operable within 10 minutes after take off when the above temperatures prevailed. The stated requirement was for one GBA for each sight in the future.[5] The Lukas-Harold Corporation and the Burroughs Adding Machine Company were selected as separate sources for manufacturing this attachment, along with special tools, jigs, fixtures, gauges, engineering data, and any assistance required from the Norden Company. Mr. Norden procured all of this material and mechanism as his own venture, and all work was to be completed within six months after award of a contract. The government obtained this complex instrument free of cost. This was in keeping with Mr. Norden's policy of developing the equipment and then reaping any profits from contracts that would later be awarded.

In 1943, a total of 900 GBA units were purchased by the Navy to be used to equip bombsights previously ordered. Procurement of large quantities had been made in the first part of 1944 for bombsights coming off the production lines. Since the GBA was considered as an accessory to the bombsight, it did not need to be specifically mentioned in any Procurement Directives, but it was added to the basic contract for bombsights.[6] Following standardization for both the Navy and Army Air Forces, the equipment was identified as the Glide Bombing Attachment Mark III, Mod 1. Most of the heavy bombardment missions were being conducted at altitudes above 20,000 feet. When air superiority was gained over Germany, heavy bombardment missions would be at lower altitudes. A new GBA attachment was developed having a range from 14,000 to 30,000 feet. Tests had not yet been completed, but results being obtained during testing were in the range required by the AAF, and the features would be comparable with the instrument having the lower range. Research by the Engineering Development Division, Wright Field, AAF, the Bureau of Ordnance, and the Norden Company personnel showed that an instrument encompassing the higher altitude could be made pending further study. The contractor was given a very high priority to work on this new development. If the 30,000 foot altitude could not be obtained, the 25,000 foot unit was acceptable until an instrument could be designed. The AAF used the present GBA for training purposes in airplanes such as the AT-11, which did not operate at an altitude in excess of 20,000 feet. Time was not lost; as soon as the 30,000 foot instrument was available it could be very easily adapted to the heavy bombardment airplanes.[7]

Testing was underway for GBA having a range from 14,000 to 30,000 foot altitude. By January 1944, tests had not been completed, but it was encouraging that a unit could be developed having a minimum of 25,000 and preferably 30,000 feet and a lower limit of 4,000 feet. In the following months, these deficiencies were soon resolved and, beginning the first part of 1944, production was started on the modified GBA. This instrument was given an extremely high priority (AA) for its production. By July 1944, the Burroughs Company was in the process of tooling up for the manufacture of the GBA and would be complete by September 1944. Due to the high priority of the attachment and its necessity by the AAF, an additional manufacturing source was added—the Manufacturers Machine and Tool Company, which was the manufacturing agent for the parts used by the Norden Company in assembling the bombsight. Burroughs assisted Manufacturers Machine by turning over tooling to permit increasing the rate of production.[8]

The Navy advised Hqs AAF, in August 1944, that quantities of GBAs were available to the AAF. The first priority was to the 73rd Bomb Wing—one each per airplane assigned and second priority—and all heavy bombardment production line airplanes (Boeing, Consolidated, etc.). The Bureau of Ordnance informed the AAF that production from their facilities was sufficient to allocate 3,287 or more units from June to December 1944.[9] No exact production quantities of GBAs were found in the examined files, however, production charts and spread sheets containing requirements versus production showed this program projected into September 1945. With WWII ending in August 1945, all production was canceled. There is little prospect that this program was carried out to termination. These charts showed a total of 13,578 units for bombers. Spares and training were projected from October 1944 to November 1945. It is not known if the previous 900 units issued to the AAF was included in the existing Navy total of 12,051, plus units produced from October 1944 to the end of the war in Germany. If this data is factual, then the previously manufactured quantity was 12,051 plus units produced from October 1944 to the end of May 1945. With the war over in Germany, many of the defense plants were either closed or cut back in production. Since this was a critically needed item by the AAF in bombing and training, it is reasonable to assume that production continued until the first part of 1945. This would bring the total of GBAs produced to 18,108 units instead of the projected 30,209. Directives from the AAF requested that each airplane be equipped with the latest GBA.[10] **Note:** Since the completion of the manuscript the author was able to obtain a GBA in excellent condition. It is one of two known to exist.

AUTOMATIC GYRO LEVELING DEVICE-MERCURY SWITCH

Abbreviated to AGLD. Many attempts and models had been made in the past to quickly and accurately level the sight gyro—which would be a boon on a bomb run—but had always ended in failure. As previously explained in another chapter, a device called the Automatic Erection System ended up in failure and had to be either removed or made inoperative. By June 1943, a model was tested extensively with and without the AGLD by the 2nd Air Force at Galveston, Texas. It was also tested by the Ninth Bombardment Group (Heavy), Orlando Air Force Base. The bombing was done in B-17s with two pilot-bombardier teams which were kept intact throughout the tests. Three hundred bombs were dropped, all by the latest modification of the C-1 Automatic Pilot. Bombing was done at 4,000, 8,000, 12,000, and 20,000 foot altitudes with half of the missions being run at 12,000 feet. In the first six months of the tests, the circular error was 20% less with the new type of erection system than with the conventional M-9 sight. The Second half of the year with the second bombardier was 25% less with the new type of erection system. However, there were failures during the tests, and it required more maintenance than the standard type.[11] This new device was developed by Minneapolis-Honeywell engineers. The AAF Armament laboratory and the manufacturer were conducting engineering studies in an attempt to correct all unsatisfactory characteristics. This new device created a great amount of interest, not only in the combat areas, but also in training of all bombing personnel. Further tests by the 2nd Air Force requested that this new system be made available on a basis not to exceed 1/3 of their number of sights. It was considered that further testing would bring out unfavorable deficiencies and improvement in bombing results would accrue. It created interest in an effort to improve bombing by inexperienced bombardiers, which was sorely lacking through the training period. By modifying only part of the fleet it would give the average combat bombardier a better chance, while the highly experienced bombardier could have a choice on the type of sight to be used. It also appeared to be a morale booster, and in addition it resulted in an improvement of bombing results by the 2nd Air Force. It also was felt that it would improve the accuracy of the average bombardier who had completed the four stages of combat crew training, which made up a large portion of bombardiers departing for combat areas.[12] It was decided that, when all commands were in agreement and when the AGLD was standardized and had progressed to a point where it could be considered suitable for factory line production, it was directed that the modification was to be accomplished in Depot Level Overhaul facilities.[13] Prior to departure for overseas combat duty, the tests conducted by the 351st and 380th Bomb Group for one month indicated highly satisfactory results. This modification did not have any moving parts, nor did it require any increase in maintenance. Unfortunately, the modification required depot level machine shop capabilities to be done by very experienced machinists and could not be done in the field due to considerable machining of the bottom half of the bombsight housing, carden, and the gyro. The Technical Order governing this modification consisted of 37 pages of detailed instructions.[14] Minneapolis Honeywell stated that they could furnish modified sights at the rate of three per day if given a high priority (Plate #2). The author is well acquainted with the AGLD. He worked on the modification for a great length of time, making many changes on the device. All changes were sent to Army Air Forces, Wright Field, Dayton, Ohio, for distribution. A Letter for this effort is on file in the author's office.

Tests were continued from altitudes of 6,000 to 12,000 feet and over by the 2nd Air Force at Colorado Springs. Combat experienced bombardiers and recent graduates of bombardier schools dropped 825 bombs using the AGLD and 822 bombs using the standard M-Series bombsight—error was 292.18 feet in favor of the AGLD. Not much in terms of feet, but of sufficient difference in placing bomb(s) in a critical area. Combat bombardiers claimed the greatest advantage was that the bombardier could level his sight gyroscope without removing his eye from the sight optics, and less trouble with poor visibility so common when bombing enemy targets in Germany. Because of this high rate of success, the AAF wanted 50% of all Norden sights in use or to be produced to be modified to include the Mercury Erection System.[15]

As usual there were the "Doubting Thomases" as with any new modification or design changes. The Minneapolis Honeywell personnel, AAF Chief Physicist, AAF Project Engineer, and the AAF Armament personnel at Wright Field continued to conduct extensive tests at Midland, Texas. Two expert bombardiers were selected and two student bombardiers approximately 50% complete in training at the Bombardier School were used in this latest test. For some unexplained reason, a bad gyro "hunt" developed in the rudder axis of the auto pilot that made manual leveling of the bubbles extremely difficult and inaccurate. This strange phenomenon was eventually corrected, but no plausible explanation was ever found, unless it was a bad case of bearings, which continually plagued the bombsight program with the lack of precision ball bearings.[16]

At the same time, there were several other contractors, including Victor Company, using other methods. These methods included a 1) liquid switch, round in shape with gaseous bubbles similar to an octant bubble containing 4 contact points. As the gyro tilted, the bubble uncovered one of the contact points, the electrical balance was disturbed, energizing the proper solenoid which then erected the gyro by means of the electromagnetic forces that were set up. 2) Spherical Gyro which employed a gyro, the fly wheel, free to pivot around the focal point of the system, and had a field coil mounted on a pendulum inside the flywheel. This system was scrapped. 3) Delco Products erection system was similar in principle to the one used in the Sperry S-1 Series bombsight and was being applied to the Norden gyro.

This system employed a rotating disc pendulum. This was later changed to a rotating ring in place of the pendulum. 4) Specialties Company system used three ball bearings mounted in a track at a constant speed with the weight equally distributed around circumference of the gyro. Simple and effective, but not combat usable. None of these proved successful under intensive testing.[17] The leader was one developed by Minneapolis Honeywell using the mercury switch.

There was a uniform and most insistent demand on the part of the combat operating units among heavy bombers for the mercury erected system for use with the Norden M-9 type bombsight. In June 1944, Mr. Robert J. Lovett, Assistant Secretary for Air, wrote to the Chief of the Air Staff concerning this modification, outlining the adamant and continuing requests from the 8th Air Force (The Mighty Eight) in the United Kingdom for this new system. Since 1943, the 8th Air Force had been trying to get modification kits up to the time that the Assistant Secretary of Air visited the war zone. Up to that time only 41 kits had been received by the combat units. The results of actual experience in the combat zone with this device were not only desirable, but necessary in that it permitted group bombardiers to effect more accurate sight leveling under war conditions and flak air disturbances. The Eighth Air Force felt that all Norden bombsights should be equipped with the AGLD modification.

The Assistant Secretary for Air, Mr. Robert J. Lovett, on his visit and investigation of complaints by the 8th Air Force, discovered that there was a uniform and most insistent demand by all of the operating units among the heavy bombers for the mercury erected system. As of June 1, 1944, the Minneapolis-Honeywell Company had orders for 250 AGLD kits, which did not cover the total quantity of 600 kits that were originally requested by the 8th Air Force to provide 15 AGLD equipped bombsights for heavy bomber groups.[18] This equipment was of the utmost importance in connection with precision bombing, which was necessary to pinpoint and take out certain targets still remaining unattacked in Germany. A letter from Assistant Secretary Lovett to the Chief of the Air Staff, in June 1944, directed that this program be given the highest priority and every effort be made to accelerate delivery of these articles. This request had the unqualified backing of General Doolittle and General Spaatz, and they had made a personal plea to obtain this equipment as fast as possible.[19]

There was apparently some argument as to whether the AGLD sight was more accurate under theoretical or ideal conditions compared to the standard bombsight. It was emphatically stated that the 8th Air Force airplanes were not concerned with theoretical conditions, but with actual, and, as they had used both, it seemed that their distinct preference was conclusive evidence as to what was needed. If there was any tendency to delay the project on the basis of statistics which showed that an unmodified sight was theoretically more accurate than one equipped with the mercury erector, any such discussions were to be postponed until after the German targets had been hit and destroyed by the equipment which the combat units had found to be a most effective weapon for this purpose.[20]

Immediate and accelerated delivery was taken to get these units to the 8th Air Force combat squadrons by the fastest means possible. The 8th was advised that their request for 250 AGLD kits were to be filled as follows: one hundred were ready by July 20, 1944, and the remaining 150 were delivered by August 20, 1944. Thereafter, the rate of production was to be over 150 kits per month. Action was taken to increase the production of the Minneapolis-Honeywell Company to meet the requirements established by the Eighth Air Force. This item was not to be standardized until a more suitable gyro was obtained. As mentioned previously, several types were under development. All other requests were filled as the kits became available from production.[21]

REFLEX OPEN OPTIC SIGHT
This attachment for the M-Series bombsight was designed as an aid in picking up a target partially obscured by smoke, haze, or clouds. The unit consisted of a set of auxiliary reflex optics directly coupled to the telescope rack in the M-Series bombsight. This device gave a greater field of vision for the bombardier, facilitating initial alignment of the bombsight telescope crosshairs on the target in a minimum amount of time. The reflex sight was rigidly attached to the bombsight housing, but the optics were not gyroscopically stabilized. This device was to be used only to establish *initial* course and range synchronization. It was not designed to be used as a bombsight, and no bombing was to be done with it. Due to minimum loss of light transmission through the reflex optics, it was particularly adaptable for night bombing and at very high altitudes when visibility of the target was limited.

Very little engineering and design data was found in the examined files. It also came into use late in the war as an aid to bombing. The use of the Reflex Optic sights was discussed by Engineering Personnel, Wright Field, Materiel Division, AAF in November 1942. No action was taken at this time due to concentration of producing critically needed bombsights for airplanes departing for combat. Because of weather conditions over Germany, Italy, and the Pacific areas, at times, the visibility over the prime targets was so poor that the bombers were directed to alternate targets. One of the main features of the June 1944 Bombing Conference brought out that the Open Optic Sight was absolutely necessary, and personnel returning from the 15th Air Force stated that their needs were urgent. [22] An experimental article was undergoing tests at the AAF Eglin Proving Ground, and preliminary reports indicated that it was ready for approval. The AAF purchased an additional 25 units for further service test by combat units. Unfortunately, to install the Open Optics Sights necessitated several changes in the bombsight itself. These major changes were of such a nature that they could be incorporated in the production lines with basic changes to already set up lines. These changes could not be

Chapter XIV: Augmenting the Mark XV, M-9 Bombsight   171

done at field level, nor by intermediate repair stations. This modification was accomplished by rear depot level facilities due to the extensive machining which was not available to service personnel in the field.[23] The Navy did not intend to use this item and it would not be installed in the production lines at the Norden plant or at any other Navy facilities. The Army and Navy facilities were now producing bombsights at a record level, and there was to be no disruption of this program. The most feasible option then was to have a reliable manufacturer other than the M-Series contractors. Using this method required the device to be in kit form and would then be identified as Government Furnished Equipment (GFE). These kits would be furnished as GFE and not as a contractor item (Plate #3).

By July, 1944, Hqs AAF directed that the procurement of the Reflex Sight be procured, since it passed all tests at the AAF Eglin Proving Ground, on the basis of one for every "M" Series bombsight made. Every effort was to be made to obtain necessary units for distribution to the combat theaters. This was to include retrofit installation on sights in service. In view of the experience of the 15th, $14^{th,}$ and 10th Air Force, plus those in Europe required immediate distribution of this equipment. Consideration was also given to the installation and assembly of the Open Sight received from the Bureau of Ordnance, which was to be accomplished by forwarding them to Depot Level Repair facilities for installation. Hqs AAF directed the Engineering Division, Wright Field, Materiel Command to standardize the new Open Reflex Sight. It was also directed to begin retroactive installation for those bombsights now in service. The plan was to establish installation at the Victor Company to incorporate the Open Reflex Sight received from Navy facilities.

The Reflex Sight was furnished as GFE for both the Victor Company and Navy facilities. The new unit was to be installed and fitted with dowlings. The serial number of the bombsight to which it had been fitted to was transferred to the name plate of the Reflex sight. A problem arose in that the Reflex Sight could not be attached to the sight to which it had been fitted for shipment, as the shipping container was not large enough to accommodate the assembled unit. Therefore, after it had been fitted, it was necessary to remove the unit and place it in the metal shipping container with the bombsight. The installation of the Reflex Sight was a very simple operation and could be accomplished by the airplane manufacturer when the bombsight was installed in the airplane, or by service personnel in the field. Top Command Headquarters issued directives for the production of this unit, and was given top priority. It was also furnished in kit form.[24] In keeping with directives issued by the Chief of the Army Air Corps, the AAF Materiel Command advised the Commanding General AAF that Reflex Open Optic Sights for M-Series bombsights be delivered at the rate of 1,000 in November 1944; 3,500 in December 1944; 5,000 in January 1945; and 7,000 in February 1945. Provisions were also made to allocate 300 to the United Kingdom; 300 to the Mediterranean area; 100 to the Fifth Air Force; 100 to the Seventh Air Force; and 50 to the CBI theater. This attachment was installed at the factory, removed when shipped (due to size, it would not fit in the bombsight shipping box), and reassembled at destination or where there were depot level facilities in the combat theaters.[25]

The W.L. Maxon Company of New York was selected as the manufacturer. Complete and accurate production quantities are not available. Production spread sheets show projected quantities of 26,000 units to be produced from November 1944 to April 1945. Since the goal of the AAF was to equip each M-Series bombsight with this unit, it appears that thousands were produced before the end of the war in August 1945. Since there was such a high requirement for this item, it is assumed that the Maxon Company produced their contract of 26,000 units.[26] The bombsight that was used to drop the first Atomic bomb by the Enola Gay was fitted with an Open Reflex Optic Sight. No actual cost was found for this device, but some correspondence in the National Archives showed an estimated cost of from $800.00 to $1,400.00 per unit.[27] This device is readily available in many military surplus outlets. Apparently, a large quantity was produced. The author has one on display.

SUPER ALTITUDE ATTACHMENT AND HIGH ALTITUDE BOMBSIGHT
To keep abreast of new bombardment aircraft and to avoid anti-aircraft fire, the C.L. Norden laboratories were engaged to develop a Super Altitude Attachment for the M-Series bombsight. This attachment took the place of the telescope motor and gear reducing mechanism by means of a flexible drive shaft providing a disc speed range, raising the effective bombing altitude. In October 1943, the Bureau of Ordnance advised the Commanding General, AAF, Engineering Division, AAF Hqs and the War Department as to the availability of a super altitude attachment for M-Series bombsights proposed by the Norden Company. The device consisted of an attachment mounted in a manner similar to the GBA, and connected to the bombsight by a flexible shaft. The minimum time of fall was not more than 45 seconds, giving a 7 second overlap compared to the present rate motor speed allowing a maximum time of fall of 52 seconds. In addition to this overlap, it was practical to disconnect the flexible shaft and fuse the rate motor for a time of fall of less than 45 seconds. The upper limit of the time of fall was designed, if the requirement existed, of not over 80 seconds. The AAF desired to procure eight units for testing and combat use.[28] Due to the cost, the AAF reduced their requirements to three units.[29] However, in ordering the eight units, the price dropped to $3,500 for the first unit, and the additional units were $1,500 each. The manufacturer of these units by the Norden Company or Norden Laboratories, all rights to this device would remain with the manufacturer and that the government would not receive any patents rights, licenses, or other rights and interests with respect to any improvements, refinements, or additions that could be later developed.[30] The units were made available within 12 weeks after awarding of the contract. Unfortunately, very little data was

found in the archives on this attachment. The Bureau of Ordnance advised Commanding General, AAF, Chief of the Air Force, AAF Materiel Division that, for three units, the price would be $7,000.

NORDEN SUPER ALTITUDE ATTACHMENT
The Norden Super Altitude Attachment was an auxiliary instrument for the purpose of extending the altitude of the M-Series bombsight from its present limit of 34,000 to 70,000 feet, made possible by driving the rate of the disc speed knob as calibrated in rpm from 65 to 114, of the bombsight disc, which was equivalent to an altitude range of 28,000 to 70,000 feet. The standard 0-150 mils of trail that could be set in the bombsight remained unchanged.[31] The tachometer connection running twice the disc speed was provided without changing the design of the bombsight itself. This modification was accomplished by gear changes with the bombsight.[32] The modified range of disc speed was 77-450 rpm, compared to 102-590 rpm, which through this modification to the disc speed it was possible to synchronize the bombsight throughout an altitude range of 2,200 to 60,000 feet, compared to 1,400 to approximately 34,500 feet previously.[33]

In order to determine which instrument was to be accepted by the AAF, comprehensive tests were undertaken by Hqs AAF Proving Ground Command, Eglin Field, Florida, in May 1944 and ended in December 1944. The task for the final decision rested with the Army Air Forces Board, Orlando, Florida. The testing consisted of bombsights from the Norden Company and Victor Company. Sights used in the tests were mechanically in perfect condition, giving constant speed readings through the increased altitude range. AAF Armament Division, Wright Field, was in charge of all flight tests. Bombardiers participating in the tests were well acclimated to high altitude operations, having made daily flights to extreme altitude operations for several months.[34] The Victor Company type M-9B high altitude bombsight was used exclusively for the Aberdeen tests. The project officer took the test equipment to Muroc Lake, California (now Edwards AFB), for flight testing. Each bombsight was installed in one of the Two B-17G airplanes being used for the tests. Concurrently with this development, the Victor Company, who was already producing M-9 sights, came out with a modification to the rate end of the M-9 Series capable of performing the same function as the Norden type.[35] In addition, a "trail spotting" device was added, which enabled the bombardier to decrease trail input in such a manner that the center of the target could be used as an aiming point and the trail of bombs would bracket the target. The Victor Company's M-9B High Altitude sight changed the trail extension from the standard 0-150 to 0-230 mils.[36]

During these tests, one hundred thirty-four releases of every AAF standard type of bomb were made at altitudes of 25,000 and 38,000 feet.[37] The average altitude was 30,000 feet, average true air speed (TAS) was 241, and a circular error of 1,250 feet for the Norden Super Altitude attachment; and for the Victor M-9B High

Altitude sight, the average altitude was 35,000 feet, average TAS was 251, and a circular error of 1,049 feet.[38] The results of the tests showed neither of the two sights suffered from the extreme low temperature. However, on the Norden unit, the bombardier had a tendency to hit the flexible shaft, causing it to bind, resulting in a different disc speed setting. The speed drum on the left side caused the same error. Normal operation of the attachment restricted the bombardier because he had to clear the attachment.[39] The Victor Company Unit did not have any of these problems, plus the fact that the M-9B high altitude sight was already being used by B-29 combat crew training. The AAF Board adopted the Victor Type M-9B as standard for extremely high altitude bombing for use by the Army Air Forces.[40]

The Victor Company, in February 1945, received a contract to develop and modify the rate end of the M-9 bombsight in which tests at Eglin Proving Ground proved that bombing could be accomplished at great heights. With the successful testing of the Victor sight, work on the Norden Super Altitude Bombing Attachment was discontinued. Under this contract, the above was one of the five new bombsights that the AAF requested Victor Company to develop under this contract. The company had sufficient personnel, space, and equipment for present bombsight contracts, but not for any additional new projects. The instruments to be developed were the Gyro-Vortex, modifying the M-9 with electrolytical automatic erection system, modifying the M-9 with the gyro vortex, and modifying one of the new bombsights with the gyro vortex. The Victor Research Division estimated it would require 18,000 square feet of space and a two story building, dust proof and humidity controlled. The cost was $265,000 for the building, laboratory equipment, machinery, etc. The Board of Directors of Victor Company authorized construction of the new building providing the War and Treasury Department would issue a certificate of necessity, and to provide accelerated amortization for the new building. The research facility would be paid for by Victor Company's own capital, without government aid.[41] The land was to be leased from Northwestern University at Evanston, Illinois. Due to the war coming to a close in Europe and with radar being installed in aircraft, the entire project was canceled.[42]

LOW ALTITUDE BOMBSIGHT ATTACHMENT (LABA)

Low altitude bombsights had been around for many years. The Mark III of WWI vintage was obtained from the RAF after the war and updated by the Navy Bureau of Ordnance by Mr. Norden. It was fairly accurate but, even with redesigning, it was not successful and eventually declared obsolete before WWII. With the advent of faster medium and light bombers, there was an urgent need for a low level bombsight. Both the RAF and the Navy were using the Norden as a low level sight, but it was not the best as it was designed for high altitudes and did not perform well. The Mark XV, Mod 3 could not be synchronized below 3,000 feet. Mr. Norden was contacted on this problem, and by 1935 had made test models available to the Naval Proving Ground. During testing, many engineering and design changes were made. By November 1936 additional difficulties were encountered in the form of "Cardanic Aberration," which disturbed the azimuth setting introducing errors of considerable magnitude. In the meantime, three production contracts were awarded to the Norden Company for 185 units. Because of these serious deficiencies, production was temporarily suspended pending further tests so that the design changes could be incorporated in the production model. All special tooling, sub-assemblies, and parts were almost completed for these contracts. This unit consisted of high precision workmanship, with all calibrations and symbols engraved on the sight. The usual excellent workmanship of the Norden Company was evident in this new attachment. The Norden Company also stated that the Bureau of Ordnance would not be charged for any parts completed that could not be used. This was in keeping with their policy. Also, production was resumed on a short notice and delivery of the entire 185 units was within 75 days after awarding of the contract.[43] The LABA device gave continuous ground speeds and dropping angles for any set of conditions. A revised LABA was sent to Aircraft Scouting Force, Patrol Squadron Two, Coco Solo, Canal Zone, in March 1938 for extensive testing. Only one bombardier was available for testing, and about eight bomb runs were made using the new attachment. The errors for radial was 204, range 87, and deflection 185, with a wind force of 26 knots, at 1,900 feet and air speed of 95 knots. Further refinements needed to be addressed before production started.[44] By November 1938, the NPG was testing and redesigning the LABA using a Mark XV, Mod 3 bombsight. A total of 86 drops were made with the Mark VII using 100 pound water filled bombs from a TBD-1 airplane (B-25) at an air speed of 130 knots. The increased air speed compensated for the poor results obtained by the NPG. The radial error for the eight flight was consistently 45.6 to 60.7 feet. The results obtained were superior with the new LABA than what was currently used by the services. This device required that more time be devoted to bombing practice. All personnel were required to conduct bombing missions at the expense of other types of training. It was decided to concentrate on training a limited number of bombardiers and confining them to this duty in connection with this specialty. This would provide a large nucleus of instructors in the event of war.[45]

With the solution of the troubles in the LABA, production was started and the orders were completed. The AAF, in 1943, placed an order for 4,000 LABA units to be delivered at the rate of 1,000 per month. In September 1943, the AAF Materiel Command again initiated production of the LABA with one low altitude device for each M-Series bombsight. The modification developed by the 8th Air Force was to be incorporated in this order. The directives from higher authority stated that all necessary action be taken to procure these instruments as soon as possible, furnishing the contractor all specifications, drawings, changes, and any other data necessary for completion of this work. Since extensive combat use of this equipment modified by the 8th Air Force worked so well, all service tests were suspended.[46]

In May 1944, the Bureau of Ordnance received another order from AAF for an additional 6,000 units similar to the 4,000 previously ordered. Delivery was 1,000 per month following the completion of the present order. The Bureau of Ordnance made sure that sufficient materials were available to furnish this quantity. In August 1944, the Norden Company advised the Bureau of Ordnance that for another order of 8,500 LABAs the cost would be $800.00, and a cost plus fixed fee of seven percent (7%) for a total of $856.00 each. Delivery was started in November 1944 with 500, 800 in December, 1,000 in January 1945, 1,250 in February, and each month thereafter until completion. Any parts that were left over from prior orders were used as government furnished equipment, and on completion of the contracts adjustments and charges were made.[47]

Production charts and spread sheets prepared in October 1944, in reviewing the bombsight progress, showed AAF requirements for the LABA units extending from October 1944 to November 1945 as 14,140, and existing production from the Navy of 11,270 units. However, with the war ending in Europe in May 1945 and Japan in August 1945 there was very little likelihood that production would continue after the end of the war as all defense plants were closed down. Assuming that production continued until May 1945, the total production (including prior purchases) would be in the neighborhood of 21,440 units.

Chapter XIV: Augmenting the Mark XV, M-9 Bombsight

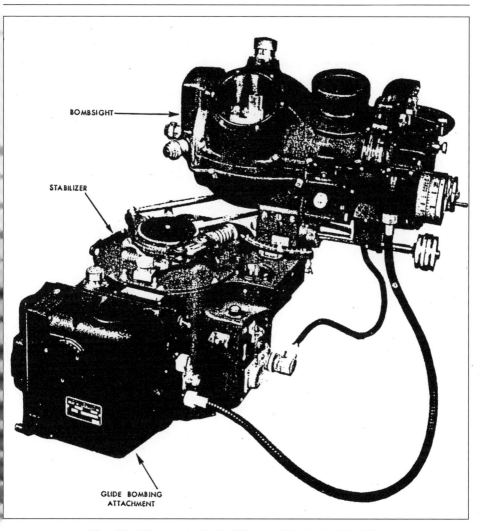

Plate #1 - GBA Attached to Stabilizer and Bombsight Mark X

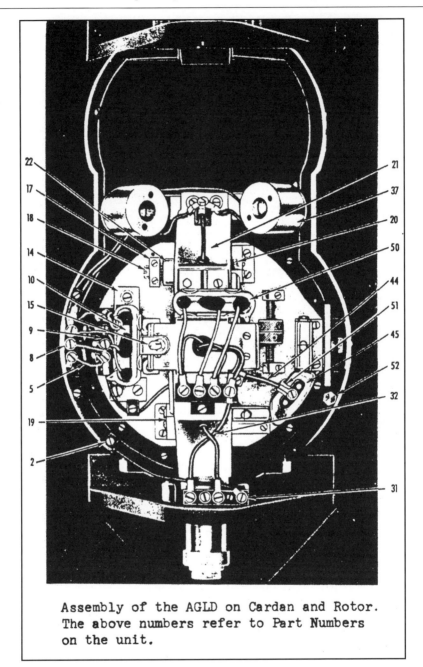

Plate #2 - Automatic Gyro Leveling Device Assembly

Chapter XIV: Augmenting the Mark XV, M-9 Bombsight

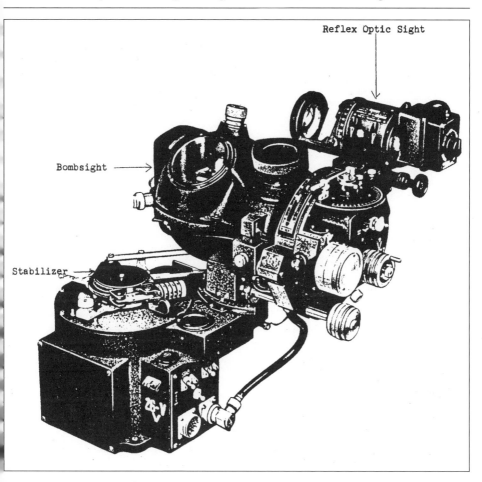

Plate #3 - Bombsight Assembly, M-9B Equipped with Type X-1 Reflex Sight

1. Glide Bombing Attachment, Mark 2, Mod, Ordnance Pamphlet No. 1116, dated March 25, 1944.
2. Memorandum from AAF Materiel Division to General Arnold, Subject: Glide Bombsight, dated March 17, 1941, National Archives, MS.
3. Ibid, page 2.
4. Letter from Bureau of Aeronautics to Bureau of Ordnance, Subject: Heating of Bombsight, Glide Bombing Attachment, and SBAE, dated October 18, 1943, National Archives.
5. Letter Bureau of Ordnance Memo for Bureau Files, Subject: Glide Bombing Attachment, Procurement Policy, dated January 22, 1944, National Archives.
6. Ibid, page 1.
7. Letter from Materiel Division, Wright Field, AAF to Chief of Bureau of Ordnance, Subject: Glide Angle Attachments for M-Series Bombsights, dated January 7, 1944, Frame 0456, National Archives.
8. Letter for Bureau of Ordnance to Procurement and Materiel, Subject: Rearrangement of Contract Schedule - Burroughs Adding Machine Company and the C.L. Norden Company, dated July 11, 1944, National Archives.
9. Letter Hqs to Commanding General, Materiel Command, Subject: Glide Bombing Attachment - M-Series Bombsight, dated August 2, 1944, Frame 0549, National Archives.
10. Headquarters AAF, Memorandum for Record, Subject: Present Bombsight Progress, dated October 7, 1943, Frame 0586, National Archives.
11. Letter from Headquarters Ninth Bombardment Group (Heavy), Orlando Air Base, Florida, to Commanding General, Wright Field, Armament Section, Subject: Mercury Erection System on M-Series Sights, dated August 26, 1943, National Archives.
12. AAF, Materiel Command, to Commanding General AAF, Subject: Report on Auto Gyro leveling Device, dated June 14, 1943, National Archives.
13. Ibid, page 2.
14. Technical Order No. 11-30-66, Bombsights — Operation and Service Instructions on Automatic Gyro Leveling Device (Mercury Switch) for M-Series Bombsights, dated December 18, 1943, National Archives.
15. Routing and Record Sheet, AAF, Subject: Minneapolis Honeywell Regulator Company Automatic Gyro Leveling Device (Mercury Switch) for M-Series Bombsights, dated December 18, 1943, National Archives.
16. Hqs AAF Letter to Commanding General AAF, Subject: Status and Comments - Automatic Gyro Leveling Device for M-Series Bombsights, dated November 13, 1943, National Archives.
17. Ibid, pages 2 & 3.
18. Letter from Office of the Assistant Secretary for Air, Memorandum for the Chief of the Air Staff, Subject: Automatic Gyro Leveling Device for Norden Bombsights, dated June 30, 1944, Frame 0537, National Archives.
19. Ibid.
20. Ibid, page 2 (Frame 0538).
21. Memorandum for the Assistant Secretary of War for Air, Subject: Recommendations made by the Assistant Secretary of War for Air, dated June 12, 1944, Frame 0543, National Archives.
22. Memorandum for the Record for the Commanding General AAF, Subject: Bombsight Development Conference - Materiel Command, dated June 5, 1944, Frame 0522, page 2, National Archives.
23. Hqs AAF, Materiel Command, CTI 546, Addendum No 14, Subject: Procurement of Reflex Open Optics Sights for "M" Series Bombsights, dated July 4, 1944, Frame 0539, National Archives.
24. Ibid, page 2.
25. Letter to Commanding General, Materiel Division, Subject: Reflex Open Optics Sight for M-Series Bombsights, dated October 23, 1944, Frame 0599, National Archives.

Chapter XIV: Augmenting the Mark XV, M-9 Bombsight   181

26 Hqs Departure News, July 1947, New Departure Division, General Motors Corporation, Worlds Greatest Producer of Ball Bearings.
27 New Departure News, July 1947, New Departure Division, General Motors Corporation, Worlds Greatest Producer of Ball bearings.
28 Memorandum from Bureau of Ordnance to Commanding General AAF, War Department, Subject: Super Altitude Attachment for M-Series Bombsight, dated October 7, 1943, National Archives.
29 Ibid, 1st Indorsement, Same Subject, dated November 5, 1943, National Archives.
30 Letter from C.L. Norden to Naval Inspector of Ordnance, Subject: Mark 15 Bombsight Super Altitude Attachment, dated July 21, 1943, National Archives.
31 Final Report on Testing of M-Series Norden Super High Altitude Attachment, dated December 14, 1944, Defense Technical Information Center, page 1.
32 AAF Board Project, Orlando, Florida, Test of M-Series Super High Altitude Bombsight, dated January 13, 1945, Defense Technical Information Center, Page 1.
33 Ibid, page 1.
34 Hqs AAF Proving Ground Command, Eglin Field, Florida, Final Report, Test of M- Series Norden Super High Altitude Bombsight, dated December 14, 1944, Defense Technical Information Center, page 6.
35 Ibid, page 6.
36 Ibid, page 5.
37 Ibid, page 6.
38 Ibid, page 7.
39 Ibid, page 9.
40 AAF Board, Orlando, Florida, AAF Board Project: Test of M-Series Norden Super High Altitude Bombsight Attachment, January 13, 1945, Defense Technical Information Center.
41 Letter from Victor Adding Machine Company to Commanding General AAF, Subject: Summary of Report to the Board of Directors of the Victor Adding Machines Company, dated February 24, 1944, Frame 04080, National Archives.
42 Letter from Aircraft Production Board to Victor Adding Machine Company, Application for Priority Assistance, dated April 21, 1944, Frame 0519, National Archives.
43 Letter from Carl L. Norden, Inc. to Bureau of Ordnance, Subject: Your letter from Bureau of Ordnance, dated November 13, 1936 and November 19, 1936, National Archives.
44 United States Fleet, Aircraft Scouting Force Patrol Squadron Two, Coco Solo, Canal Zone, to Bureau of Ordnance, Subject: Attachment for Low Altitude Bombing Mark XV bombsight, dated March 3, 1938, National Archives.
45 Letter from NPG to Bureau of Ordnance, Subject: Low Altitude Attachment to Mark XV V-2 Low Altitude Bombing Attachment (Compatron Two Letter VP2/A51/F41 of July 14, 1938) dated November 3, 1938, National Archives.
46 Letter to Commanding General from Assistant Chief of the Air Staff, Subject: Low Altitude Bombsight Attachment for M-Series Bombsight, dated September 28, 1943, National Archives.
47 Letter, 1st Indorsement from Bureau of Ordnance to Naval Inspector of Ordnance, Subject: Low Altitude Bombing Attachment — Price Quotation for, dated August 24, 1944, National Archives.
48 Headquarters Army Air Forces, Memorandum for the Record, Subject: Present Bombsight Progress, dated October 7, 1944, National Archives, MS.

Plate #1 Bureau of ordnance Pamphlet 1116, Glide Bombing Attachment Mark 2, Mod 1, dated March 25, 1944.
Plate #2 Bombsights - Operation and Service Instruction on Automatic Gyro Leveling Device for M Series Bombsights, T.O. No. 11-30-66 dated April 15, 1945.
Plate #3 Bombsights, Series M-9, and including M-9B with Type X-1 Reflex Sight, T.O. No. AN 11-5A-8, dated October 15, 1945, Revised December 29, 1947.

Photo Gallery

Carl L. Norden standing alongside gyro controlled airplane, 1931. (USN Historical Center)

Author working on cross trail assembly. (Author)

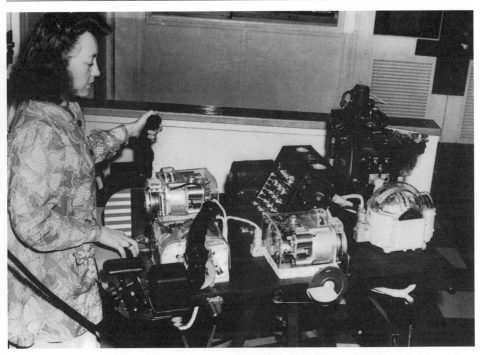
Factory worker checking a unit of the C-1 autopilot. (Courtesy of Mrs. Marge Eiseman)

Final bombsight calibration and inspection. (NASM)

Plexiglass bombsight. (Author)

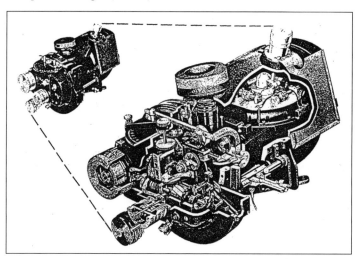

Right-Front cutaway view of bombsight. (T.O. AN 11-30-69)

Photo Gallery

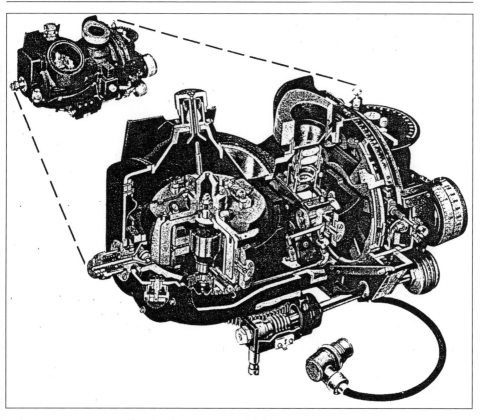

Left-Rear cutaway view of bombsight. (T.O. AN 11-30-69).

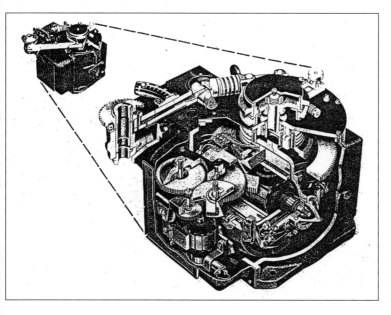

Left-Front cutaway view of bombsight. (T.O. AN 11-30-69).

186 The Legendary Secret Norden Bombsight

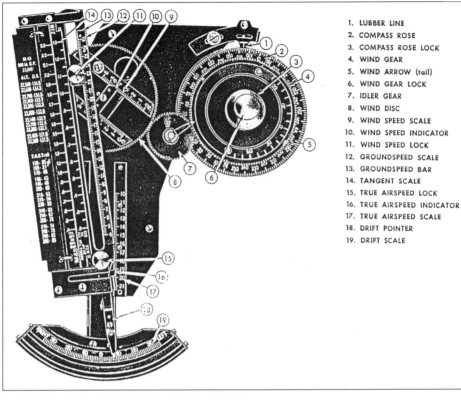

1. LUBBER LINE
2. COMPASS ROSE
3. COMPASS ROSE LOCK
4. WIND GEAR
5. WIND ARROW (tail)
6. WIND GEAR LOCK
7. IDLER GEAR
8. WIND DISC
9. WIND SPEED SCALE
10. WIND SPEED INDICATOR
11. WIND SPEED LOCK
12. GROUNDSPEED SCALE
13. GROUNDSPEED BAR
14. TANGENT SCALE
15. TRUE AIRSPEED LOCK
16. TRUE AIRSPEED INDICATOR
17. TRUE AIRSPEED SCALE
18. DRIFT POINTER
19. DRIFT SCALE

Automatic Bombing Computer. (Bombardier's Information File)

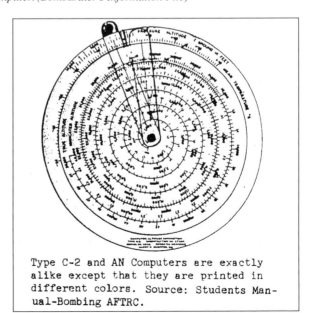

Computer, type C-2 and AN. (Student Manual-Bombing) Type C-2 and AN Computers are exactly alike except that they are printed in different colors. Source: Students Manual-Bombing AFTRC.

Type J-1 Computer Time of Run-Sighting Angle Computer. Thirty seconds on the front side and 45 seconds on reverse side. (Source: Bombardiers Information File)

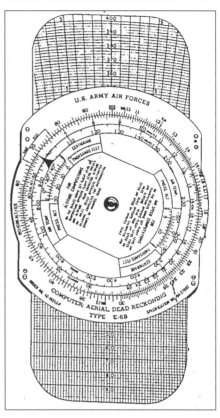

Computer, Type E-6B. (Student Manual-Bombing)

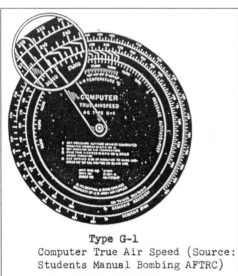

Type G-1 Computer True Air Speed (Source: Students Manual Bombing AFTRC)

Computer, Type J-1 (top), Time of Run Sighting Angle and Type C-1 (bottom), True Air Speed. (Student Manual-Bombing)

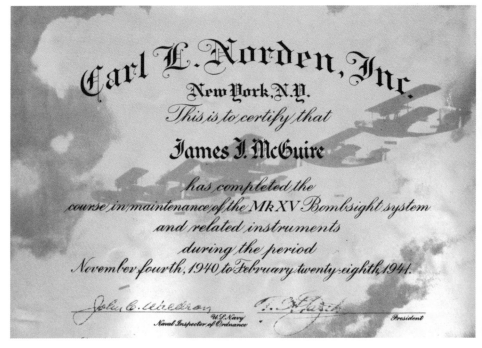

Norden bombsight Training Certificate. (Author)

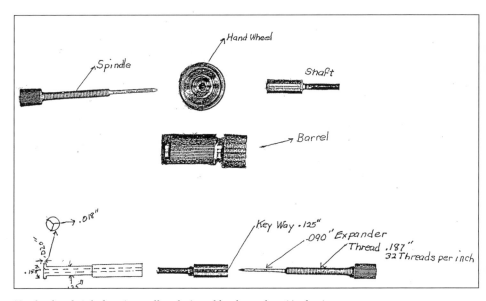

Norden bombsight bearing puller, designed by the author. (Author)

Photo Gallery

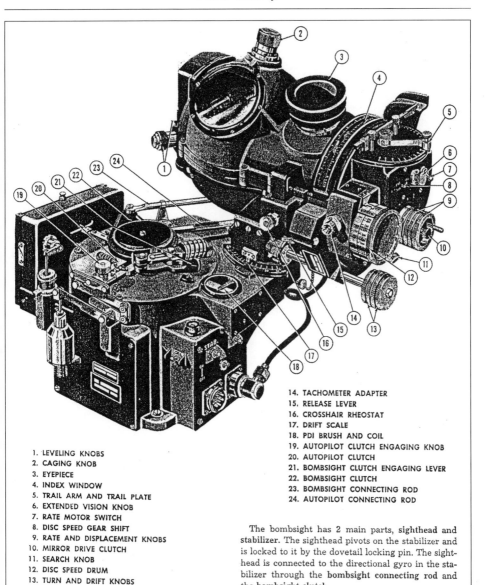

14. TACHOMETER ADAPTER
15. RELEASE LEVER
16. CROSSHAIR RHEOSTAT
17. DRIFT SCALE
18. PDI BRUSH AND COIL
19. AUTOPILOT CLUTCH ENGAGING KNOB
20. AUTOPILOT CLUTCH
21. BOMBSIGHT CLUTCH ENGAGING LEVER
22. BOMBSIGHT CLUTCH
23. BOMBSIGHT CONNECTING ROD
24. AUTOPILOT CONNECTING ROD

1. LEVELING KNOBS
2. CAGING KNOB
3. EYEPIECE
4. INDEX WINDOW
5. TRAIL ARM AND TRAIL PLATE
6. EXTENDED VISION KNOB
7. RATE MOTOR SWITCH
8. DISC SPEED GEAR SHIFT
9. RATE AND DISPLACEMENT KNOBS
10. MIRROR DRIVE CLUTCH
11. SEARCH KNOB
12. DISC SPEED DRUM
13. TURN AND DRIFT KNOBS

The bombsight has 2 main parts, sighthead and stabilizer. The sighthead pivots on the stabilizer and is locked to it by the dovetail locking pin. The sighthead is connected to the directional gyro in the stabilizer through the **bombsight connecting rod** and the **bombsight clutch.**

Explanation of operating parts for the bombsight and stabilizer. (Bombardier's Information File)

Photo Gallery

Bombardiers taking the oath to protect the bombsight. (NASM)

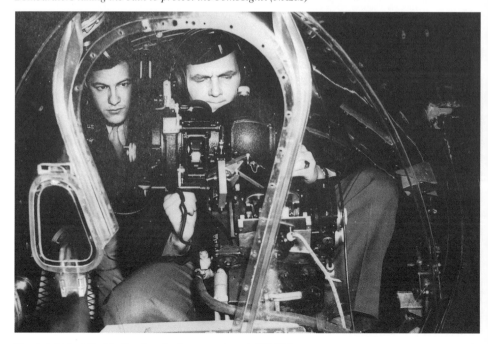

Bombsight installed in Bombardier's compartment. (NASM)

Hangar type Norden bombsight trainer. (NASM)

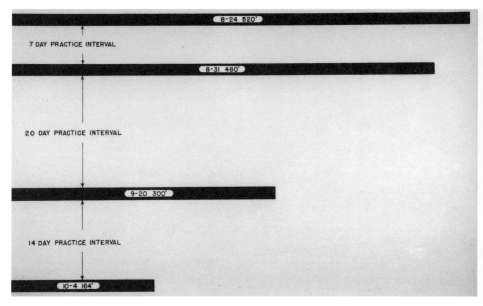

Reduction in radian error of a bombing team. (NASM)

The destruction of ball bearing factories. (NASM)

The destruction of marshalling yards for rail cars by Gen Samuel E. Anderson's 9th Bombardment Division. (NASM)

The German rail yard and supply depot at Lebach after bombing by First TAF Martin B-26. (NASM)

Bombing bridges at Kochendorf, Germany. (NASM)

Destruction of factories. (NASM)

Aerial attack of Schweinfurt. (NASM)

The bombing of Budapest, Hungary. (NASM)

Lebach prior to bombing by the First Tactical Air Force (see p. 193). (NASM)

Presentation of the Norden bombsight that dropped the first atomic bomb on Hiroshima to the Smithsonian Institution, Washington, DC. From left to right are Rear Admiral W.S. Parsons, General Carl Spaatz, A.C. Buehler, and Dr. Alexander Wetmore. (NASM)

Kit for the Bambardier and Navigator. (NASM)

General Carl Spaatz and Lieutenant Jimmy Doolittle discuss the results of the bombing raid on the oil refinery at Halle, Germany, with crews who participated in the mission. 303rd Bomb Group, England, 31 March 1945. (NASM)

A worker is shown operating the turn control, which is checked against servo m otors on the table before the instument is shipped to plane builders. (NASM)

Typical name plates used by manufacturers (Courtesy Fred Miller)

XV

Norden Bombsight Ancillary Equipment

Many common hardware items used in automotive, aircraft, railroad, government, etc., industries could not be applied to the bombsight. However, for the most part they were not acceptable for use due to the precision that was built into the Mark XV sight mandated that these items also have this precision quality.

Disc Speed Tachometer

A very essential hand held item in using the bombsight. Early models of the Mark XV had a tachometer (R.P.M.) counter installed on the bombsight case. When the sight was redesigned, the tachometer sleeve and bracket were relocated on the new "rate end." This modification was not only more accurate, but it took up less space, as the old unit was so large it covered part of the disc speed knob. In 1936, the Bureau of Ordnance, not satisfied with the ones presently in use, requested the manufacturer, Hassler Manufacturing Company, to provide two integral tachometers for testing by the NPG. The first of the two was a Hassler constant indicating chronometric counter. The instrument was large, its pointer behavior jerky, and it possessed an inherent indicating error of 1 1/2 R.P.M. over the entire range. The second instrument was an "Alpha" tachometer. This was also a constant indicating chronometric type. It contained a computing feature and an R.P.M. indicator, for indicating elapsed time in minutes. The pointer behavior of the Alpha tachometer was steadier than the Hassler, and the operating range was considerably better than the Hassler. The Alpha instrument was 1 1/2 inches thick, or one half the thickness of the Hassler. By eliminating the computer feature (which was not needed) it could even be further reduced. The face diameter for both were the same, or 3 1/8 inch. Also, provisions were made for quick and easy detachment of the integral tachometer from the bombsight case so that the tachometer and disc speed could be checked by a hand held tachometer. This unit proved very usable until a better instrument was developed years later.¹ In June 1942, price quotations were requested from American Time Products, Inc., for a Tachometer Calibration Unit, Type TCU-4, in order to assure that the tachometers in use were accurate. The cost was $900.00 for the first unit and $700.00 thereafter. It was such an urgent item that it was given an A-1 priority by the War Production Board. Delivery for the first unit was within 45 days, and ten units were furnished within 90 days.² Experimentation continued over the years, and it would be several years before a truly precision tachometer was available. In November 1943, six instruments selected at random from a lot of 120 units were shipped to the Bureau of Ordnance by the Technical Oil Tool Company. These instruments were subjected to a thorough inspection and testing by NPG. Results of the tests revealed that all of the units failed to meet specifications set by the Norden Company, Bureau of Ordnance and Aeronautics. Some of the more serious malfunctions were: the unit pointer was off at zero; the timer plunger stuck; the escapement mechanism did not function when the plunger was operated; the spacing between the drive shaft spindle and case bearing varied from .017 inch to .037 inch (specifications called from .003 to .008 inch); readings were so erratic at various speeds that the instrument could not be calibrated; this particular type of tachometer was obsolete and it did not have the accuracy and ruggedness required for the bombsight; and the end of the drive shaft spindle varied from .012 inch to .031 thousands of an inch. There were also other minor deficiencies too numerous to mention.³ The result was that all shipments to using activities were suspended until a better unit was available. In April 1944, the Bureau of Ordnance required a unit in the 0-1000 range, Type A, portable tachometer. Five units produced by the Technical Oil Tool Company were tested at NPG, revealing certain deficiencies and did not meet specifications. The Bureau of Ordnance, through the Naval Inspector of Ordnance, sent four Jaeger hand held units to the Naval Research Laboratory for tests.⁴ Extensive tests were conducted at NPG on the latest model (43D-3 Serial No. 1). The counter proved satisfactory in checking the bombsight constant (Plate #1).

Stop Watch
Plunger (stop watch) Type Revolution Counter. As newer bombsights were more accurate (M-9, M9A and B), the supporting equipment also needed to be of precision quality. As common an item as a stop watch, which was used in all kinds of events such as sports, racing, etc., was not acceptable for use on the sight. The ones that were in use were borderline. In July 1943, the Bureau of Ordnance procured a new type of stop watch from the Jaeger Watch Company, encompassing all Navy specifications. One watch was delivered in August 1943 for test purposes. The new unit was used by the Naval Inspector of Ordnance at the Norden plant to check the bombsight constant and Glide Bombing Attachment. It proved to be very suitable

for both instruments. It worked so well that an additional four counters were procured for other Naval Inspectors of Ordnance for final inspection and test. The price of the unit was $650.00.[5] With modification, the counter was adapted so it could be used on the Glide Bombing Attachment, Mark 2. This modification was made so that it would be possible to use the resetting button to stop the hands of the counter at any instant, as well as to reset them, making the control the way it should be used on stop watches. One push of the button would start the counter, another would stop the counter, and a third push would reset the hands to zero. The Jaeger tachometer became the standard hand held instrument for use with the bombsight for both the Navy and the Army Air Forces.

Total production on this item was not available in the Archives. Since there were several contractors building bombsights, it may be assumed that each NIO at the contractor's plant would need this instrument.

Anti-Glare Lenses

Filters were used to reduce glare to a minimum. Depending on the visual condition, different filters were used. It not only cut the glare and was an aid to poor visibility, but also gave better visual acuity against search lights, fog, and haze. The filters came in kits of five, with an interchangeable rubber eyepiece to fit over the telescope eyepiece. The kit contained two uncolored Polaroid filters, which cut out about 60% of the light, and the other filters could be adjusted to cut from 60% to all of the light. The Bureau of Aeronautics procured two sets of Polaroid eye glasses for examination and testing used with the Mark XV, Mod 3 bombsight. They were tested to determine if they were an aid in seeing a target in the path of the sun. Could the sight be operated while wearing the glasses heading into the sun and not heading into the sun? Could the bombardier see and read well enough inside the plane on bright and cloudy days to read tables and operate switches to obviate the necessity of taking off the glasses to read the tables and then putting them on to use the sight?

The glasses were used during acceptance bombing on bright, partly clouded, and overcast days, and were very satisfactory under all conditions. Since all the sights in use had moving optics, the use of the eye glasses was the most practical solution of the sun glare problem, since it required no fitting of filters or other changes to the bombsight. Future bombsights were handled in a simple manner by merely providing a filter to be placed in the eyepiece. It was directed that these eye glasses be issued as an accessory to all bombsights in use, and the bombsight manufacturer was to furnish a filter kit for fitting on the eyepiece for all Mark XV-Mod 4 sights in service and those under production. All sights in the future would be provided with this item.[6] The NPG comments concerning the use of Polaroid glasses was about the same as the Bureau of Aeronautics. Under actual conditions, the target was plainly visible through the sight when heading into the sun. The sight was operated with the glasses not heading into the sun and was satisfactorily operated with no difficulty. In addition, everything in the cockpit was clearly seen during bright sun and slightly clouded conditions. The glasses were of great help in looking directly into the sun glare on the water. Although the cockpit was somewhat darkened, all switches could be plainly seen without removing the glasses.[7] In August 1942, the Bureau of Ordnance requested estimates from the Polaroid Company for bombsight filter sets consisting of one variable density Polaroid unit, one Polaroid filter (single), Navy Red filter, 2 rectangle Polaroid filters, 1 Navy yellow filter, Rubber eye piece, and carrying case. The cost for one set was $20.50, with delivery commencing within 45 days after awarding of the contract, with 7,000 sets to be delivered within 30 days. A priority of AA-3 was assigned to the contract. In February 1943, an additional order was awarded to the Polaroid Corporation with a priority of AA-1.[8]

Modification to the C-1 Automatic Pilot Steering/Turn Control

The function of the manual turn control was to precess the stabilizer gyro, causing a change of course of the airplane. This motor provided only one throw of the airplane rudder, with no provision for different rates of return for the plane. Numerous complaints were being received from the operating squadrons regarding this deficiency. The NPG was requested to investigate this problem and to provide a variable rate of turn. The NPG initiated a study concerning this item. Although this was not a spectacular modification, it was vitally needed for control of the airplane. The modifications provided a means for the pilot to maneuver a properly banked turn while the airplane flew under automatic pilot without creating violent action. Early types of Turn/Steering Controls were a separate unit similar to the Bombardier remote control knob. This new design was developed by Minneapolis-Honeywell Regulator Company, which placed the turn control on the auto-pilot panel and was much easier for the pilot to operate.[9]

Experimentation on this new device started in late 1942 and, by the fall of 1943, it was ready for production. The Chief of the Air Staff issued a directive that action be taken to purchase 300 of these new units, which were to be used in the B-29 and B-32 in connection for use in Bombing Through Overcast (BTO-radar). The 58th Wing requested these units to be delivered as soon as possible. The 58th Wing stated that this equipment was most necessary in order to make their BTO equipment operate satisfactorily. Procurement was immediately authorized by the Commanding General, Hq, AAF, and directed the AAF Production Division to obtain 300 steering control units for the C-1 at the earliest possible date. The first 150 units were delivered by February 1, 1944, and the remaining 150 were delivered by June 1945 to cover the H2X radar accelerated equipment program (Plates #2 & #3).[10]

This was the other major modification on the C-1 AFCE. Not an exotic device, but an extremely useful one for the bombardier. It was developed by MH. It provided a convenient and positive means in which evasive maneuvers through the C-1 auto-pilot were accomplished from the bombardier's compartment. The turn knob assembly was installed on top of the stabilizer. The use of this turn knob permitted evasive action more easily, positively, and with greater precision than was obtained with the existing method of manual operation of the secondary clutch arm of the stabilizer. A limiting switch was designed which, at the discretion of the bombardier, was able to limit the degree of bank obtained by the use of the turn knob to 12 degrees in either direction for greater safety in formation flying.[11] Total production on this item was not available in the examined files. However, there was some reference that several thousand had been manufactured and installed on the stabilizer. The cost of the unit could not be determined, but some correspondence showed about $80.00 per set.

Formation Stick
The formation stick permits the pilot or co-pilot to maneuver an airplane quickly even when it is using the automatic pilot. It gives the pilot, with a minimum of physical effort, additional control of the airplane that was necessary during formation flying. Two formation sticks are installed in the pilot's compartment—one for the pilot and the other for the copilot. However, only one stick can be engaged at one time. A control switch button on top of each stick was provided so that if the occasion arose, transfer from one stick to another could be accomplished without disruption. The use of this device was a boon to both the pilot and co-pilot in long flights, as the unit had an arm rest and it saved the pilots considerable fatigue. Although this device was not part of the bombsight, it was essential in formation flying, especially when using the "Pathfinder" method of bombing. This unit was manufactured by Minneapolis-Honeywell and Jack & Heinz Company.[12]

Hq AAF, Chief of the Air Staff reviewed the production spread sheets in October 1944 and determined that the contemplated production, as scheduled, was sufficient to take care of all requirements for this device. Total production for these items was not available in the examined files. The spread sheet in October 1944 showed that total production for this unit, manufactured by Minneapolis-Honeywell and Jack & Heinz Company from October 1944 to April 1945, was 27,500 units in addition to existing stock. Spares and special units amounted to 1,204 and were delivered by April 1945. No detailed production costs were available in the files, but some correspondence stated that the unit was estimated at about $250 to $320 per set.

Intervalometer

The intervalometer is an instrument that provides for space intervals in bombing. The bomb release interval control is a timing device used to actuate either selective or train electrical bomb rack release mechanisms, associated equipment, and the bombsight. The instrument was designed so that spacing was provided from 0 to 750 feet depending on the model. It was located in the bombardier's compartment. The number of bomb releases and distances depended on the target. This instrument was in use many years, and very little data was found in NARA files. According to an updated Technical Order they were in use prior to 1942. Correspondence indicates that the Glen L. Martin Company installed 23 units in January 1941 in PBM-1 Navy planes. This indicates that this instrument was in wide spread use in bombardment airplanes.

In September 1943 the Bureau of Ordnance immediately advised all Commanders of Naval Fleets, Numbered Air Forces, Operational Training Commands, Fleet Air Units, Air Technical Air Training, Chief of the Army Air Forces, and Chief of Naval Operations regarding the malfunction of Bombsight Mark XV, Mod 7 — Air Corps M-9 using Intervalometer K-2 and Mark 2, Mod 1 was due to Bombsight Automatic Release Switch Assembly. This was used to lock the automatic release lever in the "armed" position until a few seconds after the bomb had been released. The lock was released by an electromagnet connected across the release circuit within the bombsight. The "feed-back" circuit was from the hot side of the intervalometer lamp through the lamp and the hold relay coil, to the safety electromagnet, which was grounded to the bombsight. Immediate orders were issued to disconnect the system until this interference was corrected. All modification of bombsights Mark XV Mods 4 to 7 (Air Force M-9) was also held in abeyance pending resolution of this problem.[13]

Extensive research and engineering studies were made and reports submitted in February 1944 by the Applied Mathematics Group, Bombing Research Group, Columbia University to the Applied Mathematics Panel of the National Defense Research Council.[14] The problem was corrected. Very little data was found on this instrument in the National Archives. The Technical Order AN-11-5E-2 was found in the Library of Congress. According to the Technical Order, it was in use as late as June 1944. Some correspondence indicated that units of this model were used up to the 1950s and beyond, however, confirmation of this could not be found. Intervalometers are readily found in military surplus stores (Plate #4).

Automatic Release Lock

The purpose of the Automatic Release Lock for the Bombsight Automatic Release Lever was to hold the bombsight Automatic Release Lever in the "armed" position when placed there by the bombardier, and to automatically release the lever from this position (after a suitable time delay) when the Bombsight Automatic Release Switch was closed. The escapement mechanism was designed to give approximately a half second interval between closing and opening of the Bombsight Automatic Release Switch. The electromagnet relay was designed for 24-28 volt D.C. on Mark XV, Mod 7, and M-9 and later series.[15]

The Naval Proving Ground conducted tests on this lever, releasing it from its "armed" position by the bombsight mechanism approximately 1,500 times in the shop. There were no variations in the interval, closing, and opening of the release switch. It also operated satisfactorily at temperatures of -35° with no effect on the equipment. Due to the small size of the control switch, a continuous effort was required by the bombardier in pushing the locking pin until caught and held by the spring loaded catch. Due to its size and location, a great possibility existed that this pin would not remain in the locked position when the bombardier, operating under hurried and stressful circumstances, or when wearing heavy gloves in sub-zero conditions, would inadvertently not lock the unit.

Automatic Release Lock had been in use since 1941 when it was introduced by the Norden Company as an assist for the bombardier. With extensive use, several deficiencies appeared and, by 1943, it was necessary to completely modify the switch. To prevent accidental release from the "armed" position, due to interference with the bombardier oxygen and size of trigger for the switch, all automatic release assemblies having the manual release pin located at the top of the unit were modified and relocated. The release pin was located on the left side of the arming bar, and larger control buttons replaced the small one. Although this was a major modification, the reworking and relocating of this unit did not take overwhelming time, but considerable effort was expended by repair shop mechanics. The author was well acquainted with this deficiency, expending many hours in instructing machinists and other mechanics on this modification. All bombsights requiring this feature were forwarded to the machine shop for rework and repair before being released to combat units.[16]

Ballistic Co-Efficient of Bombs

In November 1938, the Bureau of Ordnance issued a directive to establish new ballistic charts or tables for bombs being used. At that time the tables in use were obsolete due to new airplanes being received with greater speeds and higher service ceilings. For the new charts to be of any value, it was necessary that bomb ballistics be determined with an accuracy consistent with the accuracy of the bombsight over the entire range of speed and altitude. The project was assigned to the Naval Proving Ground at Dahlgren, Virginia.[17]

By June 1940, the Proving Ground had taken action to study and formulate plans for developing this important project. It was discovered that a considerable amount of new and improved equipment would be required to conduct bomb calibration, study of airplane trajectories, Anti-Aircraft burst ranging, etc. The layout and placement for target and theodolite (precision measuring equipment) stations were designed for a maximum of from 1,000 to 30,000 feet, and air speeds of up to 100-300 knots true air speed. In order to do this testing, additional land was acquired for the Proving Ground, which held up the program pending finalization of this transaction. The precision required for the ballistic charts required considerable preliminary engineering studies with assistance from commercial sources. Specialized equipment was also obtained. Arrangements were made for obtaining photographic theodolites from the Mitchell Camera Corporation. Provisions were made for two systems, one for work at low and moderate altitudes and air speeds, and the other for air speeds and altitudes in the order of 30,000 feet and 300 knots. Arrangements were also made with the U.S. Coast and Geodedic Survey (U.S.C. & G.S.) to assist in this project. Their expert assistance was required for the placement of lights, theodolite towers, etc. The existing Bamberg system at the Proving Ground was expanded to permit high altitude tests of all types of service bombs.

Due to the precision demanded in the use of this installation, it required that the geographical positions of the theodolite stations be determined with great care and that their elevations be accurately known. Permanent charts of the theodolite stations' target positions and references points were to be preserved on a plotting table. For future use they were engraved on brass on a scale of 400 feet to the inch, which would be used with the new cine-theodolite system. These new stations would be located at five positions approximately 38° Latitude (N) and 77° Longitude (W). The installation of this equipment would take considerable time. In order to obtain precision ballistic coefficient in forming new charts/tables, actual bomb drops were undertaken for 100 lb. water filled, 250 lb. general purpose, 500 lb. light case, 500 lb. heavy case, 1,000 lb. light case, 1,000 lb. heavy case, and 2,000 lb. general purpose bombs. This would be later expanded to include all types of bombs.[18] In order to obtain reliable data for bomb ballistic tables, the Bureau of Ordnance initiated action to obtain a bombing airplane of adequate performance to carry out this project. This also included information on the overall performance.

Wooden Tweezers

An unusual tool was developed for working on bombsights and related equipment. The very fine leads needed to be shaped, and using metal tweezers caused considerable trouble with the leads. This company produced several types for shaping leads. They were originally developed for photographic and radio work. A model was fashioned for use on the Norden Bombsight that had extremely thin and flexible leads. The standard types could not be used, the reason being that the Norden Company required long, slender shapes and soft tension. The development of this item was a great asset to the assemblers, testers, and inspectors working on gyros and other components. The delicate touch obtained by the use of wooden tweezers greatly simplified the tedious job of forming leads, besides providing good protection against fraying and short circuits. This tool was first used by employees of the Norden Company, who found them durable and extremely satisfactory in every way over the metal types.

By July 1944 these were being used by many of the bombsight contractors. There was only one company that specialized in manufacturing wooden tweezers. It was necessary to use select walnut grained wood so that the prongs were flexible and durable. The mahogany types had excellent conformal qualities and were porous and absorbed adhesives. For shaping stiff wires, maple wedges were used. They were carved to a comfortable balance. The end of the prongs lapped and fitted so that all surfaces of both sides exerted even pressure on delicate flexible leads. They were water proof and resistant to any chemical that would not harm wood. In the beginning, they were hand made, with one model requiring seven hours of work. With the acquisition of relatively inexpensive power machines, the cost dropped to $1.50 to $2.00 each depending on quantity.[19] The author has seen and used these types of tweezers. To date he has not been able to locate a pair. They were very much in demand, not only for bombsight work, but other delicate instruments being produced for the military. Note: Since the completion of the manuscript the author, through the Internet, found a pair of tweezers and special tools that were used on the sight from the son of the inventor. They are on long term display at the museum.

Illustrated Manuals or Pamphlets
Although manuals/pamphlets were available to both services, they did not keep up with the new equipment currently delivered to the squadrons. The Bureau of Ordnance had total responsibility for all engineering/design changes and publications. Normally, these changes were distributed as the changes occurred. This was one of the problems that Army Air Corps expressed discontent in their appeal to higher headquarters over the lateness of the changes. With the increase of bombsights in early 1942, these illustrated publications (pamphlets - Navy/Manuals - Air Corps) were not adequate for training or instructional purposes for concerned personnel in the field. New details of the bombsight or any development incorporated in the Mark XV Mod-7 (Navy) or M-Series (AAC) were not available to operating personnel.

In the past, the NPG prepared the respective bombsight pamphlets. Since this task was too large for the NPG because of the burgeoning workload, it was directed that commercial sources be obtained in preparing these new publications. However, any activity that was selected would be required to undergo a thorough screening and be approved due to the classification of the equipment before awarding a contract. At the same time the AAC was also in the process of preparing manuals/pamphlets for their bombsights, as well as the (SBAE/AFCE) Mark 2, Mod 1. Since the AAC was already in the process of updating their publications, they were agreeable to make these items also available to the Navy. In December 1942, it was estimated that 7,000 copies of the bombsight publications would be needed. The Army Air Corps would need 4,000 copies and the Navy 3,000 units. The cost of producing this publication was $8.00 a copy.[20]

Due to the present needs of the SBAE/AFCE pamphlet, it was given priority in cases where the work on the two pamphlets could not be done concurrently. The material was classified "Restricted" in order that they would have widespread distribution and were issued to the enlisted technicians. In February 1943, the new publications were simplified to make them more understandable to the average enlisted student of limited schooling in the present bombsight school classes. The utility of the SBAE/AFCE suffered from inadequate field maintenance to a much greater extent than the bombsight. Reports from units operating airplanes equipped with SBAE/AFCE concerning their usefulness were very favorable in direct proportion to the availability of experienced maintenance personnel.[21] In March 1943, the NPG informed the Bureau of Ordnance that the new publications were to be used for training purposes in the bombsight, SBAE/AFCE schools, and by graduates of these schools upon being assigned to duty in their respective units. They were also intended to be correlated with various motion picture films that were in the process of preparation. Due to the scope and type of work involved, it was not considered practical for service personnel to undertake this task. In March 1943, various commercial sources were available and produced writings that were submitted to them. The requirement was for 4,000 copies of each textbook and 7,000 copies of maintenance pamphlets.[22]

As the war progressed, each service developed their own manuals or publications. No further cost or quantity of these manuals or pamphlets were found in the files in NARA. Note: The author has a complete set of motion picture training films developed by the AAF.

Data Books for Bombsight (SBAE/AFCE)

The question of maintaining adequate repair records for each bombsight and SBAE/AFCE was of prime importance to the Bureau of Ordnance and the Army Air Forces. Several methods had been used in the past by both services with varying degrees of success. With the tremendous build-up of bombsights due to the bombardment program, it was mandatory that a workable system of record keeping on repair of bombsights and SBAE/AFCE be initiated. Separate log books were used in the past, but were not very successful. By April 1941, the Naval Inspector of Ordnance requested that 1,000 log books be immediately purchased, as there were no log books on hand or in stock that were to be provided with shipment. To forestall incomplete records on repair and use and to expedite delivery, it was directed that partial shipments of 100 books be forwarded upon completion.[23] This project was given a high priority to be accomplished as soon as possible. Prior to April 1944, separate log books were not used for recording data concerning bombsights and stabilizers, and that appropriate entries in the bombsight log book should be made regarding stabilizers. This practice caused considerable inconvenience in marking running time entries in the bombsight log in cases where the stabilizer was used with the SBAE/AFCE, and kept with the airplane and the bombsight was not carried in the airplane. Prior to April 1944, separate log books were not used for recording data concerning bombsights and stabilizers. This book contained information about unit operating time, inspection, last overhaul, periodic maintenance, bombing date, bench checks, etc (Plate #5). All entries were printed, including signature. In April 1944, to alleviate this cumbersome method, all service commanders were authorized to waive these practices and record the stabilizer in the SBAE/AFCE log book.[24]

To eliminate this practice, information on the bombsight, stabilizer, and SBAE/AFCE were replaced by cards. In August 1944, the title of the Data Book was changed from Bombsight-Stabilizer-Flight Gyro Data Book to Bombsight Equipment Data Book. However, the use of the old book was authorized until supply was exhausted. Stabilizer and flight gyro units were based on airplane time, and data books were no longer required. To insure that the new data cards would be a part of the bombsight, in August 1944 the Bureau of Ordnance awarded a contract to the Norden Company for 11,000 card holders to be completed at the earliest possible date. The holder was constructed of aluminum, was located on the back of the bombsight cover, and secured by screws that held the cover to the lower part of the bombsight.[25] To prevent the card from being lost, a plastic retainer cover held the log. No cost or additional quantities were found in the examined files.

The AAF, in addition to developing for bombsight and AFCE equipment, required the bombardiers to maintain a "Bombardier's Individual Log Book." In March 1942, Wright Field procured 10,000 of the booklets. At the request of the Bureau of Ordnance, in June 1942 the AAC forwarded a copy for their use for publication. The intent of this log was to provide a permanent and practical record of all bombing missions, containing name, rank, station, squadron, altitude, type airplane, type

Chapter XV: Norden Bombsight Ancillary Equipment

bomb, size, number, total bombs dropped to date, etc. For each bomb release information was required as to airspeed, drift, weather, visibility, errors in feet for range (over or short), deflection (right or left), and circular. Also, the point of impact was indicated on the bull's eye grid by a small circle, and opposite the circle the release number. The average circular error was entered at the completion of each mission. Progress charts were prepared by plotting the converted average circular errors of each mission against the number of bombs dropped and the running average circular errors of all mission against the number of bombs dropped. The booklet contained the necessary instructions, forms, graphs, and remarks to comply with this requirement. No cost or further information was found in the examined files at the Archives referring to this publication. [26]

Electrically Heated Bombsight Blanket

New airplanes flying faster and higher caused problems in the operation of the bombsight. Due to the intense cold at 15,000-20,000 feet and higher, altitudes caused the sight not to work properly. Different methods were attempted, including running the sight during flight. In order to keep moisture from forming, which would fog up the optics and damage the working parts of the sight, it was turned on and left to run. The Bureau of Ordnance requested General Electric Company (GE) to design a heating cover for the sight. By May 1942, GE had developed a suitable heating blanket. It was ready for tests and, in June 1942, a test blanket was sent to the Naval Proving Ground, Dahlgren, Virginia, to undergo intensive testing.[27]

The Chief of the Air Staff advised the Commanding General AAF that initial deliveries were made in December 1942. By January 1943, the Type A-1 bombsight heater cover was in production, and the entire output was designated for combat theaters and airplanes scheduled for early departure. When sufficient quantities were available, they were issued to bombardment and training units in the United States.[28] It was designed for either 12 or 24, and later for 26 volt application. The change from 12 to 24 volt was a simple change by servicing personnel. Instruction for the change was furnished with each blanket. The temperature was controlled by a built-in thermostat to prevent overheating. The blanket was equipped with a web strap and padlock, and it could serve both as a heater and a cover. The blanket was turned on when low temperatures were encountered. In approaching the target, the blanket was removed, and the thermal capacity of the sight was utilized to maintain satisfactory temperatures during periods of action. The extreme cold also affected units of the auto pilot. Covers were also developed for the vertical flight gyro, servo motors, and other components. The author has a bombsight blanket, and also covers for the auto pilot parts in excellent condition on display at the local air museum (Plate #6).

Plate #1 - Model 43A-2 Portable Bombsight Disk Speed Tachometer

Chapter XV: Norden Bombsight Ancillary Equipment

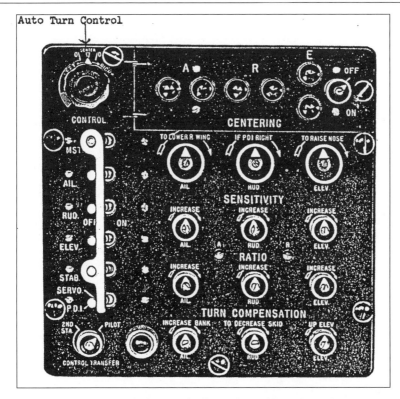

Plate #2 and #3- Autopilot Control Panel/Turn Control

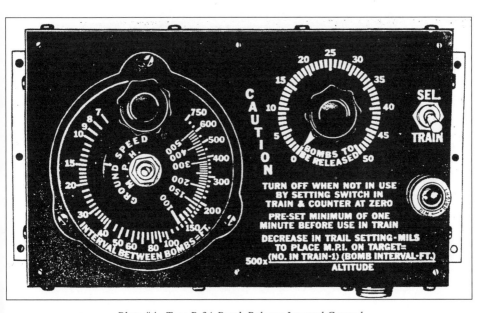

Plate #4 - Type B-3A Bomb Release Interval Control

OPERATING TIME

DATE	Operating Time HR.	MIN.	REMARKS
BROUGHT FORWARD			
5-19-45		10	BAC Bench Check
5-22-45		10	PREFLIGHT GROUND CHECK - EWT
6-4-45	1	00	Bench Check & oiled Bearings. BMC Matched with Stab. 29576
			[illegible] check
7-1-45	1	00	[illegible]
7-6	7	20	BOMBING
7-6	4	00	"
7-13-45		45	15 day [illegible] [illegible]
7-12-45	5	00	BOMBING
7-18-45	1	10	"
7-21	4	45	"
7-22-45		30	Changed Crosstrail Rack [illegible]
8-4-45	3	30	Bombing
TOTAL	28	00	

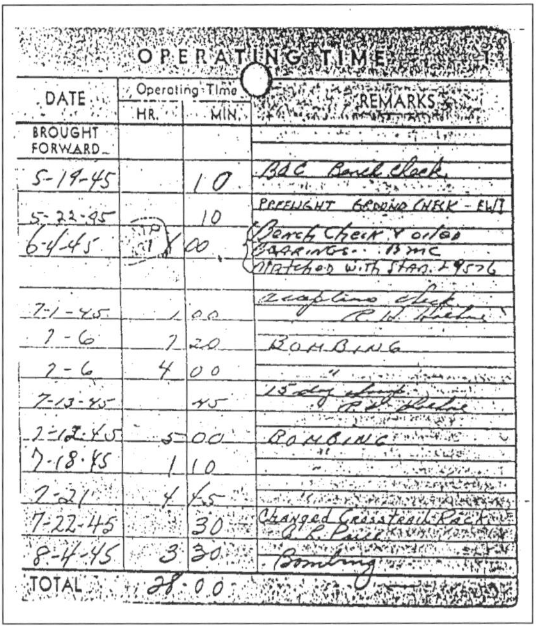

Plate #5 - Bombsight Equipment Data Book.

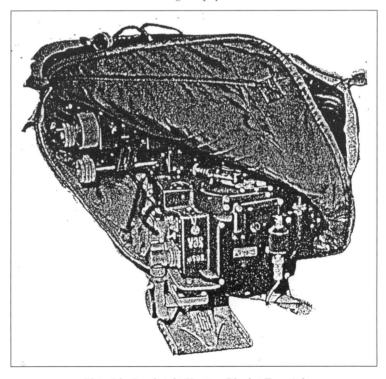

Plate #6 - Bombsight Heating Blanket Type A-1

Chapter XV: Norden Bombsight Ancillary Equipment

Notes:

1 Letter from NPG to Chief Bureau of Ordnance, Tachometers for Bombsight Mark XV, Mod 3, June 19, 1942, National Archives.
2 American Time Products, Inc., Tachometer Calibration Unit Type TCU-4, Request for price quotations, June 3, 1942, National Archives.
3 Letter Naval Inspector of Ordnance, Acceptance of Mark XV Bombsight Disk Speed Indicators Manufactured by Technical Tool Corporation, November 9, 1943, National Archives.
4 Letter, Bureau of Ordnance, Request for Authorization of Test Tachometers by Jaeger Watch Company, April 13, 1944, National Archives.
5 Jaeger Watch Company, Plunger (or stop watch) Revolution Counter, August 7, 1943, National Archives.
6 Naval Proving Ground Letter to Bureau of Ordnance, Polaroid Glasses for use with Mark XV-3 Bombsight, May 24, 1940, National Archives.
7 Letter from Naval Proving Ground to Bureau of Aeronautics, Polaroid Glasses May 23, 1940, National Archives.
8 Letter from Polaroid Corporation, Proposal for Furnishing Ray Filter Kits, February 16, 1943, National Archives.
9 AN-11-60AA-1, Technical Order, Handbook, Operation and Service Instructions, Type C-1 Automatic Pilot, June 25, 1943.
10 Hqs Army Air Forces letter/memo, Present Bombsight Program, October 7, 1944. Frame 0583.
11 Automatic Flight Control Equipment, Installation Bombardier's Turn Knob Assembly on C-1 Auto-pilot, Technical Order 00-11-60-10.
12 Bombardier's Information File, May 1945, Letter, Present Bombsight Program October 7, 1944 Frame 0583.
13 Letter from Department of the Navy to all Commanders of Navy and Army, Bombsight, Mark XV Mod 7, Modification of Intervalometers, K-2 and Mark 2 Mod 1, due to bombsight Automatic Release Switch Safety Assembly, September 24, 1943, National Archives.
14 Intervalometer Spacing, H.H. Germond, A Report Submitted by the Applied Mathematics Group by the Bombing Research Group, Columbia University, February 10, 1944.
15 Bombing Equipment and Accessories — Automatic Release Assembly — M-Series, Technical Order No. 11-5A-4, June 18, 1945.
16 Letter, Naval Proving Ground, Test of Automatic Release Mechanism for MK15 Bombsight, January 23, 1942.
17 Letter, Bureau of Ordnance, Preparation of Ballistic Tables, to Naval Proving Ground, November 15, 1939, National Archives.

[18] Letter, Naval Proving Ground to Bureau of Ordnance, Equipment for Calibration. Location of Bamberg Stations and Construction of Plotting Tables for Bomb Ranging, July 24, 1944, National Archives.
[19] Letter, George Armando Company, to Naval Inspector of Ordnance at Carl L. Norden Company, Wooden Tweezers, July 24, 1944, National Archives.
[20] Letter from U.S. Naval Proving Ground to Bureau of Ordnance, Request for Illustrated Manual or Pamphlet for Norden Bombsight, December 9, 1942, National Archives.
[21] Letter from Bureau of Ordnance to Naval Proving Ground, Illustrated Pamphlets for Bombsight Mark XV and Stabilized Bombing Approach Equipment Mark 1, February 5, 1943, National Archives.
[22] Letter from Naval Proving Ground to Chief of Bureau of Ordnance, Request for Cancellation Illustrated Manuals or Pamphlets for the Norden Bombsight and SBAE, March 9, 1943, National Archives.
[23] Letter to Chief of Bureau of Ordnance from Naval Inspector of Ordnance, Request for SBAE Log Books, April 21, 1941, National Archives.
[24] Letter from Bureau of ordnance to all Air Force and Naval Commanders, NAS and Marine Corps Air Stations, Information concerning Log Books for Bombsight and SBAE/AFCE, April 19, 1944, National Archives.
[25] Letter from Chief of Bureau of Ordnance to Carl L. Norden, Inc., 80 Lafayette Street, New York, Request for CMP-4 Transmittal Bombsight, Stabilizer and SBAE Data Card Holders, August 1, 1944, National Archives.
[26] Bombardier's Individual Log Book, Wright Field, March 19, 1942, National Archives.
[27] Letter, General Electric Company, Electrically Heated Bombsight Blanket, June 3, 1942.
[28] Letter War Department Hqs AAF to Commanding General 2nd Air Force, Subject: Low Temperature Operation of M-Series Bombsights, dated January 17, 1943, National Archives.

Plate #1 Portable Bombsight Disc Speed Tachometer, Technical Order No. 11- 5A, Library of Congress.
Plate #2 and #3 C-1 automatic Flight Control Panel, Students Manual, AFRTC Manual No. 51-340-1, dated September 1949.
Plate #4 Operation and Service Instructions, Bomb Release Interval Control, Technical Order No. AB-11-5E, June 2, 1946.
Plate #5 AAF Form No. 48, National Archives.
Plate #6 Bombsight Heating Blanket (GE) Type A-1, Bombardiers Information File, March 1945, chapter 5-7-1.

XVI

Security

The problem of providing security for the new bombsight was of paramount importance to the Navy and Army Air Forces. In the beginning, all correspondence and sights could only be picked up or received by a commissioned officer or courier. With only a few sights in the inventory, this did not present much of a problem. In 1934, General Ben Foulois stated that the NBS was the most single important secret under development in the Army Air Corps. There was only one manufacturer, and his facilities were restricted to reduce to a minimum and to preclude the possibility of disclosure to any unauthorized individual, either in the U.S. military, civilian work force or any foreign governments. Due to the very restrictive means in safeguarding this new weapon, the Navy Bureau of Ordnance removed the bombsight from "Secret" status to "Confidential".[1] Also, no visits were permitted by Army Air Corps officers to the manufacturer's plant without express permission of the Chief of the Army Air Corps. In May 1935, a directive was issued by the Bureau of Ordnance setting forth instructions to provide proper security for the NBS. Confidential material is of such a nature that its disclosure, although not endangering national security, could be prejudicial to the interest of the nation. The system of classifying the NBS as secret was awkward, cumbersome, expensive, and time consuming. In handling the NBS, the Navy would secretly notify the AAC that a quantity of NBSs were available at Dahlgren (Proving Ground), Virginia. An airplane would be dispatched from Wright Field, Dayton, Ohio, and the sights would be brought back to Wright Field and secretly locked up in a secret vault. It would be several months before the accountable Supply Officer would learn of the matter. At that time, the sights would be entered as "found on Post." With the expansion of the AAF, this method could not continue without throttling the entire program.[2]

The declassification of the NBS did not mean that the sight would not be safeguarded at all times. The design, development, testing, and production of the Mark XV was carried on by the Navy under conditions designed to insure an *UNUSUAL DEGREE OF SECRECY*. The decision to reclassify it as confidential rather than secret was made only to permit sending correspondence within continental limits of the United States by registered mail rather than by officer messenger.[3] Note: Registered mail is the safest method of sending material. Every item is handled under armed guard and locked in a vault. Every time it is transferred, the receiving person must sign for the item. Strict control is maintained on every item, including placing the material in a safe or vault. The AAC was obligated to follow the same rules as the Navy. The AAC placed the bombsight in a secret status and all correspondence relating to it. The development and production of the M-1 (Mark XV) was carried under conditions designed to insure that an *UNUSUAL DEGREE OF SECRECY* would be maintained. *Strict compliance with these instructions WERE MANDATORY ON ALL CONCERNED.* The bombardiers were supposed to take an oath to protect the NBS with their lives if need be.

The AAC advised the Navy that these precautions were outlined in U.S. Army Regulation 330-5, paragraph 6, dated 11 February 1936. It was necessary for the Chief of the Air Corps to issue regulations to cover inspection, calibration, acceptance of sights, and the distribution of spare parts, tools, storage, and receipt of each bombsight and packing case keys. The Depot Supply Officer would personally inventory all articles on the shipping ticket prior to issuance to a responsible commissioned officer. All spare parts were to be stored and treated on confidential status. The bombsight installed in the airplane was covered at all times when the airplane was on the ground, so that the sight was not visible through the bombing window. When the planes were not used in bombing, the sight was to be removed from the directional stabilizer and placed in storage. If proper storage was not available, the sight was covered, and a guard(s) established over the airplane. Absolutely no photographs were permitted to be taken of the bombsight no matter where it was installed, including photos taken of the sight through the bombardier's window. Identifying lettering as to type, manufacturer's name, bombsight assembly, and PDI was covered on the shipping box. In addition, an instructional manual in pamphlet form was furnished to each station receiving the sights. This pamphlet contained information that utmost care must be exercised to prevent unauthorized persons from obtaining access to the sight. A bombsight log book was provided for each sight, and would be posted up to date. Unserviceable sights were to be treated as serviceable and turned in to the AAF Chief, Materiel Division, Wright Field, and were given the same safeguards as a serviceable one. In the AAF, the method of shipment was as follows: The Chief of the Materiel Division outlined that domestic shipments were to be made by Railway Express. Shipment by Regular Army Air Corps airplanes, with the receipt signed by the pilot, showed the serial number of the airplane in which the shipment was made. Foreign shipments were made only by Army and Navy transports, with the sights placed in the personal custody of the Commanding Officer of troops, who personally signed the receipt. The Navy requested the sights be placed in the custody of the Commanding Officer of a Naval Man-of-War and that he personally sign the receipt, to be delivered to the Commanding Officer at destination. In all cases, each packing case was stenciled in bold letters, "**EXTREMELY DELICATE INSTRUMENTS**," with special care to prevent damage in transit (Plate #1). In no case would unserviceable bombsights be disposed of by the stations. These were disposed of by the Depot Supply Officer, Fairfield Air Depot, All Wings, Bombardment Groups, Air Bases, Training Schools, Air Base Fields, AAF Depots, Composite Groups, Advanced Flying Schools, and the Bureau of Ordnance.[4]

According to a photograph from the Smithsonian Institution which shows a canvas bag on a table flanked by two armed guards with automatic weapons, it is assumed that a Norden bombsight was in the bag while an officer administered the oath to a group of standing bombardier cadets. Caption on the photo states "Bombardier Cadets Swearing to Protect the Bombsight." There is no information as to where this was taken. The wording of the oath is not known to the author. The Oath was not found in any of the security directives or Operations Pamphlets issued by the AAF and the Bureau of Ordnance (Navy). The oath was administered in the movie, "Bombardier." Whether the wording in the movie was the actual oath is not known. Research in this area failed to reveal any information on the subject.[5]

It was the Navy's desire to avoid divulging the existence of a source of supply of the latest model of the bombsight. To accomplish this the Bureau of Ordnance required the manufacturer to include, in his contract, that no visitors be allowed in his plant except by the authority of the Chief, Bureau of Ordnance and AAF. This included commissioned officers of the Navy and AAF. A bombsight log book was provided, and the keys to the packing case and the log book were sent to the receiving activity by Registered Mail. A fully qualified officer was appointed to function as a consultant on the matter of the pamphlet, and was charged with its proper care and safeguarding. This Officer executed a certificate once every six months, July 1 and January 1, to the effect that the pamphlet was in his possession and was properly being safeguarded. There was no deviation to this rule. Information covering the maintenance, repair, adjustment, and operation of the sight was given to Regular Officer Pilots, active duty reserve officer pilots, cadet pilots, bombardiers on continuous duty, and enlisted pilots. Extreme care was used to select these personnel. Not only must they have be well qualified, but must be trustworthy and citizens of the United States.

In May 1941, the Bureau of Ordnance reclassified the SBAE/AFCE equipment to "Restricted" instead of "Confidential." This reclassification applied to the following items of SBAE/AFCE equipment: Flight Gyro, Rudder Servo Motor, Elevator Servo Motor, Banking Motor, Aileron Servo Motor, and Sector Panel Assembly. The purpose of the re-classification was to facilitate the installation of the SBAE/AFCE in aircraft at the manufacturer's plant. This action did not include the bombsight stabilizer, nor did this de-classification of the SBAE/AFCE affect the classification or status of the bombsight. This action by the Navy Bureau of Ordnance was forwarded to the AAC for dissemination to their personnel handling the sight. The Bureau of Ordnance also stated that the descriptive operations pamphlet dealing with this equipment would continue to retain its confidential classification, and that all correspondence which discussed design, characteristics, performance, and malfunctions was still classified as confidential. All service activities, training schools, manufacturers, and any other activities involved with this equipment were notified immediately. [6]

The Bureau of Ordnance notified the Army Air Corps of the change in classification of the SBAE equipment. The reclassification applied to the same equipment as the Navy's. The AAC took immediate action to disseminate this information to all activities. The descriptive pamphlet SBAE O.P. No 630 was to retain its confidential status, since it contained information on the bombsight stabilizer. Action was taken to change the classification of spare parts to restricted. Purchase orders, contracts, and correspondence concerning spare parts would have the same classification.[7]

HQS AAF, in July 1942, issued a new policy governing M-Series bombsights and Automatic Pilot Equipment (Stabilized Bombing Approach Equipment—Navy). This action was coordinated with the Bureau of Ordnance and superseded and cancelled the letter of July 27, 1938, which provisions had been in effect since that time. The bombsight unit, complete bombsight assembly, consisting of bombsight stabilizer, glide angle attachment, descriptive matter, unsatisfactory reports, correspondence, and drawings were to be treated as CONFIDENTIAL. The Automatic Flight Control Equipment pertaining to performance, malfunction, design, unsatisfactory reports, drawings, basic instructional manuals, and technical instructions were treated as CONFIDENTIAL.[8] The Flight Gyro, rudder, elevator, aileron servo motors, banking motor, sector panel, requisitions, shipping tickets, purchase orders, contracts, etc., were to be classified Restricted. Items such as pilot director indicators, tachometers, and bomb trainers, when not equipped with the bombsight, were classified Restricted. Radiograms or telegrams making reference to the Bombsight Equipment were to be properly coded. Commanding Officers were to exercise extreme care in the selection of personnel to be charged with use of this equipment. Operation and maintenance was to be given only to regular and Warrant Officer pilots, bombardier, enlisted pilots and bombardiers, selected enlisted repairmen, civilian employees, students assigned to bombardier training, and other personnel designated by the Unit Commander, as well as specially selected and trusted employees of airplane manufacturers charged with the installation of this equipment. Information was limited to the minimum.⁹

The bombsight, in the event of transfer of any aircraft, was to be transferred with the airplane. If bombsights shipped within Depot Control Areas to another, then all shipping instructions were to be furnished the consignor and, in every case, a copy was forwarded to the Chief, Field Services, AAF Wright Field (Plate #2). All markings, such as "Bombsight," etc., were obliterated or covered so that they were not visible when checked for shipment. The consignor notified the consignee by radio as to the date, method of shipment, approximate time of arrival, pilot's name, serial number, and model of the airplane bill of lading in case of express shipments, in addition to truck driver's number and any other information required to identify the shipment. In the shipment of the bombsight, code words were to be used. The following words were to be used in place of the word "Bombsight." If an item was shipped on Sunday the code word was SIDING; Monday - LATENT; Tuesday - TELESCOPE; Wednesday - MICKEY; Thursday - DUSTY; Friday - EXTERNAL; Saturday - TALON. If it was shipped by Railway Express the code was RELAX; Army truck - TRUAX; Air Transport - PINAY. In the case of shipment by Army truck, the driver and another person were to be armed during shipment, and there was *no deviation* to this rule. For shipments to Theater of Operations, or Combat Zones, no serial number record was to be maintained. No record was kept by the Chief, Field Services, Wright Field showing location and serial number of bombsights shipped to combat zones. When shipments were made by Army Transport, the bombsights were placed in the custody of the Quartermaster, and a personally signed receipt was obtained from him (Plate #3). When circumstances required expedient means, the Navy Department was requested to have the shipment placed in the custody of the Commanding Officer of a Navy transport or man-of-war who would deliver it to a commissioned officer at the destination.¹⁰

The safeguarding, storage, and maintenance facilities were charged to the Commanding Officer, who was also charged with the responsibility of proper maintenance of bombsight storage vaults and maintenance facilities under his command. He also controlled the keys and combination to bombsight storage vaults and bombsight maintenance facilities. One set of keys was provided to the commissioned Armament Officer. When the bombsight was installed in the airplane, the bombardier was to be armed with a service pistol at all times. If the bombsight was installed in the airplane when it was on the ground, the bombsight and stabilizer were to be covered at all times. On completion of a bombing mission, or when the airplane was placed in the hangar, the bombsight was removed and placed in the bombsight storage vault. When operating in the field, or where adequate storage was not available, the bombsight was left in the airplane, and sufficient guards were to be maintained to prevent unauthorized entrance. The bombsight vault was constructed with steel reinforced concrete and had heavy steel or vault type doors installed and locked, with a guard established as required. The number could vary depending on location, however, in no case was there to be less than one guard on duty at all times. If conditions were such, additional guards were to be posted to provide maximum security.[11]

The destruction of M-Series bombsights prior to forced landing over enemy territory was to fire two rounds with a .45 caliber service pistol into the rate end mechanism, and one round through the telescope. The bombsight was to be removed and thrown overboard if time permitted. This policy was ordered by General Arnold, Commander-in-Chief, Army Air Forces. The instructions applied to everyone and were strictly adhered to.

It was mandatory for any person who handled the bombsight (maintenance, operations, supply), either military or civilian personnel, to read and acknowledge by signature the Espionage Act of 1917, (Act of June 15, 1917); (Stat. 217); U.S.A. 50-31, 32; (Sec. 2181, 2182, M L 1929). Any unauthorized disclosure of data or information was a violation of this act. The author vividly remembers reading and signing this act.[12]

In December 1942, the AAF issued new sweeping directives governing M-Series bombsights and Automatic Flight Control Equipment (Stabilized Bombing Approach Equipment), Type B and C, and attachments and Low Altitude Bombing System. The development of the bombsight system had been carried under conditions to insure an *unusual degree of secrecy*; however, with the advent of war it was found necessary to reclassify this equipment. This letter superseded and supplemented the Policy letter of July 1942 concerning AFCE, SBAE, types B 1 and C 1, and all attachments. The M-Series Bombsight and Automatic Pilot types B 1 and C 1 and all attachments. The M-Series Bombsight, Automatic Pilot types B 1 and C 1, complete sets of drawings, manufacturing processes, log books of the bombsight unit, Auto Pilot B 1 and C 1 stabilizer unit, glide angle attachments of the M-Series, and performance results of the bombsight were to be treated as CONFIDENTIAL. Commanding Officers were to continue using extreme care in the selection of military, contractor, and civilian employees. Security at the various contractor plants was handled as stated above concerning the NBS. Bombsights shipped within the depot system were given shipping instructions furnished by the consignor, who then forwarded a copy to the AAF, Air Materiel Command (AMC). The consignor was to notify the receiving activity by radio or teletype. Code words were no longer needed and shipping information could be sent in the clear. The use of storage vaults continued to be used. When the airplane was on the ground, the bombsight and stabilizer were to be covered at all times. If adequate facilities did not exist for protecting the sight, it was left in the airplane with sufficient guards to prevent unauthorized entrance. At the completion of a bombing mission, the sight was not removed provided it was on a military installation. Regarding the disposition of sights in eminent danger of capture, the bombardier was to fire two rounds with a .45 caliber service pistol into the rate end mechanism and one round through the telescope, and the sight was to be thrown overboard. No mention was made about the destruct package that was enumerated in previous directives. Apparently this method of destruction was never developed and was not used.[13]

In January 1943, the Bureau of Ordnance issued another directive in the proper handling of the Mark XV bombsight, stabilizer, etc. It became evident that this equipment was not being processed in accordance with prior guidelines. Classification had been changed to Restricted from confidential for the purpose of expediting shipment and dissemination of information pertaining to this equipment, which was still considered Confidential. Information pertaining to accuracy and manufacturing data and log books after entries had been made still remained Confidential. New precautions were added for shipment, stowage, security, and packing. Sights and stabilizers, when shipped singly or together, the gyro was caged, sight strapped to base, and safety wired, and then secured in metal or wood packing box by shock proof type mounting. The packed box was enclosed in a wooden non-tip crate. In addition to Extremely Delicate Instrument, HANDLE WITH UTMOST CARE was added. Custody receipts of the equipment were required. The bombsight was to be placed in a vault, if available. If not, a suitable locked room barred to other than authorized personnel. Storage rooms were to have a temperature range of 65° to 95° F, dust proof as practical with a filter and dehumidifying unit. All other provisions remained the same and were to be followed as outlined in previous directives.

It was brought to the attention of the Navy in March 1944 that the Maintenance Division of the Bureau of Ordnance was experiencing trouble with the method of transporting sights and supplemental equipment during shipment. Bombsight material was handled as Delicate Instruments, and the Money Waybill was waived. Current directives outlined the method of shipment in detail. On the other hand, the Production Division of the Bureau of Ordnance was handling material on a Confidential basis and necessitated the use of armed guard service and handling the Money Waybill under hand to hand signature. They felt that in handling the bombsight as Confidential that greater care would be accorded these shipments. It was determined that the source of the trouble originated in that parts of the bombsight were identified as Confidential and other parts were listed as Restricted. This confusion was creating considerable hardship in the delivery of critical sights to the service activities, especially the combat units. There was no reason for this discrepancy. The original directive was clear in this area, and it was misinterpreted by the Production personnel. Accordingly, this matter was cleared up immediately.[14]

The practice of forwarding keys under separate cover by registered mail for bombsight shipping boxes to AAF and Navy consignees designated to receive the bombsight proper was discontinued June 21, 1944, for both the AAF and the Navy. The AAF initiated a system whereby all of their activities charged with handling bombsights were provided with a master set of keys capable of opening locks on all bombsight boxes. The action speeded up the processing of critically needed sights. Immediate action was taken by all activities involved with the NBS. Direct contractors operating under the jurisdiction of the activity for whom they were manufacturing sights were also to delete this requirement for keys for bombsight shipping boxes. New master keys were ordered for them immediately by either the AAF or the Navy.[15]

In July 1944, due to the great quantity of bombing airplanes that had been shot down carrying the bombsight over enemy territory, it was felt that the German Air Force had obtained complete intact bombsights from planes that survived crash landings. Accordingly, the equipment was downgraded, as new electronic gear was beginning to enter the combat theaters, such as radar bombing and AZON/RAZON radio guided bombs in the later part of 1944 to the end of WWII. No correspondence was found stating the date all security was removed on the NBS. The author's recollection is that in April 1945, the two heavy steel doors to the entrance to the Bombsight Building were removed, as was the gate leading into the shop, offices, and supply areas. The doors were replaced with a regular door, and the guards were removed and the front gate left open. It was after that the NBS began to be displayed to the public, and soon after were made available to military surplus dealers for sale to the public. Many of the bombsights on hand in the shop were dismantled by the bombsight employees, who segregated the different parts in separate containers.

Bombsight Repair Shops & Storage Facilities
With the tremendous increase of bombsights, for both the Bureau of Ordnance and AAF the problem of proper repair and storage facilities became of paramount importance due to the secrecy of the equipment and had to be addressed as soon as possible. All of the Armed services were building up to an unprecedented scale, especially the AAF expansion program caused by President Roosevelt's declaration of building a fleet of 50,000 airplanes in May 1940. To cope with this requirement, some form of standard facilities to protect the sight was mandatory. Realizing this problem, the Bureau of Ordnance issued directives for "Bombsight Shops and Storage Facilities" in July 1939. The AAF soon followed suit.

The Navy facilities were constructed at new seaplane bases on squadron requirements. All shops storage and workshop space was to be combined in one room. The room was to be constructed to provide maximum security for the bombsights. The buildings, bulkheads, deck, and ceiling were to be reinforced concrete with as few windows as possible, and those were to be barred. The vault door had a complex combination lock. The necessary power for generators, motors, and lights was obtained from commercial sources. Emergency backup power was also provided.

There was forced ventilation with provisions for dust elimination. Provisions were made to maintain necessary heating. In tropical stations the building was equipped with "dry lockers" to prevent sweating. This locker was provided with heating elements or other apparatus to reduce the moisture content of the enclosed air. Adequate lighting was required for the entire area and for the work benches and check stands. The test stand foundation consisted of a 3' x 3' reinforced concrete slab and was 6 feet deep into bed rock. This block was separated by at least 1/4 inch on all sides of the floor in order to eliminate any vibration, later called a "seismic block." The number of pads required depended on the size of the repair unit.[16]

Requirements for a bombsight overhaul shop were for one to three squadrons. This type of shop was to be located on the ground floor, in a separate building so as not to be affected by heavy machinery, trucking, or railroads and wind effect on a large building. The overhaul for one to three squadrons required precision lathes, milling machines, precision grinders, precision drill presses, and motor generator sets for both 12 and 24 volt usage. Also, the required number of bombsight test stands depending on the workload. All necessary bombsight and SBAE tools, bombsight test panels, a painting booth for repainting bombsights and SBAE, 1 set of tanks for treatment of salt water immersion, a stove for boiling parts after salt immersion, and rotor balancing equipment were included. The storage of the bombsights was in the same building. The floor, ceiling, and walls were reinforced concrete with a vault type door and combination lock. Adequate ventilation and dry lockers were mandatory. Overhaul shop requirements for four to six squadrons differed in the size of storage space and spare parts bins. The use of machine tools was greater, as a larger quantity of sights would necessarily be received for repair, and additional floor space was needed. The expansion of present shops was left up to the base commander.

The authorized Bombsight overhaul shops for the NAS in the continental U.S. did not take into consideration certain isolated bases in foreign countries (Coco Solo, Panama, Pearl harbor, etc.). The time for normal procurement of special tools peculiar to bombsight repair shops required a lengthy time frame to receive the material. These activities did not have bombsight test stands, drawings of jigs used to balance the cardan cradle, bombsight gyro, dies used in the manufacture of parts, and other items to perform complete overhaul of the bombsight. The isolated location of these bases, the time required for the normal procurement of supplies and material, plus the possibility of hampered shipments and communications in time of war, made it essential that the repair and overhaul facilities of these bases be complete and independent of mainland supply bases as practicable as possible. Pearl Harbor, Territory of Hawaii, was a huge installation, and it was at that time very isolated. For this reason, an installation of this size shop was to be completely self sufficient. Proof of this became evident in the Japanese attack on Pearl Harbor in 1941.[17]

The Army Air Forces also built similar bombsight buildings to repair their bombsights. Some of the items, such as salt water immersion, were not included by the Air Service Command, and AAF designed and built their own version of bombsight buildings. See "A Typical Bombsight Building" described by the author. The Navy and AAF bombsight shops changed very little over the years. The only difference was that after Pearl Harbor the repair facilities were greatly expanded with the addition of new personnel, but the configuration of the building remained unchanged. The interior was rearranged so that more repair people could be utilized in repairing bombsight equipment. To alleviate crowded conditions, several shifts were inaugurated to repair the additional workload.

Typical AAF Bombsight Building
A Bombsight Building was constructed by the Air Material Command, AAF, at each of its major depots for the express purpose of repairing, calibrating, and storing the NBS. As with the Navy, it was apart from all other maintenance, supply, and base facilities. The building contained about 10-15,000 square feet of shop and office space. The AAF went through a great effort to plan and construct a secured area. The Navy, at their Naval Air Stations, had already built and had in place their repair stations. The building was constructed of concrete and steel. In the AAF installation, there were no windows. The only opening was the personnel door and two large powered doors for receiving and shipping. The floor, walls, and ceiling were constructed of reinforced concrete. The perimeter of the building was fenced off with chain link fencing about 10-15 feet high, with barbed wire on the angular arms. At each corner of the fence, a radio detection system was installed (latest of its type, we were told). The personnel gate was located outside the fenced area next to a large gate for trucks to enter to deliver sights. There was only one personnel entrance door. To be admitted to the building, it was necessary to push a bell from outside the fence, which registered in the vestibule and alerted the guards. The guards could see who was entering by a port hole, and would open the gate to enter the courtyard. Once inside the court yard, which was about 40-50 feet, and to enter the vestibule, the heavy outside steel door opened and the employee entered the vestibule. The guards checked the employee against an authorized roster. When the front door was open, the inside door remained closed—they were counter balanced doors. After being cleared by the guards, the outer door was closed and the inner door would open, and you could proceed to the office, supply, or shop area. Each employee was required to wear a special badge. Later, the inside door was removed, and only half a metal door was used inside the building. This made it easier to enter and leave, and saved a lot of time. Guards were rotated on a regular basis so that we never did get acquainted with them. Inside the building contained another vault (a vault within a vault) where all the engineering drawings were kept. The office was in front of the vault. It was an aberration, where the Directorate of Maintenance and

Directorate of Supply occupied the same building. This was due to the high classification of the bombsight, and this building was the only one that was secured. Normally, Supply kept all of its material in the Depot Supply Warehouse Buildings, but, in this case, it was all in one building. Maintenance, because of the workload, was allotted the most space. Supply had their Stock Records and Shipping and Receiving near the Machine Shop. In the military, every item is issued by Supply, whether it is an atomic bomb or a round of ammunition. Shop space was assigned to the various repair activities. There was a partition from the doorway to the shop area, with a wainscot 3-4 feet high constructed of wood, and then glass to the ceiling, so there was an unobstructed view from the office. Tear down, disassembly, assembly, calibration, testing, and inspection were done in one part of the building before being sent to final inspection. This was the last phase before being forwarded to supply for recording and storage.

The machine shop was off limits to all personnel for various reasons, one being the machinery, and another the engineering drawings laying on work benches and tables. Most employees worked on one part of the sight. Machinists were called on to manufacture any required part, since none were available from the factory. At that time there was a critical shortage of bombsights. The Training Table Section was located between the machine shop and the repair activities because machine shop work was required. Next to the machine shop, the gyro balancing room was located. It also had a 3-4 foot wood wainscot with glass to the ceiling. It was close, so that if any machining was needed, machinists were readily available. The entire shop was covered with one foot square blocks of light linoleum. Janitors cleaned all areas daily, including the machine shop, and waxed weekly. No parking was permitted on the grounds. The roof contained a layer of dirt, some said about 2 feet, which made a good camouflage, as grass was growing on the top. Unfortunately, no pictures are available. The building was painted a cocoa brown, and it looked grim. Similar types of testing and calibration equipment were the same as the Navy, which had been developed through the years. Although the work force was dramatically increased after Pearl Harbor, this was accomplished by adding additional shifts. Many women employees were hired to replace the men being drafted into the services. All employees selected were carefully screened for ability and character. The vast majority were excellent in their work and, after the war, many became professional people.

Working in the bombsight building was at times somewhat difficult. Other base personnel would query as to what we were doing. Generally, the workers never left the building when reporting for work. The author got around this gracefully by stating that certain parts needed to be washed and was assigned to the wash rack. This stopped all questions. The entire work area was fully air conditioned (this equipment was also inside the building, along with an emergency backup system). Constant room temperature was maintained at all times, and the entire building was fully air-conditioned. This was the beginning of the "Clean Room." The interior was painted in a soft color and pleasing to work in. The machine shop contained some of the best precision lathes, milling machines, drill presses, etc., like the Navy, and were the best available at the time. No heavy machining was done in the shop. If the need arose, this was done in the main manufacturing machine shop by one of the bombsight machinists. A partition was placed around the machine until the job was done. Of course, this caused friction with the other machinists. Under no circumstances could we divulge what the part was for.

Very few visitors were permitted in the building. The only ones were personnel who needed or wanted bombsights repaired or were in for instructions on the repair or servicing of the system. Of course, high level civilian and military officials were admitted, such as the Base Commander, Commanding General, Chief of Maintenance, and other visiting high ranking military personnel. When General Doolittle (then Lt Col) was at the Base, he was a frequent visitor in the shop, when the B-25 airplanes were at McClellan before being loaded on the Carrier Hornet at the Alameda Naval Air Station.[18] This story, as related by Mr. Milton Stomsvick, concerning security of the Norden Bombsight will give insight as to what extremes were taken to safeguard the bombsight. We were continually reminded as to our duty to keep the information secret (See Mr. Stomsvick's story).[19]

The following story is furnished by Mr. Milton Stomsvick.

I will relate to you the importance of the security for the Norden Bombsight (NBS) and associated equipment. I was held accountable for every item and equipment.

This is my story on what happened to me concerning security on the NBS. On December 6, 1941, I was having dinner with my wife, 5 year old daughter, and a three week old son when a knock on the door interrupted our dinner. I opened the door, and there stood two armed military Sergeants with side arms. I was informed that I was to be escorted to McClellan Field. I was stunned, and naturally I objected to this and asked for an explanation. They would not give me any type of information. This upset me and the rest of the family.

I was escorted, under guard, to the offices of the two Directors, and I was informed that a NBS was missing in one of the B-17s enroute to Hawaii. This consisted of a group of 15 new planes that were from the factory that were flown into McClellan Field for outfitting before making the long journey to Hawaii. This included both gun sights and bombsights.

Director of Maintenance and Director of Supply said that they would provide me a chance to clear up this terrible situation and escape a possible 20 year sentence in a Federal Prison. The airplanes were at Hamilton Field. During this time, I was not allowed the use of a telephone or to go to my office, to physically check the bombsights, to look at any records, or to call my wife. The airplanes were scheduled to leave the next morning for the overseas flight to Hawaii.

Chapter XVI: Security

The McClellan Field Test Pilot—a Captain—was located to fly me to Hamilton Field. All of the top civilians in Supply were brought out from their homes to inventory the NBSs, and I was not to talk to any one of them. To make sure, they kept me in the air terminal waiting for the test pilot and an airplane.

About 10 p.m., we boarded a small Cessna (4 passengers) and proceeded down the runway. The airplane bucked up and down, and the Capt explained that he had never flown a small plane. The Sergeant that accompanied us took the controls away from him and straightened out the take off.

To make matters worse, the fog was so thick that we had to fly low so that we could see and follow the automobile lights. We had no assurance that we could land. Fortunately, a hole in the fog opened up, and we came in for a landing. Unfortunately, security at Hamilton Field was so tight that no lighting was provided on the flight line. Armed guards were stationed at each airplane, and it was quite an experience to approach each aircraft in the dark without being shot, as the guards were evidently briefed on the tight security for what was aboard each aircraft.

I spent the entire night going through personal baggage, .50 caliber machine guns, and associated equipment. About 7 a.m. on the morning of December 7, 1941, the Major in charge told me they were taking off. I explained to him that I had not finished checking for the lost bombsight. He said he had orders to leave. I finally persuaded him to get his orders changed, as I was threatened with a 20 year prison term if we could not find this lost bombsight.

They relented and gave me another two hours. Due to their lack of knowledge regarding the appearance between a *bombsight* and a *gun sight*, they did not know the physical difference of the two items, leading to this terrible confusion and unfounded accusation. This discrepancy was finally cleared up when it was pointed out to them that *no bombsight* was missing. I was cleared of any wrong doing, and the possibility of two Colonels being court marshaled certainly was a relief to all of us. During those days we seemed to be more fearsome of the Navy than of Japan or anyone else.

An article appeared in Readers Digest many years ago titled, "Queens Die Proudly." The author also remembers these airplanes. These were the airplanes that were involved in my adventure about the NBS and at Hamilton. The article gave a vivid description of how they were shot down in the air and on the runways. I witnessed this lack of armament while going through them in my quest to locate the missing NBS. They had no protection, and machine guns were not mounted. I sometimes think that if I could have possibly delayed their flight by two hours or more they would have had sufficient fuel to return or go to Alaska. They had no alternative but to attempt a landing even under enemy fire.

In March and April of 1942, the Bombsight Section was host and headquarters for the crews of the B-25s that were to bomb Tokyo. Lt. Col. James Doolittle was given Command Authority over all operations at McClellan Field. A very unusual aberration of normal command operations (this was my understanding of the events at that time). All of the top secret pictures of Tokyo and other classified information in their possession was entrusted to the Bombsight Officer and myself. I point this out because of subsequent events in the extreme security and secrecy of their mission. Other than the Bombsight Officer and myself, no one but the pilots involved knew of their mission. Who would ever dream that B-25 bombers would or could take off from an aircraft carrier? The pilots would joke that I could keep their NBSs that they came with. They said they were going to use clothes pins as their sights.

When they left for their mission, they asked us to witness their departure. We were told to stand in front of the Flight Hangar, and they would say goodbye from their airplanes. We stood there as they dived those B-25s at us. I swear they dropped from several thousand feet straight at us, and the tails of the planes almost touched the hangar as they pulled out of their dive and cleared the flight hangar. Even knowing that they were the cream of the pilot crop did not keep us from being scared to death. The author vividly remembers this incident. It was recently discussed with the members at the last reunion of the Bombsight Retirement group.

I now had their NBSs in my possession. To ensure continued secrecy, I was directed by our Headquarters in Dayton, Ohio, to ship the sights to a destination that had no bearing on the true nature of their mission. I guess this was in an effort to hide their mission even from me.

In 1942, I was promoted to a position in charge of Warehousing and Records. At this time, I thought that this would end my association with the NBS. It was not to be. Unfortunately, I was not aware that an employee with a name similar to mine had been apprehended sending information to the Germans.

I have never attempted to verify this information that was provided to me by an unnamed employee in the Office of Special Investigations (OSI). During my assignment as Branch Chief of Warehousing from 1942-45, I was aware that someone was tailing me whenever I left home. My wife and I would come home late on Saturday nights, and I knew that I was being followed by headlights, and we would be escorted home. They always stopped a few houses from ours. I knew my neighbors and was familiar with their vehicles. I never associated these occurrences with the NBS at that time.

I resigned my job in the fall of 1945 and bought a small restaurant from my wife's aunt in a small town of about 500 people, called Drumond, Montana. I sold the business in 1946 and returned to Sacramento. I was rehired at McClellan and was Assistant Warehouse Superintendent in the Warehouse Division.

Approximately twenty years after leaving the Bombsight Unit, I was given a run down of what happened by the OSI. Because of the similarity of my name and this employee who was caught giving the Germans information on the NBS, and since I had access to Top Secret bombsight data and supposedly had extra money to buy a restaurant, they tailed me to Montana.

Chapter XVI: Security

They knew details of my restaurant operations, i.e. installation of a stainless sink, letter from the IRS commending me on excellent record keeping, and closing my business after I sold it. I must say that I was involved with the NBS for about twenty years. Needless to say, the experiences I had will always be vivid in my memory of working for 37 years, and especially the bombsight.

The author felt that many times he was followed home. He lived in a small town about 35 miles from the work place and also operated a small deciduous fruit orchard on the outskirts, where he observed strange cars parked nearby. It was reported to the police. No information could be obtained from the local police, as all of the inquiries were never answered. However, nothing was ever heard about this, and the cars continued to park a distance from the home. This caused concern, as the author's mother was a widow and he worked various night shifts. In all of the inquiries, he never did find out the real reason for this. In a very small town, any unusual activity of this type is readily noticed. Working in the machine shop, we had access to all of the engineering drawings in order to perform our work, where other employees did not have this privilege. Perhaps having an Italian-American name with Italy on the side of Germany may have had something to do with it.

Field Repair Box
Bombsight Repair Shop (Airborne)
With thousands of sights being used by squadrons all over the world, some form of field repair for bombsights was needed. It was not practical or feasible to return the sight to a rear major repair depot for small items that could be taken care of in the field. This would cut the number of operational sights available to the bomb groups, and much time would be lost in transit. Many of the minor and intermediate repairs could be accomplished by field service personnel, thus increasing the bombing capabilities of the units.

To this end, a small portable repair shop was developed to accomplish only field repair work. No major repair was to be taken. No facilities were provided to do any major overhaul of the sight. The unit was designed by the Herman Body Works Company for the Army Air Corps. It is known that a total of four contracts were awarded for the manufacture of this item. A generating plant was provided to supply 6.3 KVA, and it could also produce 110 to 220 volts, 60 cycle alternating current. The box consisted of two sections, the lower and upper, which fit the upper box over the lower box when it was collapsed. When the upper box was raised into position, it formed a complete work shop (Plate #4). It came equipped with a dryer, heater, auxiliary power plant, air compressor, a jeweler's lathe and tools (this lathe was used due to the weight of a standard size lathe), two small benches, a calibrating stand, and a drill press (Plate #4). No air conditioning was provided in the box. In order to have some form of temperature control in the box when working on the sight, the top and bottom were insulated with 2 1/2 inches of fiberglass, and the walls with 1 1/2 inches. The box was constructed of water proof plywood. Three quarter inch plywood was used for the floor. All of the corner edges were steel reinforced inside and out. The shop was wired with outlets for 12, 24, and 110 volt, and the inside was painted a light color. Box size was 6 feet 7 inches wide by 11 feet 2 inches long. The entire weight of the box, complete with all equipment, was 4,900 pounds. The preparation and erection of the unit was not complex, and it could be assembled in a short time. This prefabricated box was self contained, fully equipped, and ready for use in any part of the world. It was designed specifically for field use.[20]

Technical Orders AN 11-30-16 and T.O. No. 19-65-2 were published under joint authority of the Commanding General Army Air Forces, the Chief of the Bureau of Aeronautics, and the Air Council of the United Kingdom. The level of distribution of Field Repair units was not available.

In January, 1943, the Ross-Gould Company published 15,400 technical orders, and the revised edition published June 1945 was 14,900.[21] Apparently this unit had wide distribution, as normally T.O.s were sent to those who have use for them. It was also used by U.S. allies. No correspondence was found in the files on this subject. It would be assumed that these would be sent to those squadrons that were not within a reasonable distance of major overhaul depots. However, it appears that it would be somewhat expensive, as specialized equipment was used. For example, the Dryer unit made by the Carrier Company was not a common item, and it weighed about 800 pounds. Other specialized equipment was furnished by Homelite Company, Westinghouse Electric Company, etc.

Plate #1 - Bombsight Shipping Box.

Chapter XVI: Security

Plate #2 - Bombsights and Stabilizer stored in vault.

Plate #3 - Bombsight and stabilizer for installation in shipping box.

Plate #4 - Box, Field Bombsight Repair Shop.

Notes:

[1] War Department, Air Corps, Materiel Division, Memorandum to the Chief of the Air Corps, Subject: Bombsights, dated Jan 18, 1936, National Archives, page 2.
[2] War Department, Office of the Chief of the Air Corps, Subject: Bombsights, Storage, Issue, Etc., and maintenance of Secrecy letter to Bureau of Ordnance, dated November 14, 1936, National Archives.
[3] Ibid, page 2.
[4] Ibid, page 6.
[5] Letter from the Chief of the Air Corps to the Chief of Bureau of Ordnance, Subject: Bombsights, Maintenance of Secrecy in Connection herewith, dated May 16, 1935, National Archives.
[6] Navy Department, Bureau of Ordnance to Chief of the Air Corps, Memorandum, Re-classification of SBAE, dated April 29, 1941, National Archives, MS.
[7] Technical Instruction, Materiel Division, Air Corps, Subject: Reclassification of SBAE, Letter to Chief of Materiel Division, Wright Field dated May 2, 1941, National Archives.
[8] Ibid, page 7.
[9] Letter from Headquarters War Department, Subject: Policy Governing M-Series and S-Series Bombsights and Automatic Flight Control Equipment (AFCE/SBAE), dated July 1, 1942, page 2, National Archives.
[10] Ibid, page 6.
[11] Ibid, page 12.
[12] Ibid, page 13.
[13] Letter War Department Hq of the Army Air Forces, Subject: Policy Concerning M- Series Bombsights and Automatic Flight Control Equipment (SBAE) Types B and C, Automatic Pilots and Attachments, such as Glide Angle Attachments, Low Altitude Bombing Attachments, dated December 11, 1942, National Archives.
[14] Letter Navy Department, Bureau of Supplies and Accounts to Chief of Bureau of Ordnance, Subject: Transportation of Bombsights and Supplemental Equipment, dated Marc 25, 1944, National Archives.
[15] Letter, Bureau of Ordnance, Commanding Officer, Naval Ordnance Plant, Indianapolis, Indiana, Subject: Authority for Discontinuance of Delivery of Keys to Naval Activities, Bombsights, Mark 15, Shipping Boxes, dated June 21, 1944, National Archives.
[16] Letter Naval Proving Ground, to Chief of Bureau of Ordnance, subject: Bombsight Shops and Storage Facilities, dated July 7, 1939, National Archives.
[17] Letter, U.S. Fleet Air Base, Pearl Harbor, Hawaii, Subject: Facilities for Bombsight Repair Shop, U.S. Fleet Air Base, Pearl Harbor, T.H., dated March 10, 1939, National Archives.
[18] From the recollections of the author who worked and was supervisor of the bombsight machine shop from 1941 to 1949. Some details may not be too clear after a lapse of almost 56 years.
[19] Mr. Milton Stomsvick was Chief of Stock Records and was responsible for the Norden Bombsight and associated equipment for Supply. However, all records and equipment were physically in the Bombsight Building. The story is printed through the courtesy of Mr. Stomsvick.
[20] Box, Field Bombsight Repair Shop, Handbook of Instruction with Parts Catalog, Technical Order, AN 11-30-16, dated November 30, 1943, Revised January 20, 1945, Library of Congress.
[21] Field Bombsight Repair Shop Box (Airborne), Operation, Service and Overhaul Instructions with Parts Catalog, Technical Order No. 19-65-2, dated June 25, 1945, Library of Congress.

Plate # 1 Bombsight Shipping Box, National Archives.
Plate # 2 Bombsight and Stabilizers stored in vault, National Archives.
Plate # 3 Bombsight and Stabilizer ready for installation in bombsight shipping box.
Plate # 4 Technical Order AN-11-30-16, Box, Field Bombsight Repair Shop, dated November 30, 1943, revised January 20, 1945.

XVII

Norden Mark XV
From Visual to Radar/Radio

With the increasing introduction of high speed and operational altitudes of bombing airplanes, the capabilities of the visual bombsight were becoming proportionally less, and greater reliance was put on radio and radar equipment as the war progressed.[1] Unfortunately, there was only fragmentary information in the National Archives on the development of this new equipment. The preponderance of research, engineering, etc., was conducted by the Radiation Laboratory at the Massachusetts Institute of Technology (MIT), Cambridge, Massachusetts, and the National Defense Research Committee (NDRC) headed by Dr. Vannever Bush, in conjunction with various universities, research centers, and companies who had modern facilities in the field of electronics. A few documents were found in Division 14, NDRC, but were of little value concerning the application of airborne radar bombing. The author was unable to review the extensive files of the Radiation Laboratory at MIT on the development of this new weapon, which was in its infancy.

It became evident that a visual bombsight could not penetrate adverse weather conditions such as haze, clouds, fog, or darkness, which were prevalent over Germany much of the time, in order to reach enemy targets. In September 1944, the AAF Office of the Assistant Chief of the Air Staff directed that the necessary project, or projects, be established for the purpose of obtaining this new equipment at the earliest possible date.[2] At the same time, this experimental program on radio controlled bombsights was discussed, and some results had been achieved. Due to the experimental status and time required for development to take this program to a production level, many detailed studies were needed. On November 10, 1942, a Bombing Equipment Conference was convened by the AAF Materiel Command, Wright Field, for the explicit purpose of determining the requirements of AAF Ad-

Chapter XVII: The Norden Mark XV - From Visual to Radar/Radio

vance Bombing Equipment. It was attended by many of the Numbered Air Forces, Training Command, Director of Bombardment, Proving Ground Command, Technical Training, Flying Training Command, Air Service Command, etc., concerning the radar bombsight.[3] The development of a "Universal Bombsight" was undertaken by Bell Telephone Laboratories and Minneapolis-Honeywell Company. Very little data was found on this item.[4] The use of radar sights for bombing through overcast (BTO) was being investigated and was assigned to Bell Laboratories, the Radiation Laboratory at MIT, and the Sperry Gyroscope Company; BTO for Anti-Submarine Research Project 297-T2 was given to the Radio Corporation of America.[5] Military characteristics for this type of bombsight were also established. Stabilization of sight optics and electronic antenna for yaw, pitch, and roll was needed with this new equipment. The maximum time of fall was 65 seconds. Trail was required to 1,200 mils, with accurate positioning features for manifold type of aerial missiles on hand and under development.[6] These developments did not come without cost. It was necessary to design simplicity, ease of operation, ease of maintenance, light in weight, compactness, and adaptability to mass production. The equipment that worked in the laboratory did not mean that it would work in the airplane under combat conditions. Much designing and engineering was expended before this new weapon would reach the combat units.[7]

Expert personnel, in June 1943, from Hq Eighth Air Force in the United Kingdom, returned to the United States from combat theaters to work on BTO devices. This new device grew from discussions with several physicists, Eighth Air Force personnel, and conference with the British. The Chief of the Air Staff was advised of the Eighth Air Force's plan to form a Pathfinder force of about one squadron reinforced, using a minimum of eight B-17s and four Liberators (B-24) equipped with the latest British devices. The goal of this plan was for these airplanes to be ready by the time fall weather set in, to lead formations of bombers to German targets for bombing with BTO. To supplement this initial squadron of Pathfinder force, the addition of at least two squadrons in strength, equipped with the latest American instruments assembled and trained in the United States, was formed. As soon as these squadrons were operational, they would be moved to the Eighth Air Force, join up with the Pathfinder force already there, and form a Pathfinder Group, which would be operational sometime in early 1944. Personnel from the Eighth Air Force were to organize, train, and obtain equipment for this American force. As soon as this group was ready to be transferred, the personnel responsible would accompany the American Group as requested by the Commanding General, Eighth Air Force.[8]

The Office of the Chief of the Air Staff, AAF, in September 1944, established principal characteristics for bombsights caused by changing conditions in the combat theaters. Emphasis was on visual radar sights that were either optical, electronic, or Universal (Optical and Electronic) to provide accurate computations of the bombing problem to insure placing all types of bombs on target. Offset bombing was also to be a feature in radar sights, which would permit use of an aiming point six nautical miles from target in range, and this same distance in deflection, being the minimum. Forward vision of optic indices provided 90° forward search.[9]

In December 1944, representatives of the Armament Laboratory (AAF), Radiation Laboratory, MIT, Bell Telephone Laboratories, and T.B. Gibbs Company met to discuss the progress and to procure needed components for the synchronous Bombing Attachment used in AN/APQ-7 radar equipment and the Norden sight. The final design of the interconnection system between the AN/APQ-7 equipment and the Norden sight was not ready to meet the combat theater demands, due to the modification of the equipment and the considerable specialized tooling required to get quantity production. The interim design presently being used consisted of two units to furnish point sighting angle information instead of a continuous sighting angle information. In the final design the sighting angle selector box was replaced by the modified Norden rate end. No equipment changes were necessary. The special control box required was used without changes in the interim and final design. The Radar Laboratory procured one each of the necessary two units for each AN/APQ-7 equipment on order. When the modified Norden rate ends reached sufficient quantities, one of the units was canceled, and the final design was adopted as a standard unit. The quantity on order was 1,660 as of January 1945.

In order to complete this complex design, the Army Air Forces Air Technical Service Command was responsible for immediately procuring one modified Norden bombsight, one modified Norden Rate End, one meter box, and all of the associated cables required to finalize a standard system.[10] By late Fall of 1944, a military requirement existed for a synchronous connecting device for automatically setting into the standard bombsight range and drift data obtained from radar sets. This speeded up the transfer of bombing responsibilities from the bombardier to the radar operator, or vice versa, necessitated by clouds or other hazy conditions encountered during a bomb run, which would have prevented completion of bombing operations by visual means.[11] Parts to be purchased by the T. B. Gibbs Corporation were expedited, as well as the high priority assigned to the project. Also, the Radiation Laboratory devoted much of its time to the possibility of modifying NOSMO for use with the Eagle AN/APQ-7 (also known as NOSMEAGLE). The 60,000 foot maximum slant range in the AN/APQ-7 radar did not provide sufficient in range offset for the AAF. However, this was easily remedied by using a larger resistor, thus raising the clamping voltage by 20 volts and extending slant range to 80,000 feet.[12]

Installation schedules were developed in December 1944 for production and delivery of this new specialized equipment. Under this program, H2X equipped B-24s would receive 247 sets beginning in December 1944 and ending in May 1945 for the various Numbered Air Forces. These sets were distributed to the Far Eastern

Air Forces, China-Burma-India Theater, 6th, 7th, and 11th Air Force. The AN/APQ-15 equipped B-17s and B-24 aircraft would receive 460 sets from December 1944 through May 1945 to be equally divided between the 8th Air Force and 15th Air Force. The AN/APQ-7 equipped B-17 and B-24 aircraft, and Training Command received a total of 305 instruments from December 1944 to May 1945. B-29 airplane requirements for combat and specialized training was 155 sets. With the expected increase in production in the coming months, all B-29 aircraft would be equipped with this specialized bombing equipment in the aerial war over Japan.[13] In the first part of 1942, H2X and AN/APS 15 sets were being tested on the bench. Research of the 8th Air Force Target Summary indicates that the use of H2X was used to attack Emden, Germany, industrial areas on September 27, 1943, through October 2, 1943, dropping a total of 1,459 tons of bombs in both raids and losing 9 airplanes in 516 sorties. Breman, Germany, was hit on the 13th and 26th of November, 1943, dropping 1,469 tons of bombs with a loss of 45 airplanes in 536 sorties.[14]

During the same time, experimentation was being conducted on visual and radar type bombing. This new type of bombing equipment was given the name NOSMO. Correspondence identified this as Norden Optical Sight Modification (NOSMO), and it also was used to identify it as synchronization of the optic bombsight with radar information (Plate #1). The AAF did not go through the Navy for this modification, and contracted directly with the T. B. Gibbs Corporation for their engineers to prepare engineering and production drawings. Emphasis was placed on the importance of this project so that its completion would satisfy the urgent need by bombardment units in combat. On October 14, 1944, engineers from T.B. Gibbs Corporation were at the Radiation Laboratory working on production drawings for NOSMO.[15] Due to the scope of the project, it was divided into two parts. One part was procured by the AAF Armament Laboratory, Wright Field, Ohio, for components that tied together the sighting angle of the Norden bombsight and the range marker of the AN/APS-15, AN/APS-15A, or AN/APS-13 radar sets. The other parts were procured by the Aircraft Radio Laboratory that determined the direction of ground track by the pulse Doppler method. In order to accomplish this program, Hq AAF appointed the Armament Laboratory to monitor this project with the view of getting this equipment into production as soon as possible. It was given a classification of secret, and it was of such importance that the project justified draft deferment for involved personnel and a priority of 2B. The MIT Radiation Laboratory was advised by the AAF that a Project on NOSMO had been set up by the National Defense Research Committee. The Air Technical Service Command provided the rate ends (computer) from the NBS and prepared the necessary holes for mounting them on the back plate of the equipment.

Due to the complexity of the equipment and the extensive engineering that was devoted to this project by all of the concerned contractors, it was not completely perfected before the end of the war. Although many sets of radar equipment were deployed in combat areas using both types, testing continued to be conducted after the end of the war as to the suitability of which unit was the best. A comprehensive study was conducted by the AAF Proving Ground Command at Eglin AFB, Florida, on the AN/APQ-7, AN/APQ-13, and Norden equipment on B-29 airplanes beginning in December 1944 and ending in September 1945. This project was initiated at the request of the AAF board. The object of this project was to determine the accuracy of the AN/APQ-7 equipment in comparison with that of the AN/APQ-13 radar bombing equipment and the Norden bombsight, and the operational suitability and dependability of the AN/APQ-7 installation in B-29 airplanes. The AN/APQ-7 radar bombing equipment was an X-Band radar set known as "Eagle," designed for high altitude blind bombing. The results of the tests showed that, in actual bombing of targets, the AN/APQ-7 radar equipment was more accurate than the AN/APQ-13 under the same conditions. The AN/APQ-7 radar installation in the B-29 airplane was simpler than the AN/APQ-13 because of fewer components.

The tests proved that the AN/APQ-7 radar bombing equipment, when installed in a B-29 airplane and used with the Norden M-9 bombsight, was suitable for bombing from altitudes of 30,000 to 35,000 feet, and gave good accuracy and greater resolution than the AN/APQ-13, enabling the radar operator to distinguish targets not distinguishable with AN/APQ-13 equipment.[16] The use of the Norden bombsight with radar would soon be dropped and, in the coming years, all bombing would be accomplished by electronic methods.

The exact number of radar sets procured by the AAF was not available in the National Archives records. However, December 1944 production schedules showed that 1,167 sets were procured by the AAF by May 1945. No direct cost was available for a set, nor all of the experimental research, design engineering, procuring component parts, and establishing production sources and testing. The era of electronic bombing was now at hand. Visual optic bombsights were still used when clear weather prevailed, but would soon be replaced. However, the Norden bombsight was not to be counted out, as it still had a place in the AAF.

The author wishes to emphasize that the information on this new equipment is extremely limited, as very little data was available for research in the National Archives and has been reported to the best of his ability. Most of the work was done by contractors, universities, MIT Radiation Laboratory, etc. The author was unable to conduct extensive research at the various repositories due to financial constraints. The above is shown so that the reader can get an insight as to the efforts that the Army Air Forces went through to find a better bombing system. These crude electronics systems led to the development of our present day sophisticated guided missiles.

AZON/RAZON Bomb

Experimentation was also being conducted on radio controlled bombs, along with other electronic projects. The AZON/RAZON devices were the forerunners of the guided missiles in use today. This experimentation started in late 1943 and continued on past WWII. This was a new concept in aerial warfare. AZON means control in azimuth only. The tail assemblies of the AZON/RAZON bomb consisted of similar, but not the same, types of radio receiver, gyro, servo motor, battery, and movable fins and flares for visual operation. RAZON means control in azimuth and range of the bomb during its flight from the airplane to the target. Early experiments with high-angle bombs were controlled in both range and azimuth by radio.[17]

To determine the response to control signals and oscillations, accentuating the amount of yaw was intentionally made, as well as to determine the amount free fall trajectory could be deviated from by the use of the Razon control assembly. These experimental tests were not primarily for accuracy, but to check the theory and mathematical data against performance during the controlled portion of flight.[18]

This very complex project was conducted under the auspices of Hq Army Air Forces, Weapons Branch, Air Technical Service Command, Gulf Research and Development Company, L.S. Schwien Engineering Company, Franklin Institute, and MIT. MIT was project engineer of controlled bombs for the National Defense Research Committee under contract with the Office of Scientific Research and Development. The types of bombs used were VB-1 and VB-3, and generally a 1,000 pound bomb. The official Army designation of VB-3 stood for "vertical bomb type No. 3" and was designated Razon. It was similar to the Azon bomb, but instead of using the cruciform type fins of the Azon bomb, either one or two octagonally shaped fins were used, and were adaptable to any standard bomb and could be installed in a similar manner as the standard fins. The Razon assembly had two gyros which controlled stabilization in flight, one in azimuth and one for rate. Control was effected by a five channel radio receiver (only four channels were used), which received a signal from the four channel transmitter located in the airplane, making the bomb release and operating on a high frequency carrier wave upon which was impressed audio frequencies. The four bands of the receiver were tuned to receive signals from corresponding bands of the transmitter. To provide the correct signal for control of the bomb, a control box with a stick similar to the control stick in the pilot's compartment was located so that it was near and convenient for the bombardier.[19]

Due to the complex nature in the concept, design, engineering, development, testing, and production, only the portion that applies to the Norden bombsight will be discussed and is extremely brief. The development of guided missiles would fill a library. Some data was available in the Library of Congress. Reports indicate that work was being conducted as early as 1943. The control of the Razon bomb in range required more instrumentation than the Azon. In order to hold the bomb so that it eclipsed the target, all errors except steering were eliminated, however, this required maneuverability of about two or three times that available in the present designs of the high-angle bomb. Preliminary experiments were sufficient to indicate required bomb maneuverability. The method selected permitted the bomb to proceed on its normal near-parabolic trajectory with range control. In order to use this general method of range control, the problem arose as to what kind of device could control it on its way to the target. The Norden bombsight contained a mechanism which provided such a motion in the tracking mirror, which closed the angle subtended by it at the time of release of the bomb and the Trail Angle of the bomb at impact in exactly the time of fall of the bomb as put into the bombsight through Disc Speed settings. Two adaptations of the NBS were built for utilizing the mirror motion, one by the contractor, involving a double cross-hair method, and the other by the Franklin Institute, involving a double mirror arrangement similar to that used in a ship's sextant.[20] This new device was called CRAB. The definition of what "CRAB" stood for is not known. The development of Crab was on a basis of a suggestion by Dr. V. Bush as to the possibility of a computer for guided bombs using the Norden bombsight.

Actually, the Crab was a simple device. Physically, it was a small, stationary mirror which was inserted between the telescope objective and the target mirror in such a manner as to intercept a portion of light normally entering the telescope and to substitute for this portion rays from a different direction. The NBS modification permitted Razon bombing with M-Series bombsights and was designed expressly for the Norden sight to guide the Razon bombs after they were released from the airplane, enabling the bombardier to view both the target and the Razon bomb on its way to the target. The function of the mirror was to super-impose the bomb image on the target image by splitting the field of the telescope into two components; one along a line of sight to the target via the target mirror, and the other along the line of sight to the bomb via the Crab 1 mirror. It could be used at any altitude, providing the time of fall for that altitude was accurately known. It was not designed for evasive action during the control period, although minor accidental deviations could be guided out. A straight run had to be maintained from time of bomb release until impact.[21] When using the bombsight, it was necessary to have the Crab mirror set at an angle corresponding to the Trail Angle of the Razon bomb at impact.[22] The normal moving mirror of the sight was used, and it simply continued to track the target until the bomb hit. For the second line of sight, a small mirror was mounted on the telescope objective so as to reflect into view a line of sight slightly rearward of the vertical. The second line of sight reflects the bomb image into the field since, in a normal drop, the apparent position of the bomb is always slightly rearward of a true vertical line extending from the airplane. The bombardier sees two fields of view, one including the area around the target, and the other including the terrain directly below the airplane.[23] In operation, the bombardier made a normal bombing run, using the NBS in the conventional manner to drop the bombs. After releasing the bomb, the bombardier followed it in the bombsight telescope and operated the range control stick. The azimuth control stick was operated by a second operator, who followed and guided the bomb as if it were an ordinary Azon bomb through a window replacing the space normally occupied by the ball turret.[24]

A new device was developed called JAG (**J**ust **A**nother **G**adget). The Jag Corrector was a device attached to the Norden bombsight, which caused a small rotation of the range rate synchronizing knob, Up, Right, or Left control when applied to the bomb. The effect of this rotation was to delay the predicted time of coincidence of the target-tracking and Crab mirror angles.[25] The JAG corrector was developed to eliminate the need for using an average time of fall.[26] The Gulf Research and Development Company and the Union Switch Company models were identical in all principal dimensions.[27] Even with various sources of difficulty, the quality of the bombings indicated that 32 hits were scored in a circle of 100 foot radius, of which 15 were in a circle of 50 foot radius, in which three were direct hits on the target "shack" of 8 x 8 feet square.[28] The hits were better in azimuth than range. If the target was assumed to be a stationary ship of 500 feet long and 80 feet wide, then 44 direct hits would have been scored.[29] In a number of drops, photographs were taken through the bombsight. These photographs recorded the bomb and target as seen by the range operator when controlling the bomb. In some drops, slit camera[30] photographs were taken of the bomb trajectories. In others, motion pictures were taken of the bomb in flight from the ground.[31] Each bomb drop was photographed with two motion picture cameras. One, a 35mm GSAP mounted in the bomb bay, and the other a hand held directed Eastman Model E with four inch telescopic lens (Plate #2).[32]

When the bombsight was set up for use with a Razon bomb, it had to have the Crab 1 mirror set at an angle corresponding to the trail angle of the Razon at impact, for a perfectly synchronized run the bombsight would show that no steering was required except for small effects due to variation in trail angle and parallax. For the Norden sight to give the proper time of fall, the trail angle setting on the sight was equal to the Crab 1 mirror angle (Plate #3).[33] The computer in the M-Series bombsight drove the mirror at the rate necessary to make the sighting angle to the target coincide with the trail angle at the end of the instrumentally developed time of fall. If the Crab 1 mirror was adjusted to produce a line of sight of the vertical by the amount of trail angle, and the bombsight was guided as to be kept on this line of sight, then the necessary conditions for hitting the target were fulfilled. To install the Crab 1 mirror, it was necessary to remove the lower telescope window of the bombsight to provide access to the telescope lens. Several operations were required for this installation. The author did not go into detail on this installation. Very briefly, the Crab 1 mirror was adjusted until it was parallel to the target mirror. The azimuth alignment was accomplished by sighting through the telescope and the trail angle.

To follow the trajectory of the bomb to its target, the standard Azon flares were used. In the Tonopah, Nevada, bombing range, drops of a red colored type were used and were very satisfactory, while at Wendover AAF Base, drops of white flares were used. On the first 19 bombs dropped at Tonopah during August and September 1944 a single octagonal fin was used. Eleven of the 19 bombs were steered successfully, and the remaining eight failed due to malfunction of the control equipment. Eighteen more bombs were built for a third set of experiments conducted at Wendover, Utah, in January 1945. Fourteen of the 18 bombs were steered successfully, the remaining four being caused by equipment failure. This model could be used at any altitude, providing the time of fall for the altitude was accurately known. In order to use this new equipment, it was necessary to modify the bombardier panel to include two additional switches. One was for the electrical arming of the flare, which was attached to the rear of the Azon assembly. The other was for the pre-warm up of the gyros and radio receiver of the Razon using the airplane's power source.

There was no information as to whether these new weapons were deployed before the end of the war, as they were still in experimental stages. No cost was found for contract designing, testing, and production. The Air University offered an Extension Course on Guided Missiles. This publication describes the various types of guided missiles in the Army Air Forces' inventory. It explained their use, propulsion methods, fuel range, and complete description on performance on each weapon, including Azon/Razon. The publication was dated 1948.[34] The age of electronic bombing was now a reality, and the visual sights were relegated as museum pieces.

Chapter XVII: The Norden Mark XV - From Visual to Radar/Radio 245

osmo attachments to Norden bombsight (above) are a center ro meter just above eyepiece and box shown just below rate head

Radar position with Nosmo (below) includes an auxiliary Norden rate head at right and a Doppler Drift control box to the left

Plate #1

Plate #2

Plate #3

Chapter XVII: The Norden Mark XV - From Visual to Radar/Radio

Notes:

[1] Letter, Hqs AAF from Assistant Chief of the Air Staff, Subject: Statement of Military Requirements, Requirements Division, dated September 25, 1944, Frame No. 0581, National Archives.
[2] Letter, from Assistant Chief of the Air Staff to Air Technical Services Command Subject: Military Characteristics for Bombsights, dated September 27, 1944, Frame No. 0582, National Archives.
[3] Army Air Forces, Materiel Center, Memorandum Report, Subject: Bombing Equipment Conference Held at Wright Field, dated November 10, 1942, Page 1, National Archives.
[4] Ibid, page 2 of Appendix No. 1.
[5] Ibid, page 3 of Appendix No. 1.
[6] Letter Hqs AAF from Chief, Requirements Division, Assistant Chief of the Air Staff, Subject: Statement of Military Requirements, dated September 7, 1944, Frame No. 0579, National Archives.
[7] Letter, War Department, Hqs AAF, Subject: Military Characteristics for Bombsights, dated September 7, 1944, Frame No. 0580, National Archives.
[8] Headquarters, Eighth Air Force Letter to Chief of the Air Staff, from General Eaker, No subject, dated June 26, 1943, National Archives.
[9] Letter, from Office of the Chief of the Air Staff to Hqs AAF, Subject: Military Requirement Characteristics for Bombsights, dated September 7, 1944, Frame No. 0578, National Archives.
[10] AAF Materiel Command, Routing and Record Sheet to Engineering Division, Subject: Procurement Characteristics for Bombsights, dated December 18, 1944, National Archives.
[11] Letter, from Assistant Chief of the Air Staff, Requirements Division to Hq AAF, Subject: Statement of Military Requirements, dated September 25, 1944, Frame No. 0581, National Archives.
[12] Letter, from Chief Engineering Division to Air Technical Service Command to Director, Radiation Laboratory and MIT, Subject: Project NOSMO, dated October 25, 1944, Frame No. 0593, National Archives.
[13] Ibid, page 2.
[14] Eighth Air Force Bomb Tonnage Dropped on German Targets, from 8th Air Force Target Summary from August 17, 1942 thru May 8, 1945.
[15] Air Materiel Command Routing and Record Sheet, Subject: N.D.R.C. Project on NOSMO dated October 14, 1944, Frame 0591, National Archives.
[16] Headquarters Army Air Forces Proving Ground Command, Eglin Field, Florida, Final Report on Comparative Bombing Tests on the AN/APQ-7, AN/APQ-13 and Norden Equipment on B-29 Aircraft, dated September 18, 1945, from Defense Technical Information.
[17] Hqs Army Air Forces, Air Technical Service Command, Memorandum Report on : RAZON Experimental Tests with Crab 1 Modification to M-Series Bombsight, dated February 20, 1945, page 2, National Archives.
[18] Ibis, page 1.
[19] Ibid, page 2.
[20] Gulf Research & Development Company, Experimental Investigations in Connection with High-Angle Dirigible Bombs, RAZON (VB-3), Progress Report April 1, 1945 Supplement #10, page 44,45. Permission granted for use by Chevron Corporation, Library of Congress.
[21] Hqs Army Air Forces, Air Technical Service Command, Memorandum Report on: RAZON Experimental tests with CRAB 1 Modification to M-Series Bombsight, dated February 20, 1945, page 5, National Archives.
[22] Gulf Research and Development Company, Experimental Investigations in Connection with High-Angle Dirigible Bombs, RAZON (VB-3), Progress Report April 1, 1945, Supplement #10, page 48, Library of Congress. Permission granted for use by Chevron Corporation.

[23] Ibid, page 46.
[24] Ibid, page 13.
[25] Gulf Research and Development Company, Experimental Investigations in Connection with High-Angle Dirigible Bombs, RAZON (VB-3), Progress Report September 1, 1945, Supplement #11, page 2, Library of Congress. Permission granted for use by Chevron Corporation,
[26] Ibid, page 20.
[27] Ibid, page 6.
[28] Ibid, page 11.
[29] Gulf Research and Development Company, Experimental Investigations in Connection with High-Angle Dirigible bombs, RAZON (VB-3), Progress Report April 1, 1945, Supplemental #10, page 16, Library of Congress, Permission granted for use by Chevron Corp.
[30] The slit camera is a camera designed to photograph on a single film the trace of the flare of the bomb while in flight.
[31] Gulf Research and Development Company, Experimental Investigations in Connection with High-Angle Dirigible bombs, RAZON (VB-3), Progress Report April 1, 1945, Supplement #11, page 6, Library of Congress. Permission granted for use by Chevron Corporation.
[32] Ibid, page 6.
[33] Gulf Research and Development Company, Experimental Investigations in Connection with High-Angle Dirigible bombs, RAZON (VB-3), Progress Report April 1, 1945, Supplement #10, page 48, Library of Congress. Permission granted for use by Chevron Corporation.
[34] Guided Missiles, Hqs Air University, Air Technical School, Tyndall Air Force Base, Panama City, Florida, 1948, Edition.

> Plate #1 NOSMO attachments to Norden Bombsight and Radar position with NOSMO. Source Radar Magazine. Date not known.
> Plate #2 Camera mounted on the Norden Bombsight to photograph behavior of falling bomb.
> Plate #3 Shows the Crab attachment mounted on the bombsight.

XVIII

Unknown Facts about the Norden Bombsight Mark XV

The development, production, and use of the NBS have been the subjects of much discussion. There is no question that many unexpected and unforeseen engineering and manufacturing problems were encountered beyond the control of any one individual or contractor. The first critical item was the unusually close tolerances that were demanded in the manufacture of some 2,300 parts making up the sight. In the beginning, all designing, testing, and modifications were conducted by the Bureau of Ordnance, Naval Proving Ground and the Norden Company, starting with the Mark XI, and then progressing to the Mark XV. From 1932 to 1941, the Norden Company produced all of the NBSs for both the navy and AAF. In 1939, with war raging in Europe and the AAF building up to their authorized bombardment capability as directed by President Roosevelt in May 1940, the NBS was already in extremely short supply. The question of whether additional plants should be built was a moot question. In order for the Bureau of Ordnance to satisfy the Navy and AAF requirements, the Navy authorized the building of a new ordnance plant in Indianapolis, Indiana. This new plant was scheduled to be completed and operational by early 1943. The first bombsight was accepted by the Navy on April 1, 1943.[1] The plant was completed but, again, trained mechanics and assembly personnel were sorely lacking. It was not possible to train personnel in such a short time to work to exacting standards, and the entire project was seriously threatened. To alleviate this bottleneck, training schools were set up to train new personnel in the techniques of precision work. Many times the idea of having critical tolerances was largely criticized, as stated in previous chapters. When the tolerances were not adhered to, the accuracy was affected and was duly reported by the alarmed combat units. In order to increase production, the standards were loosened, and the end result was an inferior product. The author spent considerable time retraining experienced machinists to work to tolerances never before attempted. It was no longer "it is good enough." Tolerances that were acceptable in other types of manufacturing were rejected for use in the NBS. A classic example was the critical shortage of precision anti-friction ball bearings. What was acceptable in all types of industry could not be used. It can readily be understood that "sloppy" work in manufacture resulted in inaccurate bombing. Therefore, all NBS manufacturers were immediately requested by the Navy to critically review their inspection procedures and rigidly follow the engineering specifications. This had an immediate effect on the quality of sights being received by the combat units.

The Navy Bureau of Ordnance was cognizant of the requirements of the AAF. Although they were accused of not providing sufficient sights, they made every effort to supply these needed items. The AAF had one bombsight contractor—the Victor Adding Machine Company—and they encountered similar problems that plagued all of the other manufacturers. American industry was not yet ready to mass produce precision equipment on a large scale. The NBS was not a static weapon where a continuous production run could be initiated and production of parts would be indefinite. There were many changes that required modifications during production. Even with the establishment of Modifications Centers (these were facilities set up to incorporate the latest changes after they had left the contractor's plant), the flow of changes to the sight could not be kept up. The production of the NBS program lagged badly. With all of the pressure brought on by the higher headquarters, and especially the Navy, production could only increase as to the availability of parts. Training new personnel (many were lost to the draft), etc., did little to increase production. In 1943, the critical shortage of precision ball bearings brought the assembly of bombsights almost to a complete halt at the various bombsight manufacturing plants. This was fast becoming an intolerable situation for the combat units who were waiting to receive the bombsight in order to conduct aerial bombing of enemy targets.

However, once the manufacturing contractors were able to obtain the necessary tooling, engineering drawings, trained personnel, and assembly lines set up, and critical materials, the flow of complete sights and spare parts became a flood. The same situation faced the airplane manufacturers. Once they were set up with the necessary essentials, the production of airplanes was phenomenal. Once again, American industry proved its mettle. By late 1944, the NBS production had accelerated to such proportions that it was necessary to cut back on production. It was the beginning of closing bombsight manufacturing plants and diverting them to other types of war work. Admittedly, the AAF had very many more bombardment airplanes than the Navy and, because of this, they received about 80% or more of all the bombsight production. That meant that the Navy received about 7,920 sights to the AAF's total of 45,463 units, which was consistent with very heavy, heavy, and medium bomber production. It is equivocal, since the exact quantity of bomb-

Chapter XVIII: Unknown Facts about the Norden Bombsight Mark XV

sights has never been established for the Victor Company. Research of files at the National Archives revealed no quantities that the Victor Company produced during WWII.

It has been authenticated by Dr. Alexander Wetmore, Secretary of the Smithsonian Institution, Mr. Carl Buehler, Vice President of the Victor Adding Machine Company, and Major Thomas Ferebee, bombardier of the Enola Gay, that the Victor Company built the Norden Bombsight Serial No. V-4120 which was used to drop the first Atomic Bomb on Hiroshima, Japan, on August 6, 1945. His instructions were to release the bomb using visual sight, although the Enola Gay was also equipped with a radar bombing system. It was to be used as a back up system to the Norden should it be necessary to go BTO. The author, in his research, found another Victor Company sight with the serial number 4982 at the Planes of Fame Museum in Chino, California. The Victor Company was awarded a $60 million contract to build bombsights. William Darby, in his book about the history of the Victor Adding Machine Company, titled, "It All Adds Up," did not show how many sights were built. It has been mentioned by various sources that Victor produced between 6,000 and 7,000 units. This would be compatible with other contractors, and using the weighted average of $7,560 per sight would equate to this quantity, allowing about 10% for spares.

There is a mistaken fact that the NBS was only an instrument and an adjunct to the airplane. The truth is that it was a complete weapon system. Much had been said of the sight and automatic pilot. The automatic pilot consisted of several units vital to the operation of the sight and auto pilot, which is rarely mentioned. The manufacture of the NBS itself was a small part of the overall program. Once the sight had proven successful, means were immediately taken to develop facilities to train personnel in its use. It would be of no value if it could not be deployed and used at a moment's notice during an emergency. Prior to Pearl Harbor, most training was done by bombardment squadrons. After the war started this was not possible, as all available airplanes were dispatched to the various war zones. With the tremendous expansion in the various types of bombardment airplanes—very heavy, heavy, medium, and light—would require either Norden, Sperry, or Estoppey bombsights. Training bases were established, airplanes were built to specifications for training bombardiers, repair stations or depots were constructed, and bombers built around the NBS, instead of the other way around. A total of 17 training bases throughout the western United States were constructed by the Army Air Forces to train an estimated 45,000 to 50,000 bombardiers. The number of Naval Air Stations built is not known. The AT series airplane built by Beech Aircraft Corporation were used by the AAF to train bombardiers in the actual use of the bombsight in dropping bombs. It was called the AT-11 "Kansan," and several thousand were built. Also, the AT-13 and AT-14 were engineered to take the M-Series bombsight and C-1 automatic pilot. No cost was found in the construction of all the air bases, buildings, land acquisition, housing, messing, training, base support, etc. Hundreds of millions of dollars were spent in this effort.

After construction of training bases, repair stations were needed to repair, test, calibrate, and store the bombsights for safe keeping. To accomplish this goal, the AAF established the depot repair system, and the Navy established the Naval Air Stations. The AAF Air Materiel Command, prior to and during the war, built huge repair depots to provide a source for maintenance, supply, procurement, etc., for AAF airplanes. At each depot, a bombsight building or secured vault was constructed for the sole purpose of repairing and storing bombsights and related equipment. The author is well acquainted with one of these buildings where he was a supervisor. The description of a typical bombsight building is shown in another chapter. The cost of these buildings was not available. By the time that all of these facilities were built, many more millions of dollars were spent.

In September 1942, the Bureau of Ordnance granted authority to make training films on the operation of the Norden bombsight. The production of these films was for instructional purposes in aviation schools as an aid to train bombardier cadets. A series of motion picture training films were designed and produced by the AAF Training Film Production Laboratories, Wright Field, AAF. One group consisted of five parts, starting with the theory of bombing and progressing to the more complex phases in the use of the sight. The author has all of these training film series on file. Very little information was found on the application of radar technology to airplanes. Many contracts were awarded to MIT, NDRC, universities and other research centers for applying the NBS to the new radar technology. It has been estimated that several billion (between 3-5 billion) dollars was devoted to this new weapon of war. Assuming that if only 10% of this amount was spent in trying to adapt the NBS to radar, it would be hundreds of millions of dollars. Considerable experimenting was done under the auspices of the NDRC and MIT Radiation Laboratory with the AZON/RAZON radio guided bombs.

To assist the bombardier, computers were developed. They were not the type that are in use today. They were made from plastic, with bombing information inscribed on both sides. The E-6B dead reckoning computer was designed to solve quickly most of the important problems and equations which arise in aerial navigation and bombing. Its function is separated into those involving solution of equations and ratios, and those requiring a graphical solution of the triangles of velocities. The former are solved on the front, or slide rule side, the latter on the back, or wind vector side. On the slide rule side, four basic problems are solved: true air speed; true altitude, nautical units converted to statute, and vice versa; time, space, and distance. The first three of these problems are solved on specially arranged scales, while the last utilizes logarithmic scales exactly like those of an ordinary slide rule.[3] It had a stationary outer scale, and an identical movable inner scale on a rotating disc. The outer scale usually represented units of measure (miles, gallons,

Chapter XVIII: Unknown Facts about the Norden Bombsight Mark XV

etc.), and the inner represented units of time (hours and minutes). These scales were standard logarithmic scales, and were used to solve any problems in multiplication or division.[4]

To assist the bombardier, additional computers were developed. The C-2 and the AN are exactly alike, except that they are printed in different colors. The black figures are the same on both computers. The C-2 has red figures, and the AN has light fluorescent figures. The inside disc has two spiral logarithmic altitude scales. The outer side of the spiral is the true altitude scale, with light figures to find the bombing altitude. The inside spiral shows the indicated altitude with black figures for setting the pressure altitude above the target. It was designed on the basis of the Smithsonian Institution Atmosphere Tables.[5] The G-1 Computer True Airspeed was used to obtain the true air speed when flying at high altitudes and airspeeds. This computer made the necessary correction for compressibility of air.[6] Computer Type J-1 Time of Run-Sighting Angle provided in determining the time of run and locating the start point (Sight Angle) for a short bombing run of 30 or 45 seconds. This was an aid for the bombardier in quickly obtaining the sighting angle.[7]

Not all the bombsights were discarded after WWII. Some continued to be repaired, calibrated, and upgraded, then placed in storage for future use. It was also used during the Korean War for bombing enemy targets. In 1949, on a test basis, the NBS was installed in the YB-49 "Flying Wing" built by Northrop Aviation Company. However, the airplane was so unsteady that it made the bombardier sick and so could not conduct a bombing sequence. It was felt that a new automatic pilot was needed. Accordingly, Honeywell, Inc., was requested to design a new auto pilot, and it was called "Little Herbert." The new unit was better, but still not good enough for operational use.[8] No further information could be obtained from either Northrup or Honeywell Corporate Historian. The Northrup Company Historian was contacted and stated that he could not find any information in the files on hand, but he would contact the author after the move was completed. After repeated contacts with this office over several years, no further information was received.

With the war raging in Europe, and especially the bombing of the British Islands by the Germans, much thought was given to the type and serviceability of the Nazi bombsights. In the fall of 1941 the Bureau of Ordnance was able to obtain three sights that had been removed from Nazi planes shot down over England. Two sights were in excellent condition and could be operated, and the other one was slightly damaged. Examinations of these sights revealed that they were very elaborate mechanically, had excellent optical systems made by Zeiss Optical, and the manufacturing effort involved in making them was quite comparable to that involved in making the Norden Mark XV. Two of the sights did not have stabilization. One sight, the image of a reticule, was stabilized in the horizontal plane by means of a pendulous gyro, but had no stabilization in azimuth. The pilot directing system was quite primitive, comparable to the Mark III (from WWI) developed in the 1920s. The mechanical computing system was complex and consisted of a multiplicity of gears, cams, differentials, flexible drives, and a speed variator. There was no resemblance between the German and Norden bombsight. The American sights were far superior in gyro stabilization, pilot directing in azimuth (SBAE/AFCE), and elegance of the computing system. Apparently, the Nazis did not use any of the plans that were stolen from the Norden Company. The American auto pilot included a stabilizer for the bombsight, which provided the method of stabilization in azimuth and for pilot directing. This feature was not found in any of the three German sights.[9]

The Bureau of Ordnance and Norden Company were cognizant of ballistic wind effect on bomb trajectory as far back as 1934. In an effort to isolate the effect on non-uniform winds on a bomb trajectory, thirty new Mark XV sights were used for the study. All bombs were dropped by the same bombardier. The bombs used were Mark VII water filled 100 pound practice bombs. The air speed was 70 knots. A total of 188 drops comprised the sample, 153 from 6,000 feet, 27 from 7,000 feet, and 8 from 8,000 feet in altitude. The ballistic wind was computed from pilot balloon soundings taken near the times the bombs were dropped, and the vector difference between the ballistic wind and the wind at the bombing altitude gave the force and direction of the effective wind, called "W," acting on the bomb while falling. To find the force and direction of the ballistic wind, the average value of the absolute wind force in the 1,000 foot zone was multiplied by the appropriate wind weighing factor. By graphical means, the weighted wind vectors were added vectorially, and the closing leg of the polygon was formed, representing the ballistic winds. Of the 188 bombs, 108 had positive errors and 74 had negative errors; the mean error was 1.6 to 6.9 mils. The conclusion from these data was that the effect on the bomb trajectory of ordinary values on non-uniform winds was negligible, at any rate for the range of conditions represented in the Proving Ground tests. This was the beginning of making new studies as speeds of airplanes increased, and it was a continuing effort to update bombing charts.[10]

The bombsight even caught the fancy of Hollywood. A full length movie was made with Pat O'Brien and Randolph Scott, titled "Bombardier." Of course, the Norden bombsight was not used. It only made reference to it. It was a contest between dive bombing and horizontal bombing. Of course, the B-17 out performs the dive bomber. The bombardier trainers shown in the movie were authentic and were used by the AAF to train bombardier cadets. Two books were written about the bombardier. One is by Henry B. Lent, titled, "Bombardier—Tom Dixon Wins His Wings With the Bomber Command." The other is "Bombs Away, The Story of a Bomber Team," by John Steinbeck. Both books deal with the use of the Norden bombsight.

Considerable time and expense was devoted to the Norden Mark 16 and Mark 17 bombsights—the Aiming Angle, or Dive Sight. It had been discussed since 1938

Chapter XVIII: Unknown Facts about the Norden Bombsight Mark XV   255

as to the advisability of developing such a sight. The Naval Proving Ground was directed to test both sights. In the latest tests, in May 1944, bombs were released with the completed Mark 17 bombsight mechanism. Three pilots took part in these dives, and 47 usable points of impact were obtained. After seven years of development work, this instrument had not shown any clear cut promise of becoming operational. The idea of using the Norden bombsight principles in a dive sight had merit. However, these principles had not been embodied in the dive sights. It would require more engineering and testing, but since so many years had been devoted to this instrument, the likelihood of producing a satisfactory dive sight was not in the near future. It was thought that the Mark 17 could be better constructed, or have a two-man dive bombsight, so that synchronization and piloting could be performed by separate individuals. Due to the pressure in producing needed Norden bombsights, the production effort was geared to this effort. Norden Company was not the only ones experimenting with dive bombsights. The Specialities Company had also developed a dive sight that was undergoing tests at the NPG, and its preliminary trails showed considerable promise. With the war winding down and other electronic types being developed, they did not reach the production stage.[11]There were many items that have never been mentioned that caused undue problems with the bombsight. One of them was bombsight mounts. Vibration was beginning to cause trouble with the operation of the sight. In order to combat this, the Bureau of Ordnance, in January 1938, awarded a contract to manufacture eight anti-vibration mounts for the Mark XV at a cost of $43.75 each. Drawings were completed by the NPG for a flexible mounting to effect vibrating isolation between the bombsight and the airplane. It was a stabilized base support with a trunnion cap, right hand trunnion support, and trunnion type base support with rubber shock rings. This type was used until about June 15, 1943.[12] A new type was then developed, and a contract was awarded to L.P. Stuart, Inc., for bombsight mounts on a test basis. The cost was $64.50 each, and delivery was 20-25 per week. The B-6 mount was replaced with the B-7. The B-6 was unsatisfactory and was to be replaced as soon as possible. Production aircraft were to have this new mount incorporated at the factory.[13] No further information on the B-7 type mount was found in the Archives except for the above.

Much had been said about the bearing trouble with the Mark XV bombsight. However, very few know how extensive and intensive the investigation was into this matter, and that it was of long standing (as far back as 1939). In 1939, the Norden Company ran an endurance test on gyro rotor bearings for 5 days attempting to duplicate the bearing failure as they appeared in the bombsight. The bearings were run for 52 hours before failure without lubrication. This failure appeared to be the same as those experienced by the services. SKF Industries (bearing manufacturer) agreed with the Norden Company that bearing failure resulted from lack of lubrication (once in 15 hours). The Norden Company felt that self aligning types of ball bearings were the best. To determine if this was true, the Norden Company submitted bearings supplied by SKF for calibration and testing to the Experimental Station, Annapolis, to find whether SKF was actually supplying bearings of high quality. The Bureau of Ordnance indicated that future contracts were to incorporate a 1,000 hour guarantee against bearing failure.[14]

The type of bearings being used had a special type of ball retainer ring made from Dowmetal. The experience with this type was limited.[15] Dimensional tests were conducted by the U.S. Naval Engineering Experiment Station in accordance with Bureau of Engineering specifications on twelve types of bearings. It was found that some were 0.0001 to 0.00015 of an inch too large, but the hardness and crush tests were within specifications. They also stated that self aligning bearings were not justified for this instrument.[16] The question of bearings had been studied for a number of years and had never been satisfactorily solved. Every type of manufacture was tried. The first bearings were provided with steel retainers, sometimes solid, and other times with spring retainers. Later the retainer was made of material like "Micarta." This did not work and was changed back to steel retainers. Since this was satisfactory, it was changed back to Dowmetal. By 1941, bearings were being provided with textilite, which was similar to micarta. It would not be until well into the war that a new company (Barden Company) was built by the Navy to provide precision bearings for use in the Norden bombsight. The development of precision bearings is discussed in another chapter. SKF, Norden Company, and the Bureau of Ordnance conducted laboratory tests of textilite and bronze retainer rings to determine if they would stand up better than the Dowmetal retainers. No follow-up information was found in the archives.[17]

Very few precision instruments lasted as long as the Norden bombsight. Normally, they are modified, redesigned, or discarded. It was of durable construction and sound engineering. If the sight was properly handled and care was taken not to damage any part, it would perform as required. Unfortunately, the author was unable to contact former Norden Company officials, engineers, employees, or Navy personnel. No attempt was made to contact the Norden family, as the author respects their privacy. At the end of the war all plants producing war material, including bombsight manufacturers, were closed. After WWII it became Norden Laboratories, Inc. It later merged with the Ketay Corporation and was called the Norden-Ketay Corporation. In 1958, United Aircraft Corporation formed the Norden Division when it acquired the company. This new company produced instruments and systems, such as digital converters, airborne radar, data handling systems, navigation equipment, etc. United Aircraft Corporation was acquired by United Technologies Corporation in 1978, and continued to be the Norden Division with the new company. The famous name of Norden continued to live.

Although this was not part of the NBS program, it is interesting that the AAF was concerned about the training of bombardiers. Only a few words will be said

Chapter XVIII: Unknown Facts about the Norden Bombsight Mark XV 257

about this. The plans for a tremendous expansion of bombardier training paralleled the plans for pilot training. Over 47,000 bombardiers had been trained in the AAF Training Command by September 1945. This figure includes about 3,000 navigator graduates who were also trained as bombardiers, and a number of enlisted men trained as D-8 bombardiers. A training of this magnitude required a tremendous number of aerial instructors. The relatively few graduates of 1940-41 were quickly absorbed, and instructors were recruited from new graduates. Bad flying conditions frequently interfered with training at Barksdale and Ellington Bombardier Schools in 1941. It was decided that these schools were to be reestablished in arid or semi-arid regions where good flying weather was at a maximum.[18]

The Chief of the Air Corps requested the Personnel Section of the Adjutant General's Office to conduct a study of the selection of applicants for training as bombardiers and navigators after an excessive number failed to qualify for Bombardier Third Class.[19] They were selected from flying cadets who failed to qualify for pilot rating. In order to cover this aspect, the AAF established the Aviation Psychology Research Program On Bombardiers and Navigators. The result was the establishment of a Bombardier-Navigator Section in conjunction with the Trade Test Division of the Air Corps Technical Training Command at Chanute Field.[20] Report #9, covering from 1941 through 1946, was issued, titled Psychological Research on Bombardier Training dated January 10, 1946. A complete study was made on the psychological aspects of cadets using either the Norden or Sperry bombsight. The program objective covered all sorts of tests, rating scales, cadet aerial training, etc. In these studies the psychological characteristics of the bombardier student were developed. The report consisted of hundreds of pages of detailed information that cannot be reproduced here due to the volume.

To promote interest in the training of bombardiers, the Naval Air Technical Training center, Jacksonville, Florida, devised a plan for using the Norden Bombsight and calling it the "Axis Exterminator #1." This equipment was assigned to the two outstanding student bombardiers of each class, and the name of the highest scorer was added to the list of "Axis Exterminator #1" in the display case at the Bombsight School. Two bombardiers from each class were selected. The competition for this privilege was extremely keen, and it is believed that better results were obtained in the flying phase after this competition than any of its predecessors. The first class to use this equipment was in August 1943. Not much is known about this program, however, the author found an ex-Navy bombardier instructor who participated in this competition and stated that it was very unique and an excellent training aid. The Bureau of Ordnance, C.L. Norden Company, and its employees wholeheartedly supported this program and donated three "Axis Exterminators" for the school's use. It is not known whether the Army Air Forces had a similar program. No further information on this program was found in the National Archives.[21]

It seems that the Norden Bombsight caught the fancy of the other types of news media. Sometime in 1945, four pages of Carl L. Norden were portrayed in comic book form, and it also was in color. It started with the heading "Inspiring Inventions" and went on to explain the various phases of Norden's achievements. It showed him working for Sperry, then developing the arresting gear for aircraft carriers, using the bombsight during WWII, and his receiving the Holley medal. According to the donor of documents, it was obtained from Ripley's "Believe it or Not." The original copies are in the author's files.[22]

Another unknown fact of the Norden Bombsight that has never been highly publicized was its use for precision map making and photo reconnaissance. There was no information or data on this topic in the National Archives. Some sketchy information was obtained from members of the Bombsight Employees Retirement Group, who knew of Air Force personnel who were involved in this type of project. It is not known when, where, or how extensive this practice was.

Chapter XVIII: Unknown Facts about the Norden Bombsight Mark XV

Notes:

1. Message from the Bureau of Ordnance to Employees of Lukas-Harold Corporation, Naval Ordnance Plant, Indianapolis, Indiana, dated April 1, 1943, National Archives.
2. Hqs Commander-in-Chief, Navy Department, Memorandum for General Marshall, Subject: BurOrd Table of Mark XV Bombsight Production and Allocation, dated November 29, 1942. National Archives
3. AAF School of Applied Science, Orlando, Florida, Combat Operations Section, AAF Historical Office, Subject: E-6B Computer, 250-4 180-P, dated June 1945.
4. Students Manual - Bombing, Air Training Command, AFTRC Manual No. 52-340, dated September 1, 1949, page 2-2.
5. Ibid, page 2-14. AAF Aviation Psychology Program, Report #9, page 147.
6. Ibid, page 2-13.
7. Bombardiers Information File, dated March 1945.
8. Letter from Brigadier General Robert L. Cardenas, USAF Retired, no subject, dated December 13, 1991.
9. Memorandum for the Secretary of the Navy, from Bureau of Ordnance, Subject: Report of Examination of Nazi Bombsights by Bureau of Ordnance, dated December 13, 1940, National Archives.
10. Letter, Naval Proving Ground to Bureau of Ordnance, Subject: Ballistic Wind Effect on Bomb Trajectory, dated July 24, 1934, National Archives.
11. Naval Proving Ground Report, Results of tests of the Mark 16 and 17 Bombsight — Norden Aiming Angle or Dive Sight, dated May 23, 1944.
12. Bureau of Ordnance, Project Order, Ordnance Stores, no subject, dated January 8, 1938, National Archives.
13. Bureau of Aeronautics to Wright Field, AAF, Subject: PBJ Type Aircraft, Replacement of B-6 Bombsight Mount with B-7, dated June 10, 1944, national Archives.
14. Letter, Bureau of Ordnance, to Chief of Bureau, Subject: Conference with C.L. Norden Company and SKF Representative Regarding Ball Bearing Failures in Bombsights and SBAE, dated August 22, 1939, National Archives.
15. Ibid.
16. U.S. Naval Engineering Experiment Station, Annapolis, Maryland to Bureau of Ordnance, Subject: Report on types for Ball Bearings for Gyro Rotors, dated September 28, 1939, National Archives.
17. Letter, Bureau of Ordnance, Memorandum for Chief of the Bureau, Subject: Conference Between "Ma" Section, SKF Company and Norden Company at Plant of SKF in Philadelphia, dated August 30, 1939, National Archives.
18. Army Air Forces Aviation Psychology Program Research Report, Report No. 9, Psychological Research on Bombardier Training, Midland, Texas dated June 10, 1946, pages 5 and 6.
19. Ibid, page 13.
20. Ibid, page 14.
21. Letter, Chief of the Bureau of Ordnance to the Naval Air Technical Training Center, Subject: Utilization of "Axis Exterminator # 1" dated August 6, 1943, National Archives.
22. Ripley's comic strip in the author's files.

XIX

Mark XV Bombsight Training Mobile and Stationary Devices

The Bureau of Ordnance regarded the bombing teacher or trainer as essential equipment for ground training of bombardiers in the fundamentals of bombsight operation. In 1928, the operating squadrons were using the D-4 bombsight (Estoppey). They were given this sight as an emergency measure, since it was the only one that was available, except for the Wimperis and Michelin. The squadrons were aware that the Mark XI was ready for production and would soon be replacing the D-4 sight. The Navy encouraged service personnel to develop their own training devices. One of the earliest was an ingenious device designed by Squadron V, "A" U.S.S. *Jason*, Asiatic Fleet. In December 1928, their plans were forwarded to the Bureau of Ordnance for approval. The device was known as the Price Bombing Car. Briefly stated, the apparatus consisted of a car similar, in certain respects, to the cockpit of an airplane and mounted on a specially constructed platform which moved on three ball bearing casters and could move in any desired direction. A bombing run was accomplished by the bombardier entering the car and positioning himself at the bombsight, while the pilot guided the unit. The speed of the car was so regulated that it gave the same effect as would be observed by a bombardier in a plane. Cross wind was obtained by moving the target right or left, the effect of up or down wind approach was obtained by increasing or decreasing the speed of the car proportional to the velocity of the wind selected. The bombardier manipulated the sight and directed the pilot through the Pilot Director as would be done in the airplane.[1] The Bureau of Ordnance forwarded the design to the Bureau of Aeronautics for evaluation. They felt that the device was basically sound. It was returned to the *Jason* Commanding Officer for any additional comments or changes. Although this simple apparatus did not fulfill all of the conditions in actual flight, it proved of considerable value while in use. It was also felt that much improvement could be made by installing a motor for propulsion of the car and developing a dotter gear in order that hits and misses could be recorded. The Bureau of Ordnance was very much in favor of using bombing trainers and felt that training should be done on the ground using some type of training device. It was pointed out that the more complex Mark XI would be coming out soon and would require considerably more ground training than any sight in service.[2]

The need for a training device was also requested by the Commander Naval Air Station (NAS), Coco Solo, Canal Zone in early January 1929. They also designed a trainer for their use. This trainer was manufactured from old bicycle parts, and two small steading wheels from a child's toy wagon. For motive power, a starter from a Ford automobile was used to drive a sprocket through an inertia airplane engine starter for proper gearing. The motor driven sprocket was connected by a driving chain to the larger sprocket on the front wheel. Electric current was provided by a 12 volt storage battery. It was used by three squadrons on a daily basis. The entire device was constructed on a scale of 1000 to 1 and simulated actual bombing from 5,000 feet, and the relative speed of the bombsight towards the target was 65 knots. The speed of the target was controllable between 10-20 knots. The target consisted of an electric motor, gears, pulleys, and the target, which was drawn by a chalk line over a distance of 40 feet.[3]

The Bureau of Aeronautics, realizing the pressing need in February 1929, recommended the purchase of one Vickers-Bygrave bombing teacher for trail. At about the same time, the Bureau of Ordnance submitted drawings for a training device to the C.L. Norden Company. In the meantime, the Bureau of Aeronautics recommended that four units from the Norden Company be purchased and issued to various units for trial, pending the design from Vickers-Bygrave Company.[4] By 1930, the Bureau of Ordnance was proceeding with designs of a bombing teacher which would replace the "home made" ones being used by the fleet activities.[5] The Bureau of Ordnance up to this time had not decided on any type of bomber trainer. The Coco Solo device was simple yet excellent in providing the necessary skills to train bombardiers. Since no trainers for the Mark XI were available, Aircraft, Battle Force, Torpedo, and Bombing Plane Squadron Two-B personnel designed and built in January 1932 which provided excellent advantage in training forty-four bombsight operators. In the limited time it was impossible to train all of them entirely in the air. It was a simple affair, but advanced over the Coco Solo model. This was the beginning in a development of bomb trainers that were used before and during WWII. It consisted of a triangular shaped chassis built of angle iron and large enough to accommodate the bombsight, battery, pilot, and the bombardier. The frame was 64 inches wide in front, 15 inches wide in the rear, and decked with heavy flooring and cross bracing to prevent weaving. The frame was mounted in front to a front axle assembly of a model T Ford, shortened so that the thread was 48 inches. The drive consisted of a rear wheel and rear frame section of a motorcycle, which was welded to another steel plate providing six ball bearings for the drive unit. This mobile training stand had many advantages. It simulated air conditions to a remarkable degree. It also permitted the following: instructor to talk to students while bombing; permitted several students, while bombing, to observe a bomb run, noting any mistakes and listening to discussion; and cross wind bombing training and necessary pilot directing (the most difficult problem for beginners) saved wear and tear on bombsights, overhaul time, and trouble shooting. Also, training of large number of bombardiers saved operating costs in fuel and deterioration of the airplane and bombsights.[6]

In keeping with encouraging service personnel to show their resourcefulness in developing devices, NAS, Pensacola, Florida, in May 1932, built a bomb training machine for students on how to bomb with the Mark XI sight in the classroom. This new device was more accurate and realistic than any trainers in existence. The new device provided a means for exactly reproducing a bombing run on a miniature basis. It was very much more advanced over the ones that were in use. Again, it was largely constructed from scrap material. Briefly, this new device had a bombsight platform, a pilot for controlling the device, including a Pilot Director Indicator (PDI), the same as in the airplane. The bombardier operated the sight exactly as in the air. The target was placed five feet below the level of the sight platform. The "ocean" was made to move under the bombing plane in such a manner that the relative motion between the target and the bombsight was the same as if in an actual bombing run that was being made in the air. The "ocean" was a platform 9 feet square which moved on a track mounted on another platform. The pantograph (mechanical means for indicating speeds along a track) to indicate the proper speed, friction drives for moving the platform at indicated speeds, and a plumb bob controlling device for dropping the plumb bob for the point of impact. The friction drive gears were set to move the ocean at the speeds indicated by the pantograph. Any desired altitude could be represented on the machine, from 5,000 to 10,000 feet.[7] The advantages of an accurate machine for training bombers on the ground are too obvious and numerous to mention. Commander, U.S. Fleet Battle Force, U.S.S. *Saratoga* stated that this trainer was too complex to be built by them. In a letter to the Chief of Naval Operations, he strongly recommended that this type of trainer be purchased and made available to all training centers who had the Mark XI bombsight. The device used by VT Squadron Two-B was simple in construction and required no crew other than the pilot and bombardier to operate it. It was encouraged that this model be easily converted to use with the Mark XV sight when placed in service.[8]

In February 1933, the Bureau of Ordnance, recognizing the need for training bombardiers, encouraged operating units to construct training devices on their own. Due to the Depression, funds were scarce for new equipment and, accordingly, the operating units were told to construct their own training devices within their resources. To encourage this, a liberal allowance of practice bombs was supplied for training in the air. Plans, photos, and instructions of trainers developed by VT Squadron Two-B and Pansacola NAS were forwarded to the various service activities, especially Pearl Harbor and Hickam AFB. The Bureau of Ordnance requisitioned material not readily available in order to assist and complete the trainer program.[9] Unfortunately, the training of bombardiers was not a priority item during these years. The Bureau of Ordnance realized the importance of bomber training and directed that personnel at Pearl Harbor take necessary action to replace bombing teachers that were not serviceable, by improved trainer utilizing facilities and material available at the station.[10]

The buildup of the U.S. military forces to counteract the growing menace in Europe, and the requirement for bombsights, increased dramatically. The bomb trainers being used by the services (AAF and Navy) were inefficient in training bombardiers on the advanced Mark XV bombsight. With the expansion in both the AAF and Navy, new trainers were badly needed. Bombardier training was no longer "business as usual." Training of bombardiers became an important part of the bombsight program. In September 1940, the Bureau of Ordnance requested the Naval Aircraft Factory to design and make workable drawings for a new trainer.[11] Due to the secrecy of the bombsight, all designing was done except for the certain dimensions having to do with the installation of the secret bombsight. The Bureau of Ordnance took the best features and developed an entirely new training device. This resulted in a longer time to finish an acceptable set of drawings than was originally anticipated. A delay was encountered in the manufacture due to other work being conducted by the Naval Aircraft Factory. It was recommended that the trainers be procured commercially. The first contract for 50 training stands for training bombardiers was awarded in June 1940. The cost of one unit was $2,500.00. Later, another contract was awarded for 34 more.[12] It is certain that many more of these trainers were obtained to train an estimated 50,000 bombardiers at various training bases throughout the U.S. These types of trainers were the ones shown in the movie "Bombardier." The increase in trainers also meant more bombsights would be allocated to those having the new trainers. This would further impact the bombsight shortage. These types of trainers were used by AAF during WWII with modifications, but the concept was the same. The examined files made no mention of procuring additional units. It must be assumed that many more were purchased, since both were used by the Army and Navy. The exact number or cost actually produced is not known.

7A - 3 Trainer

A new concept in training bombardiers was developed by the Bureau of Aeronautics, in the development and procurement of synthetic training equipment. This project was given to the International Business Machines Corporation for production. The first engineering studies were conducted in December 1942, at the Banana River NAS. No date was available as to when they were furnished to the training activities. Prior to this, bombing was taught by the use of a moving vehicle equipped with a bombsight located several feet above the hangar floor. Target recognition could not be taught, and it did not simulate actual flying conditions. This new device was designated the 7A-3 Trainer, consisting of a replica of a bomber compartment (B-24). Every condition of actual bombing procedure was reproduced so accurately that the student needed no period of adjustment when actually bombing with an airplane. A projection device produced a moving image of the ground or target area in exact proportion to the speed and altitude of the airplane; it simulated drift in the same manner that a plane is affected by cross winds. Very accurate photographic maps of actual target areas, taken at various bombing altitudes, were used in the projector to recreate the proper illusion of altitudes. This type of trainer saved time, maintenance on airplanes, fuel, practice bombs, ground crew time, etc., and gave the student a better picture of the bombing problem. No production quantities or costs were found in the examined files.[13]

A-6 Photographic Trainer

Earlier models were the A-2 and A-5. The A-6, or McKaba trainer, was not used for full cadet classes until the spring or summer of 1945. This trainer was designed for intermediate and advanced training of bombardiers. It was a simplified and less expensive version of the A-5, although it was developed separately. The unit consisted of a bombardier's compartment mockup located over a screen upon which a traveling terrain was projected. It contained rack control, switches, and instruments as in the airplane which must be placed in correct position to enable bomb release. The Instructor, as the pilot, had control to set drift, altitude, temperature, trail, heading, airspeed, and reset. Provisions made for interphone, machine gun noise, motor noise, vibration, and anti-aircraft gun flashes. It had up to 80 degrees of forward vision. The point of impact was shown by a flash of light projected on a screen. Again, very little and fragmented data was available. Quantities produced were not found (T.O. 11-65-12, Jan 25, 1944).

Visual Link Trainer

In November 1942, the Bureau of Ordnance looked into the possibility of modifying Navy owned Visual Link Trainers stored at Binghamton, N.Y., to provide installation of AFCE/SBAE and Stabilizers. They would include all the features for AFCE/SBAE, instrument training would carry the project beyond the possibility of the present visual trainer, and would require the development of a completely new type that would take at least a year to a year and a half to modify or redesign the new trainer. The material required for this new project would have to be sanctioned by the WPB and JAC. A standardized trainer had been agreed on by the working

Chapter XIX: Mark XV Bombsight Training

committee of the JAC. This unit had features that could satisfy all of the services (Army, Navy, British, and RCAF). For changes from regular production of trainers to the standardized trainer approval was required from the JAC. To alleviate this condition, the equipment would be installed by the stations at the ultimate destination. The Navy agreed that detailed and assembly drawings for the mount arrangements, pulleys, cables, etc., would be supplied by the Bureau of Ordnance to speed up the project.

The manufacturer estimated that the first six units would be delivered in about six weeks. Additional quantities were made at the rate of one to three per week depending on the quantities involved. The price was $1,000 per unit.[14] In November 1942, the Bureau of Aeronautics contacted the Bureau of Ordnance, stating that a basic model had been developed to accommodate SBAE/AFCE using the Norden Mark XV, Sperry S-1, and the Honeywell control equipment. The Bureau of Ordnance issued a contract for 21 of the modified Link Trainers. Plans were drawn up and forwarded to Link Aviation Devices, Inc., for modification and allocation of 25 units, four of which were sent to Special Devices Section. All plans and specifications were at Link Aviation Devices, Inc., and production of these items was completed in the first part of 1942.[15] In September 1944, the existing contract with Link Aviation Devices was increased by 55 units (various models) to the existing contract of 25 units. The first items on the contract were in production and were delivered the early part of 1945. Total Production is not known, as there was no data in the National Archives. A total of 80 modified Visual Link Trainers were built.[16]

Training Devices
Mock up
In addition to the mobile and stationary devices, there were classroom training aids. One of the most effective was the AFCE/SBAE training table. This table had all of the essential bombsight and automatic pilot control equipment installed on a table so that pilots, bombardiers, etc., could see the entire function. Mrs. Marge (Gorman) Eiseman is shown performing final inspection of the training table before being released to the operating units less the bombsight.[17] This particular table was the latest updated version. These tables were provided to bases throughout the U.S. and were manufactured by the AAF at the Sacramento Air Depot (S.A.D.). The Bureau of Aeronautics procured 38 demonstration tables from the AAF manufactured by S.A.D. (mock up stands) used by the AAF in displaying their AFCE equipment for training. It is a simplified version of the actual setup of Norden equipment in a bombing airplane. All essential controls are provided minus the instruments, which would be furnished by the receiving activities. These training tables were manufactured by the AAF at the Sacramento Air Depot where the author worked. The training section was near the machine shop, because much of the work on the tables was done by machinists. Twenty units of this total were used in

the bombsight and SBAE schools located at the NPG, Dahlgren, Virginia, and 18 were sent to NAS Jacksonville, Florida.[18]

Plate #1 shows all of the components: 1. bombsight placement; 2. Directional Stabilizer; 3. Vertical flight gyro; 4. Aileron servo motor; 5. Elevator servo motor; 6. Rudder servo motor; 7. Formation stick-left; 8. Formation Stick-right; 9. Rotay Invertor; 10. Auto Pilot control panel and 11. Junction Box (Electrical). The original scale model of the training table developed in the 1930s was donated to the local air museum by Mr. John Torbert, III, son of the inventor, who has been a life long friend of the author.

The Plexiglass Bombsight

One of the most explicit training aids ever developed for the Norden Bombsight (NBS) was the plastic covers for the sight. Depot Repair facilities (AAF) were not geared for production of bombsights. This was left to the individual contractors, since the Depot's primary and only function was repair, however, an exception was made in this case. The idea for the clear training covers was advanced by the 2nd and 4th Air Force at Galveston, Texas. Other stations over the country had turned down the project, saying that with limited facilities and lack of specialist personnel, it "could not be done." They did not want to get involved in this project. Replacing for classroom use the regular metal bombsight covers, which only had two small indicator windows and the rate end had none, would be a boon in training bombardiers. The two units only weighed 25 ounces.

The project was approved by Hq Air Technical Service Command, Dayton, Ohio. This idea of manufacturing these covers was discussed at great length with the author and other personnel. In the beginning, it was believed that it could not be done. It was mandatory that all work was to be done in the Bombsight Shop (building), both in the machine shop and repair and assembly shops, since the sight was still classified "secret." The only other personnel involved was the Plexiglass Department. It was immediately necessary to clear one of their best technicians to work with the bombsight group in developing molds. In order to cast the plastic, a new high pressure mold was designed and built in order to form the 1/4" material. There was no previous precedent on accomplishing this type of work. Before any commitments could be made as to whether it could be done, considerable time and research was spent between the author and the Plexiglass shop. Machining of plastic on this scope had never been attempted where such critical tolerances were required. All measurements were strictly adhered to in accordance with engineering specifications. There were no deviations. The upper most question was "was it possible to machine plastic to such tolerances?" After much experimenting, it appeared that it could be done. Watch-like measurements were necessary for the moving parts. The author designed and supervised the entire project from beginning to end. Before any work was attempted, it was necessary to devise patterns, jigs, and fixtures to hold the covers while machining, as well as changes in the blue prints to allow for the use of plastics. Sufficient materials were available so that it was not necessary to obtain permission from the JAC or WPB.

The first part to be worked on was the top cover. This was the easiest part, since there was less machining to be done. Assistance was obtained from the plastic manufacturer, Rohm & Hass, after several failures to produce a workable cover. More work was needed to remove "wrinkles" in the casting. After several molds were made, perfect covers were obtained. The failure rate on this part was low due to the few parts that needed to be machined, and tolerances were not as critical as those in the "rate end."

With experience gained from the successful top cover, attention was turned to the rate end. From engineering drawings of the regular metal housing, five different models were made. On the surface, it appeared simple. This was very deceiving, as there were many dowel pins that lined up the motor and holes for the turn and drift knobs, screw holes to hold various parts, mirror drive clutch, disc speed, gear shafts, extended vision knobs, etc. An attempt was made to use templates, but they were not accurate and could not be used. All the holes in the cover had to be precisely located and machined, or any binding would cause friction in the moving parts, resulting in the sight performing erratically, and it could not be calibrated to the required stringent specifications. Friction in the gyro would, in turn, affect parts of the sight. On installation, the covers split and broke.

To remedy this problem and to make the rate end covers, stiffeners in the casting were inserted in the area under the Disc Speed drum due to cracking. This was solved by cutting out a pie shaped opening and putting in a thicker piece of glass. The Plexiglass Shop glued the parts together. About 50% were discarded, as they did not hold, bubbles formed, etc. Machining this part required the most skilled mechanics in the shop. It was a difficult task to machine plastic to such exacting tolerances, as the plastic was not stable and heating caused erroneous measurements. This was overcome by the use of air jets. The discard rate was about 25%. Machinists were able to achieve tolerances of .00054 inches (five ten thousands of an inch). Three hundred hand and machine operations were required to complete a rear cover. Each cover had to be tailor made for its bombsight.

Two hundred and seventy-five hours for the production of the first experimental cover was reduced to 100 hours. To assemble the eight component parts of the two covers to the bombsight by assembly and repair personnel required an average of 80 hours, including testing and calibration. The Trail Plate and the Disc Speed Drum were also made of plastic, so that the operator could see what happened when an adjustment was made.

The bombsight workers were rewarded when the first successful covers were installed. They were the first to see the action of the gyro, telescope, and the "rate end" (computer). This was the first time that any of the working parts were ever seen in actual operation through a "window." It was a significant advancement in bombardier training.

A total of 15 covers were made. Unprecedented demands were received from many sources. However, so much time was used in producing these units and regular repair work was so impacted that a complete halt was ordered on authorizations for any further production. The units were manufactured in 1943, but the news releases were not authorized until April 1945. Note: the above information is from the author's files, recollections, and news media articles. None of the plexiglass sights are available. A rumor surfaced that a person had one in his possession, but the author was unable to locate the individual, although extensive research was conducted. They were not the plexiglass sight, but only static photos of the various working parts of the NBS. The author has, in his possession, black and white photos that clearly show all the working parts of the bombsight. Unfortunately, they do not reproduce well.

Classroom Training

With the AAF and the Navy procuring bombsights at an increasing pace, the need for training personnel to repair, maintain, and calibrate the bombsight and automatic pilot was becoming of primary importance to all of the services. The Army Air Corps did not have an effective and experienced repair capability to conduct major overhauls on this equipment. It became very evident in June 1936, when the AAC had 23 sights badly in need of repair and modernization. Since they were unable to do this work, they contacted the Bureau of Ordnance to make arrangements with the Norden Company to perform the overhaul and obtain costs. The sights were returned to the manufacturer to be brought to the latest configuration.[19] In July 1936, the Naval Proving Ground issued new instructions concerning the calibration of bombing equipment. Again, the AC could not perform this work. The calibration of this equipment was intended for use in the test rooms of the bombsight shops, which the AC did not have. The methods of setting up these operating instructions are not described here and have been eliminated for the sake of brevity, as they covered hundreds of pages of detailed instructions.[20]

Many problems were encountered with the policy of returning the sights to the contractor for repair and modification, especially when the sights were located at American bases is foreign countries. Much delay was encountered in the return of the sights, leaving the operating squadrons without bombsights. To alleviate this problem the Chief of the Army Air Corps established a course of instruction for the first bombsight school at Chanute Field in September 1936. The course of instruction was for eight weeks' duration with four students. The students underwent strict scrutiny and were especially selected and recommended after a background check was completed. In addition, the student had to be of undoubted loyalty, a high school graduate, non-commissioned officer, with at least six years service, graduate of an

Chapter XIX: Mark XV Bombsight Training

armorer course or three years in a bombardment organization, and an air mechanic with at least six years of service. The qualifications of all the candidates were forwarded to the Commandant of The Air Corps Technical School for approval.[21] This was the beginning of training schools for bombsights and auto pilot.

With the development of AFCE/SBAE in November 1940, the bombsight school added a training course for this equipment lasting four weeks in duration, or 160 hours for eight men. The qualifications for this course were the same as for the bombsight repair course. One requirement consisted of a detailed study of the stabilizer unit and bombsight operation on the Norden Trainer prior to entering this four week course in AFCE/SBAE[22], covering all phases of repair, maintenance, and calibration. After Pearl Harbor, the training of both the bombsight and AFCE/SBAE was intensified. To augment this training, the Bombsight Maintenance Division, Department of Fire Control, Lowry Field, Denver, Colorado, prepared a complete Syllabi of Instruction for bombsight maintenance covering every aspect of the system.

In the beginning, course of instruction included 1st, 2nd, 3rd, and 4th Echelons. However, 3rd and 4th echelon was discontinued in February 1944. With the ever increasing complexity of the new equipment being received, the course of instruction was lengthened. The bombsight course of instruction was designed to train AAF personnel in 1st and 2nd echelon on the D 8 Estoppey, Norden M-Series bombsights, Norden Glide Bombing Attachment, and Honeywell B-1 and C-1 Auto Pilot.[23] The course embodied instruction in operation, maintenance, inspection, adjustment, and calibration of the equipment. Instruction also provided in the use of special tools, jigs, and fixtures.

Classes were limited to enrollment. Due to the volume of material on training, only a brief outline and description is shown, and the prerequisites required for selection and duration of classes. Students were selected in accordance with existing regulations set up by the classification offices of their induction and basic training centers. Requirements were high. A GCT of 120, and a Math score of 120, or an equivalent background of mechanical and electrical experience and ability was required.[24] Duration of the course was 96 academic days, and the average class size was 36 students. The training program was divided into five blocks. Briefly, they are: Block #1 consisted of 72 hours of instruction on manual skills, tools, DC electricity, and Norden Stabilizer; Block #2 included 216 hours on bombing problems, the Norden sight, the D-8 sight, and the Automatic Bombing Computer; Block #3 was 36 hours on the Glide Bombing Attachment; Block #4 was 36 hours on the Bombsight Box, Field Repair, and care and maintenance of the Norden Trainer; and Block #5 included calibration and maintenance of the C-1 Automatic pilot, 216 hours.[25] A two week course was also given on manual skills, tools, DC electricity, supply, technical orders, and the Norden Stabilizer.[26] The course for the C-1 Auto Pilot, due to the complexity of the equipment, was increased to 48 academic days.

The students were given detailed, intensive and comprehensive instructions in all phases of the equipment being studied. They were required to attend classes six days a week. Each week, a new phase on the Norden was studied. For example, the first week was devoted to bombing problems, theory of bombing, inspection, handling, care of equipment, detailed operation of the rate end, etc. The sixth day was devoted to three hours of practical examination, 1 1/2 hours of written tests, and 1 1/2 hours of examination on criticism of practical examination. Also included was laboratory work. The series of training films developed for the AAF was also part of the curriculum. The final test was the installation and alignment of the Norden sight and stabilizer in the plane. Lesson plans, subjects to be taken daily, and hours attached to each are not shown because of length and repetition that would be too detailed and would consume many pages of information not relevant to the subject.

In addition to this school at Lowry, a course of instruction was offered by the Minneapolis Honeywell Regulator Company beginning August 28, 1943, at their plant in Minneapolis, Minnesota, on the AFCE/SBAE Mark 2, Mod 1. This was the latest on the automatic pilot. Both military and civilian personnel attended these classes. This was mostly attended by civilian personnel. There were no government quarters or messing facilities for military personnel.[27]

The Norden Company also conducted courses of instruction similar to the type at Lowry Air Force Base, but it was more detailed in the workings of the Mark XV. This course of instruction was approximately 18 weeks and covered the maintenance and repair of the Mark XV bombsight system and related instruments. Classes were held to a minimum at the factory in New York. No information on this training was available. It is not known if this course was continued after Pearl Harbor. The author knows only two persons that attended these classes, and they were before December 7, 1941.

Radar Classroom Training

With the use of radar in bombing, it was only natural that service personnel needed to be trained to keep this new equipment in operating condition. By July 1944, various types of radar equipment were being used by the AAF in the combat theaters. In order to keep this equipment in operating condition, radar technicians were urgently required. The Seventh Air Force Communications School inaugurated a class for an Airborne Low Altitude Bombsight Radar. Equipment to be studied was the AN/APQ-5 (Low Altitude Bombing). This radar equipment installed in airplanes was used with the SCR-717, Airborne Search and Navigation Radar to perform blind bombing at low altitudes. It was designed to operate up to 2,000 feet of altitude, and it weighed about 100 pounds. The course was prepared to train radar mechanics to perform maintenance and repair work on the Low Altitude Radar Bombsight, covering theory, operation, use of test equipment, and maintenance. The class was for 80 hours, or two weeks in duration. Forty hours was devoted to

Chapter XIX: Mark XV Bombsight Training

familiarization of the set, study of functional diagrams, and detailed explanation on each part of the electrical circuits. Six hours was spent on the various controls and interpretation of the picture on the indicator. Thirty-four hours was assigned to laboratory instruction, performing all turning and adjustments using the test equipment that would be expected to have under field conditions. Measurements were made on voltage, resistance, and waveform measurements on mockups.[28] Classes were limited to only qualified personnel who had a background in radar equipment and met other specific qualifications.

There were probably many other training schools on various new types of equipment, but the author is not aware of them. No other information on training was found in the National Archives files.

Field Services Training

With the ever increasing number of Norden instruments being used by the Military Services (Army, Navy, and Allies), the Bureau of Ordnance and the Army Air Forces that Field Services support was badly needed, especially in the combat zones. The Norden Company was the first to start this type of comprehensive service. To accomplish this goal, a school was started in October 1942 with a nucleus of 17 students, followed by an additional 12 students, and another class was to start in the first week of January 1943. After that, a continuing supply of factory trained personnel continued until the end of the war. The exact quantity is not known. The schedule of instruction was eight months of intensive factory and classroom instruction. Upon graduation, personnel were sent out in the field under the direct supervision of the present Field Service Engineers that were in the area. Those fully qualified were assigned to areas of activity, both in the U.S. and combat theaters, as soon as possible. Twenty service personnel were assigned to overseas duty under a contract awarded by the Army Air Forces. It was necessary to do this, as these civilian personnel were to be protected and recognized by the Commanding Authority in the areas that they were assigned to.[29] These personnel became known as "Technical Representatives," or "Tech Rep" for short, and the services relied very heavily on them. They had all of the latest information on the equipment being used, and they assisted either combat personnel or students in every way possible. A first hand narrative of a "Tech Rep" is described in another chapter.

Mobile Trainer

The hangar, or outside type of trainer, required considerable space so that it could maneuver towards the target. Different models were manufactured. All of them accomplished the same thing. Normally, the trainer was self-propelled (battery or other means) and could be guided by the pilot or bombardier. On some, the front axle was larger than the rear, or had only one wheel in the rear. The front axle was about 6-8 feet in width, and the rear axle was about 2-3 feet less in width, giving a tapering effect, while on others the front and rear axles were the same size and had inflated tires. All were about 12 feet tall and were rigidly reinforced with heavy steel cross bars laterally and vertically for stability when moving across the floor. The top level, or "perch," was of sufficient size to accommodate a bombardier and instructor. Immediately behind the bombardier was the pilot's compartment with all of the necessary controls for the unit. The bombsight and stabilizer was mounted on the perch with the student having easy access to it. Space was allotted for the placement of various instruments for a bomb run.

Using this method solved the bombing problem without going into the air. The purpose of this type of trainer was to familiarize the student with the operation of the bombsight. The motion of the trainer across the floor simulated the airplane in flight. The four wheeled electric driven box, on which the target was placed, simulated the wind. The bomb used on the trainer was an electrically operated sort of plumb bob. When activated by the bombardier at the proper instant, it would register either a miss or hit. Although these were not as sophisticated as the new trainers that were being developed and deployed, they were replacing the older ones in service. They were used to train thousands of student bombardiers as late as October 1944.

Notes:

[1] Letter V.T. Five "A", Aircraft Squadron, U.S.S. Jason, Flagship to Bureau of Ordnance, Subject: Device for Training in Bombing, dated July 10, 1928, National Archives.
[2] Aircraft Squadrons U.S. Asiatic Fleet, U.S. Jason, letter to Bureau of Aeronautics, subject: Devices for training i bombing, dated December 21, 1928, National Archives.
[3] United States Naval Air Station, Coco Solo Canal Zone, letter to Bureau of Aeronautics, subject: Device for ground training of bombers, dated January 3, 1929, National Archives.
[4] United States Asiatic Fleet, Manila, P.I. to Bureau of Aeronautics, subject: Devices for training in bombing, 10th Indorsement, dated February 18, 1929, National Archives.
[5] No basic letter, 1st Indorsement from Chief of Bureau of Aeronautics to Bureau of Ordnance, subject: Device for ground trainers of bombers, dated January 21, 1930, National Archives.
[6] U.S. Fleet, Aircraft, Battle Force, Torpedo and Bombing Plane Squadron Two, to Chief of Bureau of Ordnance, subject: Stand for Training Mark XI Bombsight Operators, dated January 5, 1932, National Archives.
[7] U.S. Naval Air Station, Pensacola, Florida to Chief of Bureau of Aeronautics, subject: Bombing training device for Mark XI Bombsight, dated July 26, 1932, National Archives.
[8] U.S. Fleet, Aircraft, Battle Force, U.S.S. Saratoga, to Chief of Naval Operations subject: Bombing Training Device for Mark XI Bombsight, dated July 26, 1932, national Archives.
[9] Navy Department, Bureau of Ordnance to Aircraft Commander of Squadrons and Attending Craft, Fleet Air Base, Pearl Harbor, T.H., Training Device for Horizontal Bombing (Bombing Teacher), dated February 2, 1933, National Archives.
[10] Ibid, page 2.

11   U.S. Navy Yard, Naval Aircraft Factory to Bureau of Ordnance, subject: Bombing Trainers, Estimate of Cost to Manufacture, dated June 21, 1940, National Archives.
12   U.S. Navy Yard, Naval Aircraft Factory to Bureau of Ordnance, subject: Manufacture of 50 Bombing Trainers, dated September 12, 1940, National Archives.
13   Development and Application of the 7A-3 Bombing Trainer, no date.
14   Letter from Link Aviation Devices, Inc., Subject: Conference on AFC Link, dated November 4, 1942.
15   Navy Department, Bureau of Aeronautics, to Chief of the Bureau of Ordnance, Subject: Link Trainers Modified by Installation of SBAE and Stabilizer, dated November 30, 1942, National Archives.
16   Navy Department, Bureau of Aeronautics, Subject: G.F.E. for 7-1 Automatic Flight Control Trainers, Contract No.(s) 2755; dated September 14, 1944, National Archives.
17   Courtesy of Mrs. Eisman collection.
18   U.S. Naval Proving Ground, Dahlgren, Virginia, letter subject: Request to Procure 38 demonstration (mock up) stands as now used by the Army in displaying their AFCE (Navy SBAE), dated November 16, 1942, National Archives.
19   War Department, Chief of the Air Corps, to Bureau of Ordnance, Subject: Navy Department contract Number 46343, dated June 23, 1936, National Archives.
20   Letter Naval Proving Ground, Dahlgren, Virginia to Bureau of Ordnance, Subject: Bombsight Calibration Equipment Instructions, dated July 1, 1936, National Archives.
21   Letter Hq Chanute Field, to Chief of the Air Corps, Subject: Inauguration of a Course of Instruction at the Air Corps Technical School in Bombsight Maintenance dated September 15, 1936, National Archives.
22   Automatic Flight Control Equipment Schedule for a Calls of Eight Men, June 1936, Page 165 Master Study, National Archives.
23   Department of Fire Control, Lowry Field, Denver, Colorado, Bombsight Maintenance School, First and Second Echelon Maintenance, Complete Syllabi for instruction, dated February 12, 1944.
24   Ibid., page III.
25   Ibid., page IV.
26   Ibid., page 3.
27   Letter to Bureau of Ordnance, Second Indorsement, Subject: Request for Course of Instruction in Minneapolis Honeywell SBAE Mark 2, Mod 1, dated July 30, 1943, National Archives.
28   Seventh Air Force Communications School, Syllabus, Subject: Field Service School, dated December 7, 1942, National Archives.
29   Letter Carl L. Norden to Bureau of Ordnance, Subject: Field Service School, dated December 7, 1942, National Archives.

XX

Myths - Spider Webs, Human Hair, Etc. Etc.

For some unknown reason myths abound around the Norden sight. The story of using black widow spider webs for reticles (crosshairs) in the bombsight has been around for as long as the bombsight itself. Extensive research into this myth has not proved that it was ever used. It is a known fact spider silk has tremendous tensile strength—stronger than steel. It is 0.0001 of an inch in diameter and has a breaking point of about 60,000 lbs. If a person has observed black widow spider webs, they can readily see that the texture is somewhat course and stringy. The reticles in the sight were required to be absolutely uniform. If the webs were used they would have had to be processed to acceptable tolerances required for this extremely important part of the sight. Perhaps Mr. Norden, in the early years when experimenting with the sight, may have used or thought of using spider webs. No records or data have been found to support this myth. In Mr. Norden's patent filed in 1932, it specifically states that the reticles (crosshairs) were etched. This seems a more plausible explanation than the use of any type of spider webs.

Some credence for this myth may have come from an article on Professor Albright in the July 1942 issue of Flying Magazine, where it mentioned spider silk was being used for bombsights.[1] Dr. Albright was a well known physicist and the author of several books on Physics and a paper on extracting spider silk. During WWII he was a professor of Physics at the Case School of Applied Science, Cleveland, Ohio (now Case Western Reserve University). He also had a side line business of extracting spider silk from the Miranda Aurentia, which is a large golden garden spider variety. He and his brother developed a system of extracting about a hundred feet of silk from the spider. The author traced Professor Albright from Case Western Reserve University to the University of Rhode Island, where he was the Chairman of the Department of Physics at Rhode Island State College at

Chapter XX: Myths - Spider Webs, Human Hair, Etc. Etc.

Kingston, Rhode Island. He was nominated for a Whitney Visiting Professors Program—John Hay Whitney Foundation, New York City, New York, in 1955 where the trail abruptly stopped.[2] Many years later an article appeared in the July 1992 issue of the Discoverer Magazine concerning the use of black widow spider silk. The article was written by an employee of the Unisys Corp, which during WWII was the home of the Burroughs Adding Machine Company in Detroit and nearby Plymouth, Michigan. During WWII Burroughs made over 6,000 bombsights. The author contacted the person who wrote the article. He had been able to find a living relative of Professor Albright. Regrettably, he passed away many years before.[3] The species were checked, and the black widow spider is of the Latrodectus species and not the same spider.

In the first year of WWII, the Minneapolis Honeywell Regulator Company at their Wabash plant were able to produce over 55,000 good reticles for military use. Prior to Honeywell producing reticles (crosshairs), it required a pantograph machine (engraving) costing $15,000, and etched only 10 reticles at a time with a 50% rejection rate. Honeywell personnel worked on this problem and developed a simplified pantograph which cost $600, achieved a higher output, and a rejection rate of only 25-30 percent.[4] Why would the Norden Company want to waste critical time on extracting spider silk, then processing it when perfectly etched reticles were readily available? The facts do not support this myth. This story has been cropping up for the past 50-60 years, and is continually repeated. However, the lore of the spider webs continues unabated.

A real favorite and well known name for the Mark XV sight was the "Pickle Barrel Bombsight." It is not known who actually coined this name. Some have attributed it to Carl Norden, others to Theodore H. Barth, and some to the Army Air Corps officers. None of these could be identified or verified as actual. There was no reference of this name in the examined files. It has been reported that it grew from a highly publicized Air Corps test of the sight in which Norden-equipped bombers hit the target dead center. A jest was started in which they did not state which "pickle" to hit. There was no information in the National Archives that covered this saying, therefore, it must be assumed that it got started by word of mouth or through the media. Of course, such sayings sounded good, and perhaps it was a morale booster for the bombardier and the general public. The advent of electronics being used in the automatic pilot for the first time went a long way toward making the sight more accurate. Before and during WWII many articles appeared in several national magazines and other media about the superiority of the Norden bombsight. In most instances they were well written. In the years before Pearl Harbor many of these magazines purported to be able to describe how the bombsight worked. Some were fairly accurate, but they did not have the solution to the bombing problem. Of course, they were not given any data by the government, as it was classified secret at the time.

The author did not attempt to determine which was the best sight being used by the services. Insufficient information and data precluded making any assumption or determination. Some studies were conducted during the war, but many were not available for research. Those that were available were not complete, and much of the information and data in the enclosures was incomplete or missing. The purpose of this book is not to determine which was the most accurate bombsight, but to chronicle the history of the Norden bombsight. There were supporters for both bombsights (Norden and Sperry). Each sight had its good and bad points. In order to make a comparison, all bombsights—Norden, Sperry, and Estoppey—would have had to be compared under controlled conditions. Since most bombsights used the same principle and were gyroscopically controlled, considerable time would be required to determine which was best. During war, time is of the essence, and military planners use the material that is available to them at the time.

Another well known story is the use of human hair for bombsights. Human hair has been used for reticles in telescopes, gun sights, weather, and other optical instruments for many years. Several years ago a rather humorous article appeared about a young lady donating her 34 inch tresses for bombsight crosshairs. It appears that the Institute of Technology, Washington, D.C., made an appeal for women's hair that had never been bleached, treated with chemicals, or hot irons. The author tried to locate this agency, but it was no longer in operation. In 1942 the young lady graciously donated her hair for use by the U.S. Government. Years later a story was written about her ultimate sacrifice. The story multiplied to where it ended up by her receiving a letter from the White House signed by President Reagan congratulating her for donating her hair for use in the Norden bombsight. She was also honored by the Colorado Aviation Historical Society for this auspicious honor. Note: On patent number 2,428,678, filed May 28, 1930, by Carl L. Norden and Theodore H. Barth, in the description phase, column 33 states in part, "Fore and aft line and athwartship crosslines are engraved on the reticle." A good case here is that if a story is repeated often enough it tends to become fact. Actually, the Bureau of Ordnance received many letters from patriotic women wanting to donate their hair to the military for war use. The author, as well as several others knowledgeable on the Norden bombsight, took issue on this. It is doubtful that this or any other book will correct this myth. Recently, the author was queried about this. In order not to offend the questioner, it was politely stated that we could find no evidence that spider silk or human hair was used in bombsights. We could not find any facts or data to support these myths.

A catchy name used in calling the bombsight was the "Blue Ox." In one magazine a bombardier is shown carrying a bombsight to an awaiting B-17 airplane naming the bombsight the "Blue Ox." How it arrived at this name remains a mystery. It was not an official name for the Norden sight, and it was never used in AAF and Navy correspondence. It may have been given this name by personnel in various combat theaters as a code word, but it appears to be tied in with the bombardier and the Flying Fortress. The author, after extensive research, was unable to find any facts on this.

A favorite was calling the bombsight "the Football." This could have some truth in it, as the Mark XI was somewhat shaped like a football (see Chapter 6). The configuration of the Mark XV does not look like or is it shaped like a football. In our research only once was the bombsight referred to by this name. This was in reviewing tens of thousands of pages of correspondence, letters, memos, reports, telephone conversations, etc. Normally, in official correspondence it was referred to and called by the Navy (Bureau of Ordnance and Bureau of Aeronautics) as the Mark 15, Mod-, and the Army Air Forces as the M-series bombsight. This name could have been used by military personnel for identification or security reasons. When information regarding the sight was encrypted (coded), no mention was made of a football (see Chapter 16). It may have been used, but it was not an official designation.

The producers of the television series, "Unsolved Mysteries" also produces a few times a year, "A Story Within a Story," and contacted three persons knowledgeable on the Norden sight, including the author, concerning the use of human hair for reticles (crosshairs). No information or disposition on this subject is known. There was no foundation or basis for this story concerning the Norden sight. The War Department advertised for blonde, female hair that had a soft texture. It was to be used in meteorological work. It was found that human hair was better than any substitute available. Prior to Pearl Harbor, most of the blonde hair came from Europe. With the war in Europe it was necessary to develop a domestic supply. It was used by the Army, Navy, and Weather Bureau for weather reporting.

Another name used was the "Cork and Jug." This one made some sense, as the crosstrail assembly fitted into the directional stabilizer. The "Cork" was the bombsight and the "Jug" was the Directional Stabilizer, hence the name. There may be other names concerning the sight, but the author is unaware of them.

In the late 1930s, and prior to Pearl Harbor, there were newspaper stories that somehow, over the years, the bombsight acquired some mystical powers. Some of the articles stated that a device had been perfected to give bombers greater accuracy and to protect them from anti-aircraft fire. This was in reference to the automatic pilot, which takes over the controls during the last few seconds during a bomb run. There was also a joke that the bombsight should be commissioned, and the bombardier could then be sent home, which he undoubtedly would have appreciated. These and other stories regarding the bombsight abound and made the regular rounds.

There is a somewhat controversial story concerning use of explosives on the bombsight when capture was imminent. Although this is not a myth, it is not truthful. Few, if any, have ever seen one installed on the bombsight. Extensive investigation did not reveal the kind of explosives that was used, or instructions on its installation. The British Government, through the Ambassador to the United States, suggested that in order to obtain the Norden bombsight, a destruct package could be installed on the instrument and would be open to inspection by the American Naval Attache in London.[5] The Navy Department, Bureau of Ordnance in Circular Letter No. V27-42, states in part: "Attention is also invited to Demolition Outfit Mark 5 Mod 1 for use with the Mark 15 Bombsight. Procurement of this item had been initiated, and instructions concerning its installation and use was to be provided at the time it was to be issued to the services."[6] However, a letter from the Naval Proving Ground under sub-heading of Demolition, states: "In connection with demolition, it was considered desirable (if such steps had not already been taken) to provide each sight with some form of demolition unit which would insure complete destruction of both sight and stabilizer."[7]

Discussions with many ex-bombardiers (Army Air Forces and Navy) revealed that they were not aware that a destruct package was being provided for the NBS. Stories continue to persist on this in the media.[8] It appears that if a destruct package was provided for the sight, the question remains as to who would install the device, where it would be placed, the method of detonation, and type of explosives. None of this information was found in the archives.

There is one item that has proved to be very elusive. Some magazines and articles state that a Lt. F. I. Entwhistle, USN, later Rear Admiral, was the co-inventor of the Mark XV bombsight. He was assigned to the Bureau of Ordnance in 1929. It was during this period that the Mark XV was being developed and tested. Some correspondence shows that he worked with Carl Norden and Theodore H. Barth on the development of this new sight when he was assigned to the Bureau of Ordnance. There was considerable activity between the Norden Company and the Bureau of Ordnance and Bureau of Aeronautics in engineering this new concept of a bombsight. In Sept. 1931 he was transferred from the Bureau of Ordnance to other assignments. Research of patents filed on the bombsight in May 1930 for the Mark XI and April 1932 for the Mark XV bombsight showed Mr. Norden and Mr. Barth as inventors, and no mention is made of a co-inventor. Perhaps military personnel were not allowed to patent their ideas or have their names associated with an invention. However, if he was actually a co-inventor he should be given credit for his efforts. Also, there were many early day naval officers that were involved with the Mark XI and XV who were never mentioned in the development of the Mark XI and XV.

In the early part of 1942 the Bombsight machine shop at McClellan Air Force Base was requested to design and construct a number of simple bombsights. The best recollection by the author was about 25 units, as it has been 56 years since the job was done. All design and manufacture was accomplished in the building by the author and fellow employees, since this was the most secure building on the base.

Chapter XX: Myths - Spider Webs, Human Hair, Etc. Etc.

No information was given, except that they were to be made as soon as possible. The finished device was a quadrant (quarter of a circle) made from 1/4" to 3/8" sheet metal, and the edge engraved from 0° to 90° and bolted on a larger metal base plate for easy installation in the airplane. Provisions were made so that the telescope could be locked at any desired position (degree), and was sufficiently tilted so that a target could be seen from the bombardier's compartment. The telescope was of normal power (2x power). The author inquired as to their intended use, but received no answer or explanation, only a stony silence. The B-25s that were used in the Tokyo, Japan, raid were stationed at McClellan Air Force Base outside of Sacramento, CA, prior to transfer to the aircraft carrier *Hornet* at the Alameda Naval Air Station. We are absolutely sure that the Norden bombsight was not used on this raid (see chapter XVI). It was interesting to watch the airplanes practicing takeoffs and landings. We could not understand at the time what they were trying to accomplish with this type of training. After the Tokyo raid it all became very clear. There are many who claim to have made the bombsights that were used by the Doolittle Raiders. Lt. Col. Doolittle (later General) was a frequent visitor in the Bombsight Building.[9] Perhaps, there were several types of bombsights made, and the best ones that suited their needs were used. Many attempts were made to determine if those manufactured in the author's shop were used, but no definite determination was ever made. This is one of the items that will never be known. In keeping with the author's philosophy of reporting true facts, unless there is actual proof, the item is labeled as unknown or with no actual proof.

Notes:

[1] Flying Magazine, July 1942.
[2] University of Rhode Island, Library Archives Division
[3] Carbon copy of letter sent to Mr. Carroll J. Watkins from John A. Bleecker, III, Staff Systems Engineer, Unisys Corp, dated Oct. 12, 1992.
[4] Honeywell, Inc. 100th Year Centennial Paper, page 21.
[5] Telegram to Marquess of Lothian, British Ambassador to the U.S., Dec. 15, 1939.
[6] Letter from Naval Proving Ground, Dahlgren, VA to Chief of Bureau of Ordnance, Subject: Mark 15 Bombsight Information on Handling of, dated Nov. 14, 1942 Nat. Archives.
[7] Navy Department, Bureau of Ordnance Circular Letter No. V-27-42, subject: Destruction of Bombsight Equipment When Capture is Imminent, dated Oct. 23, 1942, Nat. Archives.
[8] U. S. Naval Proving Ground, Dahlgren, VA to the Chief of Bureau of Ordnance, Subject: Information on Handling of Mark 15 Bombsights, dated Nov. 14, 1942 Nat Archives.
[9] From the authors recollections when he was assigned to the Bombsight Building from 1941-1949.

XXI

The Honeywell C-1 Automatic Pilot

The story of the Norden Bombsight cannot be complete without the C-1 Automatic Pilot, which was an integral part and played a very important role in the success of daylight precision bombing. Below are a few words about this device that was developed by Minneapolis Honeywell Regulator Company (now Honeywell, Inc.). The auto pilot was developed by the C.L. Norden Company in 1936. It was called the Stabilized Bombing Approach Equipment (SBAE), and was the forerunner of the auto pilot used in WWII. The AAF was experiencing considerable trouble with the Norden auto pilot. The AAF Materiel Command, Wright Field, Dayton, Ohio, asked Honeywell to look into the problem. The evolution of this famous Automatic Flight Control Equipment (AFCE) developed by Honeywell for the AAF resulted in the use of electronics in airplanes for the first time. The design, development, and production of the C-1 would require volumes. Those must have been hectic days. Through the courtesy and permission of Honeywell, Inc., the following information from their 100th year Centennial Paper will give the readers a short but interesting insight into its development.

Researchers Mr. Gille and Mr. Sigford developed a phase-discriminating amplifier, a new concept which could control direction and torque on a motor by providing instantaneous polarity and amplitude changes. The motor could now position dampers or valves for a more precise control of temperatures. Honeywell engineers quickly tried several applications of the concept, among them control of cabin temperature. In early 1941, Honeywell engineers (Mr. McGoldrick) accepted an invitation from the AAF Wright Field, Materiel Command at Dayton, Ohio, to demonstrate their electronic devices, and their engineers gave them their "dog and pony show," which involved the cabin temperature control demonstrator, remote position demonstrator, and some pressure switches. They received some "oohs" and

"Ahs," but that was it. They did not hear from the AAF Wright Field Photo Laboratory for over a month about using their remote positioner to keep the aerial camera level. The heavy Fairchild camera was located in crowded quarters in the airplane's tail and had to be positioned manually. In a few days, Honeywell engineers adapted a remote positioning device to level the aerial camera. The device was such a success that further demonstrations were requested by Photo personnel, and they called in some people from the Armament Laboratory. In May 1941, Honeywell personnel were called back to Wright Field (AAF) and asked if they could make an electronic remote controller. Honeywell stated that this could be done. The AAF Officers took the engineers into a basement vault where equipment measuring about four by six feet was covered by a white sheet. Mr. Whempner recalled what happened: "An AAF Officer just pulled back the edge of the sheet and asked, `Can you put some Potentiometers on these?' pointing to a servo motor and gyro. Again we said we could, so we were issued a blind specification which gave us a vague drawing of a control." What they had shown was the servo motor and the gyro of the automatic pilot for a bomber. Adding potentiometers or variable resistors to the gyro would allow the system to move the control surfaces smoothly. They were told that their system was to help the country's most advanced bomber, the B-17, or "Flying Fortress." As Mr. Gil Taylor later wrote, "Our existing system (remote position demonstrator) did not have a chance. Our pee-wee servo of some 15 watts total input was expected to move the massive rudder structure of a B-17B in flight." Undeterred, Honeywellers, in just 10 days, built a system large enough to handle the job. They were not yet told that the ultimate and most important function of the system was to provide a stable platform for the highly secret Norden Bombsight. The Norden auto pilot did not work well. The device caused airplanes to make exaggerated and jerky moves, and smooth moves were essential. The Norden Company's automatic pilot was connected to control surfaces with cables which ran the entire length of the airplane. Because of different co-efficients of expansion of metals used in aircraft construction, the bronze cables went slack in the very cold high altitudes, making accurate control impossible.

The development of a precision high altitude sight was of the highest priority for the AAF, and one of the most hush-hush projects of the war. Ken Covington, a B-17 bombardier who joined Honeywell after the war, said an accurate bombsight was a terrific advantage. Mr. Harrison, research director of the Brown Division of Honeywell, had designed and patented a self-balancing potentiometer in 1939 whose principles were exactly the same as those that the Honeywell team had developed. Both systems could transfer motion from one point to another. Mr. Harrison gave the Honeywell team details about adjusting the plane's wing flaps for steering. The C-1 needed an inverse feedback system which would signal a stopping position without overshooting. The best available system was a Leeds and Northrup patent, which Brown was using for potentiometers under license. The C-1 used this feedback system. Engineer Upton made the first demonstrator model, and Mr. Gille, Mr. Kutzler, Mr. Fedder, and Mr. Borrell took it to Wright Field, Dayton, Ohio, in early June 1941. AAF personnel attached it to the mockup, which was still behind the secret white sheets. Honeywell personnel were not allowed to watch the demonstration. But, after the system worked perfectly for five minutes, the AAF personnel removed it from the mock-up and asked the Honeywell team to install it right away on the pitch axis of a B-17B for a test flight. The use of this new electronic device on the auto pilot made it capable of performing over 300 flight corrections per minute, or 5 per second, unheard of up to that time. It was the best in existence up to that time and was used many years after WWII. Eventually, all auto pilots (Norden and Honeywell) were equipped with this new device.

The AAF test pilot did not have much confidence in this new system and was uneasy about it, Mr. Whempner recalled. The test pilot brought along a pair of bolt cutters and gave them to Mr. Borell. Working for five minutes on a mock-up was one thing, but having that contraption controlling his airplane was something else. He was definitely skeptical. Mr. Borell was put in the tail and was ordered that if the thing did not work he was to cut the cables immediately if something went wrong. The bolt cutters were not needed. The success of this first flight prompted the AAF to ask for a system to control all three axis of the airplane (pitch, roll, and yaw). Honeywell engineers used the basic Norden Automatic Pilot design, but exchanged its problem cables for an electronic system that gave excellent performance and reliability. The AFF designated the Honeywell system C-1 auto pilot, then issued a contract to perfect the system—quickly. The contract gave the Minneapolis Honeywell blanket permission to go ahead and spend money to develop the system. The C-1 system, for which Mr. Gille received the patent, was accepted by the AAF in October 1941, just five months after Honeywell received its original blind specifications. Many times engineers, model makers, and machinists would work all night so it would be flown to Wright Field for testing. In those days, jobs were not clearly defined, and all worked together to get the job done. When Honeywell engineers were at Wright Field and found something that needed to be changed on the auto pilot, they would phone that night from the hotel room and get someone back in Minneapolis to tell them exactly what it was that they wanted. Retired engineer Mr. Bower remembered, "One of the first test flights with the bombsight was Thanksgiving Day, 1941. I was so surprised when I called. After all, it was a holiday. This was before the war. The big airplane was loaded with feed sacks containing lime, which would burst on impact and scatter so it could be seen from the air. On the first run, the bombing was so accurate that the AAF personnel figured it must have been just luck. The plane was landed, reloaded with another set of feed sacks, and this entire load was delivered on target."

The AAF called for Honeywell to build 23 flight control system modification kits. At the time, Mr. Whempner recalled that the AAF had only a few dozen B-17 airplanes. In July, the AAF loaned Honeywell a B-17 to test the C-1. The first test flight of the entire autopilot was on this airplane in Minneapolis in August, 1941. Honeywell had to borrow pilots from Minneapolis based Northwest Airlines to make the test flight aboard the loaned B-17. When the AAF wanted to do its tests, however, another B-17 had to be flown in because Honeywell's plane was not equipped with the secret Norden bombsight.

A production contract was signed after only 60 hours of flight testing. Much of the wartime research for electronic projects was done in two "Ghost Rooms" at the main plant in Minneapolis. The rooms drew their unusual nicknames from the fact that many of the systems produced in them were covered by white sheets when transferred elsewhere. The Projects handled in these areas were so secret that employees in Ghost Room #1 did not know what was going on in Ghost Room #2 and vice versa. Later, both rooms were consolidated. Honeywell would go on to supply the AAF with over 35,000 C-1s for all United States and Allied heavy and medium bombers.

The following is an interesting item, "As a publicity stunt at a New York City trade show, we set up a gyro to measure the rate of movement of the Empire State Building where the show was taking place. People were really impressed — all but the manager of the building. He told us to get that "junk" out of there. `With this kind of publicity, we'll lose all our tenants.'" Reported by George Smith, Minneapolis, 1939-1974 and reprinted from the 100th year centennial of Honeywell, Inc. This must have provoked many a laugh.

The Minneapolis Honeywell system consisted of the following components: 1. Autopilot Control Panel; 2. Servo Unit (3 each per plane); 3. Vertical Flight Gyro; 4. Autopilot Rotary Inverter; 5. Autopilot Amplifier; 6. Directional Stabilizer; 7. Directional Panel; 8. Directional Arm Lock; 9. Remote Turn Control and 10. Bombardier's Turn Knob. Later, a Formation Stick and Steering Motor were added.

This first hand account on the C-1 auto pilot was graciously provided by Mr. Sherman Booen, about his experience as a Technical Representative (affectionately known as a "Tech Rep") with the 12th AF in Africa and 15th Air Force in Italy. Mr. Booen was a Field Engineer for Honeywell from 1942 to 1945. The following is his account:

The C-1 autopilot consisted of a flight gyro that controlled the aileron and elevator axis, while the directional axis (rudder) was controlled by the stabilizer that was part of the bomb sight. The basic circuit caused the servo motor, connected to the flight cables, to run until the circuit was balanced. All was operated on either 28 volts or 110/400 cycles. The flight gyro turned at very high (app. 20,000 RPM) speed, and was kept in a vertical position by a set of "bales," surrounding the top of the gyro (like a top) that caused the gyro to react at 90 degrees, and eventually back to vertical.

The primary object of the C-1 was precision high altitude bombing. The autopilot was connected to the bomb sight, via the stabilizer, and the airplane followed the corrections of the bomb sight, without under or over shoot. In otherwords, the bombardier was guiding the aircraft when on the bombing run, and it was very precise.

My job in the field was to assist military personnel in maintaining the equipment, instructing air crews in the use of the C-1 in accurate bombing, and in normal auto pilot cruising. As the war progressed, a lot of my work overseas was instructing crews in the use of keeping the aircraft flying when normal control cables were severed. The servos were on the end of the cable, and the electrical wires ran along a different part of the fuselage, so if manual control cables were severed, it was sometimes possible to maintain control of the elevator, rudder, and ailerons via the C-1, providing it was intact.

Numerous aircraft and crews were saved from destruction when this situation occurred, and the crew was able to land safely. I personally witnessed several such instances. My job on the overseas assignment was also to instruct crews in the use of the system on bombing missions. I would fly with the lead aircraft a day before the mission to assure that the C-1 and the Bomb sight were working properly. We would fly at 30,000 feet, full load, and I would trim the system up to my satisfaction...Aircraft at 30,000 feet shrink a few inches, and the C-1 was adjusted in sensitivity and amount of control...all of this was done on a 12 knob panel that was the heart of the C-1...it even had an "up elevator" control to be adjusted in banks...

I made a lot of test flights in Italy reference landing a B-17 on autopilot, when the control cables were disabled...It was pretty tricky, especially with a tail wheel airplane, but it worked. I was called to the control tower for several aircraft that came back with wounded, and cables destroyed...and was able to help them make a good landing. Some of the C-1 units, towards the end of the war, had a "formation stick," so to speak, in other words, he flew the C-1 and the C-1 flew the airplane. It worked very well, and took a lot of strain off the pilots. The C-1 was standard equipment in the AT-11, B-17, B-24, and B-29.

Many lead pilots did not like the idea of the bombardier guiding the airplane on a bomb run, and preferred to follow the PDI (pilot direction indicator), which was not as accurate. But other experienced pilots just turned it over to the bomb sight, and bombing was very accurate.

While in Italy, we built a C-1 autopilot into a link trainer, and it served as a simulator. All crews were required to get a check out in that system, so we had classes running all day each day. It worked well, but was a real rube goldberg...but it did work, and saved a lot of money, as we did not have to have airplanes.

With Norden connected to the C-1, it was the most accurate high altitude bombing. There was a saying going around that a bombardier could hit a pickle barrel from 30,000 feet. Well, it was not that accurate. You can't even see a pickle barrel from 30,000 feet, but I have made runs and put the practice bombs within 500 feet of dead center, which is pretty good.

When the flight gyro was turning up, it produced a lot of kinetic energy. There is a story that went around about the guy who got it turning up full speed, put it in a suitcase, and asked the hotel man to carry the baggage into the hotel. When he tried to turn, the suitcase went the other way—I don't know whether that is true or not.

The author has heard this and other stories concerning the gyro's behavior.

XXII

Cancellation of the Sperry Sight

The enormous task of supporting tens of thousands of several different types of bombsights used by the AAF was creating a monumental logistics and maintenance problem. To maintain three different types of sights in the combat field was not only uneconomical, but created a massive and difficult supply situation. The Navy did not have this problem since they only used the Norden bombsight. Due to insufficient production of Norden M-Series bombsights, the outlook for this equipment from the Navy was very poor. The AAF decided to use the Sperry equipment in some of their heavy airplanes (B-24) since the Norden was not available. This resulted in the setting up of the B-24E airplane program and all models thereafter with the Sperry S-1 bombsight and the A-5 automatic pilot, together with the B-32 program and such airplanes as would be required by the AAF Training Command.

Troubles experienced with the first Sperry S-1 sight and A-5 auto pilot caused a retrenchment of the program, deleting the B-32 airplane. The M- Series production was still far from adequate to meet the AAF requirements, and the Navy could not insure suitable production. The AAF had previously established their requirements with the Navy for the bombsight. A review of the bombsight facilities was held by the Materiel Command, AAF, Bureau of Ordnance and WPB, which resulted in the cancellation or transfer to the Navy of all Army Air Forces facilities, with the exception of the Victor Company. The current availability of sights was not affected, because production would not peak until the middle of 1944 or later. This left only the Norden and Sperry sight, which the AAF was dependent on.[1]

To obtain rapid and conclusive data as to suitability, a small group of airplanes was allocated to the Second Air Force for service test purposes between the Norden and Sperry bombsights. The first tests that were reported by the 2nd Air Force during the period of April-July 1943 indicated that the S-1 sight and A-5 auto pilot were operationally unsatisfactory, and the Second Air force issued orders that this equipment would not be used. The Assistant Chief of the Air Staff, Training, was disturbed by the possible elimination of Sperry equipment since arrangements were made to equip 200 AT-11 airplanes with the Sperry S-1 and A-5 auto pilot. Half of the order was received, and the rest was in process of delivery. With the possible abandonment of Sperry equipment the airplanes were useless to the bombardier schools until modified for the Norden sight.[2] In August 1943, the Chief of The Air Staff directed that a Board of Officers composed from various Commands and high level departmental heads to consider the unsatisfactory bombsight situation. In keeping with this directive, the Chief of the Army Air Forces appointed a Board of Officers to review all known factors affecting the operational suitability, and the future program for the Norden M-Series bombsight, associated Minneapolis-Honeywell C-1 automatic pilot, and the Sperry S-1 series bombsight and associated A-5 automatic pilot, which were the only precision bombing equipment available to the Air Force. The Officers that were appointed were specialists in their field.[3]

The Board of Officers met in Sept. 1943 to consider precision bombsights and associated automatic pilot equipment and make recommendations as to standardization. The task of conducting comparison tests and analysis was accomplished by the 2nd Air force, AAF, Office of the Chief of the Air Staff, Operations, Materiel Maintenance, AF Proving Ground, and testimony from expert witnesses from the Materiel Command, Wright Field, Consultant to the Air Forces, Bombing Project Officer, Production Engineering Wright Field, Training Command and representatives from the various Combat Theaters, etc.[4]

This high level Board reviewed many documents and studies that were submitted to them by many organizations using all types of bombsights. Of the nine Exhibits that were furnished the Officers Board, only one was available for review in the National Archives. As is always the case, the attachments were invariably withdrawn, and the pertinent data in these documents was not found. The missing documents covered the entire range of bombing: Chronological Progress Report: A-5 auto pilot; Requirements vs Availability: Bombsights; Technical Report: Modification of the S-series Bombsight; Procurement Data-S-1 Bombsights; Second Air Force Bombing Records; Adaptability for Combat Bombing: Comparison, Accelerated Service Test: A-5 Auto Pilot; Maintenance Reports: Bombsights; and Comparison of Accuracy: Bombsights. The entire range of both bombsights was reviewed.

On Aug. 20, 1943, the Assistant Chief of The Air Staff-Operations, Equipment, and Requirements advised the Chief of The Air Staff that analysis of the bombing accuracy and adaptability charts showed that Norden bombing equipment was more accurate and suitable than Sperry equipment for combat operations. Many problems of maintenance, supply, training, and logistics were encountered due to the two types of equipment. Accordingly, the President of the Board of Officers met with the complete Board on 15 Sept. 1943 to review the data that had been collected on the two systems. The conclusions were that although testing was not made under controlled conditions and the results were not conclusive, in comparing the two sights it did indicate the accuracy of the two systems: Inherent design—similar accuracy.[5] Long bombing runs, manual control—were considered similar, although some tests showed Norden superiority, and long bomb runs on auto pilot seemed similar, although some data showed Norden superiority. Experience on Sperry automatic pilot was very limited due to installation and functional difficulties in the A-5 auto pilot. On short runs no comparison was available due to the lack of necessary design features in the Sperry units, and even with proposed improvement would not be as accurate as the Norden. Sperry equipment was in the U. K., but no combat information was reported by the squadrons.[6] In the areas of visibility, target search capabilities, and orientation of the line of sight the S-1 was definitely inferior to the Norden. Light transmission through the S-1 sight was definitely less than the Norden due to the inherent design of the optics, which in turn would require redesign of the sight itself. This feature was especially important under conditions of poor visibility, especially in the European and Pacific areas, where smoke and haze often restricted clear view of the target. The operational design features of the Norden were far superior at the present, but the Sperry equipment was capable of being improved to a suitable standard in most of the features. However, the improved versions would not be available in adequate quantities in the combat theaters for at least six months. In the areas of serviceability and reliability the Norden sight and the C-1 auto pilot had fewer malfunctions and required less maintenance per operating hour than the Sperry equipment. The training of Sperry bombardiers and Sperry maintenance personnel for all echelons was inadequate in both quantity and quality, largely due to the lack of sufficient and suitable Sperry equipment for training. The supply and distribution of spare parts and maintenance equipment was inadequate and largely underdeveloped.[7] It was estimated that it would take about five years before reaching self-sufficiency and overall experience would reach the level of the Norden equipment.[8] Regarding voltage requirements in both sensitivity to small variations in voltage and maintenance requirements of the airplane electrical system, the Norden sight and the C-1 auto pilot was superior. The voltage and current requirements were extremely critical, and any small variation resulted in malfunction. This deficiency was critical in combat due to the operation of gun turrets and limited maintenance under combat conditions.[9] It was brought out that the Sperry optical system admitted approximately twenty percent less light than the Norden. This excessive loss of light would be a severe handicap to bombing accuracy in the European and Pacific Theaters of operation. Accuracy in night bombing with the Sperry was severely handicapped due to this deficiency.[10] There were several modifications (9 in all) that were essential in order to make the Sperry bombsight and auto pilot suitable for combat use, such as tangent scale for sighting angle and dropping angle, degree scales for sighting angle and dropping angle, trigger lock trail angle and fall of time lock, automatic cross trail setting, improved visibility through optical system, protection of the optical system from oil, foreign particles and moisture, installation of a locking mechanism on the front window of the sight to keep dirt and moisture out of the sight and permit an easy removal of the window when fogging occurs, and provide suitable turn control for the pilot.[11]

The representatives of the Sperry Company and the AAF Materiel Command, Wright Field estimated that these changes would take from four to six months to be developed before being installed in the bombsights presently in production. An additional four to six months would be required for training bombardier and maintenance personnel. It would take approximately one year before it would equal the Norden equipment in combat.[12]

The problems of programming and maintaining an adequate flow of the correct proportion of personnel and materiel for both systems, plus the duplication of personnel in production, maintenance, operation, and spares was wasteful and undesirable, even if the two systems were comparable in every necessary feature.[13] With the deletion of the Sperry, the problem still remained as to the adequacy of suitable sources for development and production of Norden bombsights and C-1 auto pilot to forestall any deficiencies, since the Navy controlled approximately 90% of available equipment.[14] A memorandum to the Air Staff AAF, from the Assistant Chief of the Air Staff, Operations, outlined the plan to adopt one bombsight and auto pilot for combat operations. The relative accuracy and adaptability of the Sperry and the Norden bombing equipment was investigated in order to determine the advantages and disadvantages of each type for combat operations. Analysis of the bombing results obtained by the 2nd Air Force during the period from April-July 1943 were shown in the exhibits. Again, all of the Exhibits except one were withdrawn, and a detailed comparison could not be made. On the basis of information obtained in the examined files, the recommendation was to have one type of bombsight and automatic pilot adopted for use in all heavy bombardment type aircraft, and that the M-Series bombsight and C-1 auto pilot were to be adopted as standard equipment. However, this presented problems of obtaining sufficient quantities of Norden sights to fill the gap without Sperry units. This would require readjustment by increasing Norden production from Naval facilities and also increasing AAF allocations. Sperry production would stop when a sufficient number of S-1 sights and A-5 auto pilots were accumulated for the B-24 airplanes that could not be Norden equipped. The change over to one type of bombing equipment was to occur at the earliest practicable date.[15] Every effort was made to improve the Sperry bombsight and the A-5 auto pilot so that adequate quantities of the improved version would be procured in order that maximum bombing effort would be maintained prior to standardization of Norden equipment.

In Oct. 1943 the AAF took action to have all B-24 airplanes engineered for installation of the M-Series sights and C-1 auto pilots. Also, until such time that airplanes were coming off the production lines suitable for installation of Norden equipment, the Sperry S-1 sight and A-5 auto pilot would continue to be installed. Sperry contracts for sights and auto pilot over and above the equipment required for this installation were canceled. It was also intended that Sperry equipment was not to be replaced on airplanes already delivered. A careful review was necessary of the Norden equipment production to ensure the availability for installation in the B-24 type airplanes (E and H model) as soon as they were suitable for the changeover.[16] In addition to the installation of this equipment, the Sperry Company established a program for sufficient spare parts for their units presently in use and that Sperry units continued to be manufactured in order to support airplanes in training and combat theaters.

The Sperry Gyroscope Company, on being notified that their contract was to be canceled, sent a delegation to Washington to visit each Board Member for the purpose of discussing the Board of Officers' proceedings and decisions. They also contacted the AAF Chief of The Air Staff and stated that a fair test had not been given to the equipment incorporating all of the latest changes. The result was that Hqs AAF, on 23 Oct. 1943, directed that cancellation of Sperry contracts was to be held in abeyance pending complete reports from an operational theater based on the latest types of Sperry equipment. Many changes had been accomplished since they had been reviewed by the Board of Officers. The AAF Materiel Command arranged with the International Business Machine Company to build 150 of the S-1 bombsight with all of the recommended changes as far as they were able, and ten were to be available as quickly as possible for transfer to the 8th Air Force in the U. K. This was accomplished, and the sights were delivered by air, together with officer personnel who were fully acquainted with their operation. On 21 Nov. 1943 a cable report from the United Kingdom stated that all of the latest modifications were a marked improvement over the original equipment, but the troubles were basic in the bombsight and automatic pilot. These results confirmed the previous recommendations to standardize the Norden system. Upon receipt of the cable reports from the U.K., the Deputy Chief of The Air Staff, AAF issued instructions to AMC Materiel Command to proceed immediately with the cancellation of the Sperry S-1 and A-5 auto pilot as directed by the Board of Officers.[17] A meeting was held in Oct. 1943 to review bombsight facilities and production schedules versus Army and Navy requirements. The meeting was attended by the AAF, Navy, and the War Production Board of high level personnel. The prime purpose of this meeting was to resummarize requirements of both services in comparison with existing facilities in view of the Sperry equipment cancellation, which required an increase in facilities for the additional manufacture of M-Series bombing equipment and C-1 auto pilot, and to explore the possibility of utilizing the facilities vacated by the Sperry

Chapter XXII: Cancellation of the Sperry Sight

program for other high priority war work.[18] The Air Service Command took the necessary action to insure that Norden equipment was available for installation in B-24 airplanes at the time these airplanes were delivered from production.

The question remained as to what should be done with all of the Sperry equipment. The plan was for any surplus Sperry sights and auto pilot were to be used in Lend Lease bombardment aircraft instead of Norden equipment. The Sperry Company would be out of production by May 1944. Over the years the AAF had invested tens of millions of dollars for Sperry material. The AAF had procured 6,719 Sperry bombsights and A-5 auto pilots. Of this number, 5,140 were installed in airplanes, leaving a balance of 1,579 surplus units for use as spares, replacements, and for ground training. It was estimated that without actual trail installation it would take approximately 250-350 manhours per airplane to remove the Sperry sight and auto pilot and replace them with Norden material. If the kits were made in the United States and forwarded to the operating units, the manhours required would be less, however, the AAF Air Service Command felt that it would be too big a job for the operating squadrons to perform this work. With facilities in the United States, the best the AAF could do in converting the B-24s in training units would be from 30-50 airplanes per month. The kits required the procurement of a considerable quantity of brackets, fabrication of a new electrical harness, bombsight mounts, and many other sundry items. To obtain and assemble these parts without interrupting other activities required time, and the first kits would not be available for three to five months. The 8th Air Force had the bulk of the Sperry equipped airplanes, were using "bombleader tactics," and whenever possible assigning a Norden equipped airplane as the lead airplane. It was not warranted for all the planes to be equipped with the Norden system.[19]

The AAF directed that the program of modifying as expeditiously as possible all Sperry equipped airplanes in the Continental U.S. for training (approximately 500 B-24Hs and 200 AT-11s). No action was taken to modify B-24Hs in the 8th and 15th Air Force.[20] The installation of the conversion kits for the changeover from Sperry to Norden equipment was not intended to be done in the combat theaters. Sperry equipped airplanes were to be replaced as soon as possible. They were brought back to the U.S. for use in four engine and gunnery schools, and in training bombardiers. The conversion of these airplanes to Norden equipment was accomplished at the AMC depots, which were far more efficient than attempting to convert them in the theaters by the use of conversion kits.[21] By March 1944, M-Series production was sufficient to sustain installation of Norden bombsights in production aircraft, and approximately 200 kits were available per month. It was directed that the estimated 200 kits per month were divided between training and overseas allocated Sperry equipped B-24 airplanes.[22]

The Deputy Chief of The Air Staff felt that the surplus Sperry sights could be given with Russians B-25s, and they could make their own installation. However, on investigation it was discovered that attrition, spare parts, etc., would not leave sufficient sights for use in the existing fleet of aircraft being deployed in the combat theaters. There was a surplus of the D-8 sights (Estoppey), and the use of the T-1 (British Mark 14), which was available for installation in these airplanes. There were enough sights available to fulfill these requirements for the Russians. To re-engineer and install Sperry bombsight equipment in the Russian B-25 airplanes entailed a considerable amount of work and installation manhours. All principal contracts (except for some repair and rework contracts) on the Sperry bombsight were completed by May 1944. By the end of May 1944 all bomber production was swung over to Norden installation.[23]

Note: The author did not attempt to determine which bombsight was the best. Information and comparison data on the Norden and Sperry equipment was extracted from the official files located in the National Archives. The cancellation of the Sperry equipment was an emotional one. The primary purpose of a repair shop was to repair bombsights and related equipment, regardless of make or model. In repairing, assembling, testing, machining, etc., the process was the same.

Notes:

[1] Memorandum for Major General O. P. Echols, from Assistant Chief of The Air Staff, Materiel, Maintenance and Distribution Div, Subject: Bombsight History Leading Up to Decision to Cancel Sperry S-1 and A-5 Bombing Equipment, dated 25 Nov. 1943, Nat Archives.
[2] Routing and Record Sheet, Headquarters AAF, Subject: Bombardier Training on Sperry Bombsight, dated 13 Aug. 1943, Nat. Archives.
[3] Board of Officers meeting at the call of the President on 15 Sept. 1943, No Subject, Nat. Archives, MS.
[4] Ibid, page 1.
[5] Ibid, page 2
[6] Ibid, page 2
[7] Ibid, page 3
[8] Ibid, page 3
[9] Ibid, page 2
[10] War Department, Hqs Army Forces, Memorandum for the Chief of The Air Force, Subject: Bombsights, dated 20 Aug. 1943, Nat. Archives, MS
[11] Ibid, page 2
[12] Ibid, page 2
[13] Letter: Bombing Accuracy 2nd Air Force (Exhibit J), Memorandum to Colonel Garland: Sperry Bombsight (Exhibit K), date 15 Sept. 1943, Nat. Archives, MS
[14] Ibid, page 2
[15] CTI-586, Addendum No. 8, Installation and Allocation of Bombsights, Production Division, 15 Oct. 1943, Nat. Archives, MS
[16] Ibid, page 2
[17] Memo for Maj Gen O. P. Nickols, Subject: History Leading Up To Decision To Cancel Sperry S-1 and A-5 Bombing Equipment, dated 25 Nov. 1943, Nat Archives.
[18] Bombsight Facilities and Requirements Meeting 13 and 14 Oct. 1943, with Chairman, Industrial Facility Committee, War Production Board, dated 10 Oct. 1943, Nat. Archives.
[19] Letter from Deputy Chief of Staff, Subject: Possibility of Putting the Sperry S-1 bombsight in Russian B-25's to the Chief of The Air Staff, dated 14 Feb. 1944, Frame 0507 Nat. Archives.
[20] Letter from Assistant Chief of The Air Staff, Materiel to Assistant Chief of The Air Staff, Operation, Subject: Possibility of Putting the Sperry S-1 in Russian B-25's, dated 9 Mar. 1944, Frame No. 0503, Nat. Archives.
[21] Letter from Assistant Chief of The Air Staff-Materiel and Maintenance to Assistant Chief of Staff-Operations, Commitments and Requirements, Subject: Possibility of Putting the Sperry S-1 bombsight in Russian B-25's dated 3 Mar. 1944, Frame No 0505, Nat. Archives.
[22] Letter from Assistant Chief of The Air Staff to Chief of The Air Staff, Subject: Possibility of Putting the Sperry S-1 Bombsight in Russian B-25's, dated 6 Mar. 1944, Frame No. 0504, Nat. Archives.
[23] Memorandum for the Chief of The Air Staff from the Deputy Chief of The Air Staff- Materiel, Subject: Possibility of Putting the Sperry Bombsight in Russian B-25's, no date, Frame 0509, Nat. Archives.

XXIII

Foreign Interests

As to be expected, foreign governments would be inevitably drawn to the bombsight. They were the British, French, Germans, and the Japanese. Nazi Germany took a different approach. This section is discussed in another chapter. The French were eager to trade their new machine gun for it. The British Royal Air Force (RAF) was vitally interested in obtaining the complete system, including the automatic pilot. In April of 1939, many representatives were invited to attend an Air Demonstration at the United States Infantry School Fort Benning, Columbus, Georgia, sponsored by the Army Air Corps. Those in attendance included the RAF Attaché. The air bombardment part consisted of high altitude precision daylight bombing. In this demonstration, four B-17s were to be used. A target was drawn on the ground in the shape of a "battleship." It was 600 feet long and 105 feet wide at the center. Horizontal bombing was to be done from 12,000 to 15,000 feet under excellent weather conditions. The first B-17 dropped 300 pound bombs before any of the spectators even saw the airplane. All six bombs hit the center of the target. The second B-17 was spotted before the three 600 pound bombs hit the target. This was not as good as the first attack, as two bombs hit the bow, with the other missing the target. The third B-17, carrying two 1100 pound bombs, hit the center of the deck, one just aft and the other forward of the mast. The fourth B-17 was carrying a 2000 pound bomb. The day was so clear that the bomb could be seen leaving the bomb rack at 12,000 feet and could be followed to the point of impact. It hit forward of the bow, and was sufficient to do major damage to the ship. This was indeed remarkable accuracy in bombing.

The next part of the air show was bombardment tactics. This demonstration was to be conducted by 12 B-18-A airplanes. Each B-18-A carried twelve 100 pound high explosive bombs. The target was a square 600 yards by 600 yards. The airplanes approached the target at 10,000 feet in a sub-formation of three airplanes about three miles apart. The leader of each formation was the bombardier, who dropped the first salvo, the two airplanes releasing their bombs when the first plane dropped his. The bombs, with one exception, all fell within the marked area. These were excellent results for this type of bombing. It was also a remarkable display of accurate bombing from high altitudes. All airplanes were using the Norden Bombsight.[1]

With the war in Europe caused by Hitler, the British wanted the bombsight for their bombers. There was much correspondence from the British Embassy Air Attaché in Washington, D.C., to the Air Ministry in London about the Norden Bombsight. It even went to such extremes that the British Ambassador had lunch with President Roosevelt. At that time, he discussed the subject of the sight with the President. The British were willing to exchange some of their equipment. The President stated that he would give the matter further consideration.[2] The RAF was willing to devise a destruct package to fit on the bombsight so that it would not fall into enemy hands. They already had a destroying device for their airplanes, and it could readily be adapted to the sight without much trouble.

The situation in the United Kingdom had grown so desperate that the Prime Minister of Great Britain appealed personally to the President of the United States. Normally, this is not done, as such requests are forwarded through the Foreign Office. The full text of the letter to the United States and the State Department reply is shown in Plate #1 and Plate #2.[3]

The RAF also tried to use the Norden Bombsight in the Mosquito Mark XX in 1944. Many bombing flights were made with varying results. However, it was found that the sights could not be used with this airplane for several reasons, one being that the Mosquito was somewhat unstable, problems with the sight gyro, etc. Bombing was very poor and inconclusive. Under these conditions, it was decided not to use it. It is not known why such poor results were achieved. The Norden sight required a long runup, and this required weather conditions of little or no clouds or haze. Good weather conditions in the United Kingdom are not found in the fall or winter.

In December 1939, the British Ambassador asked President Roosevelt, again, if it was possible to arrange for the exchange of the Norden Sight for ASDIC (submarine detection device), and stated that the self-destroying device would be open to inspection by the American Naval Attaché in London. The President said that he would look once more into the matter, but evidently regarded the self-destroying device as vital. He felt that it would be safer if the Royal Navy did not have the sight so that the enemy could not get it. The President felt the risk was too great that the sight could fall into enemy hands.[4]

In April 1951, the Norden Bombsight was supplied as part of the B-29 airplane (known to the RAF as the Washingtons). They were being evaluated by the RAF, and they found that they could not be used for off-set bombing. It was necessary to modify the airplane to correct the problem. The modifications were to be in two phases: 1) Modification of the cross trail slide and bracket with manual settings, 2) the introduction of a new computer, which would make heading changes either manually of automatically. Trails were conducted and were considered sufficiently satisfactory to justify continuation, though only 15 bombs were dropped. However, the use of radar assisted sights was now becoming standard equipment in bombing, while the visual sights were less used and finally phased out of the inventory.[5]

Notes:

[1] Letter British Air Commission in Washington to Ministry of Aircraft Production, Subject: Norden Bombsight and Auto Pilot, April 18, 1939.
[2] Letter from Air Attaché, British Embassy to the Director of Air Intelligence Air Ministry, Subject: U.S. Bombsight and Platform, dated December 20, 1939, No. 1397.
[3] Reprinted from Franklin D. Roosevelt and Foreign Affairs. Second Series: January 1937-August 1939 edited by Donald B. Schewe by permission of University Publications of America. Plate #1.
[4] Telegram from the Marquess of Lothian (Washington) concerning the Norden Bombsight, dated December 14, 1939, No. 860.
[5] Extract from File AIR 14/3965, Modifications to the Norden Bombsight by the RAF.

This page is unclassified

G 2/2060-1130/155

WAR DEPARTMENT
War Department General Staff
Military Intelligence Division, G.-?
Washington, D.C.

March 8, 1940
MEMORANDUM TO THE CHIEF OF THE AIR CORPS
Subject: British Attempt to Obtain Norden Bomb Sight.

The following information is extracted from Report No. 40-493, Military Attaché, London, February 12, 1940, subject, Comments on Current Events No. 94:

At luncheon yesterday, in the country, in the presence of Captain Kirk, U.S.N. Naval Attaché, and Major McDonald, a very attractive young man named Congrieve, who has a civil post in the Air Ministry, brought up the bomb sight question again. It was rather startling to hear from him that one of the high members of the Air Staff had quite seriously instructed him that as he has a most attractive American wife (she is), and as he himself has a pleasing personality, he should see what he could do where others had failed in getting the Norden bombsight released to Great Britain.

For the Acting A.C. of S., G-2

Lieut. Colonel, General Staff
Executive Officer, G-2

10 Downing Street,
Whitehall.

25th August, 1939.

My dear Mr. President,

    The Secretary of State for Air informs me that the United States Navy Department have developed a new type of automatic bombsight known as the Norden bombsight, and I understand that this sight (together perhaps with a similar development of the United States Army Air Corps) is the most efficient instrument of its kind in existence. We are therefore most anxious to obtain details of the sight and have inquired urgently through our Air Attache in Washington whether they can be given to us. For reasons which I can readily imagine the Air Authorities of the United States have not felt able to accede to our request.

    In normal times I should not make a direct approach to you in such a matter, but in the present grave situation I venture to ask you whether you could help us to obtain the information we desire. I make this urgent personal request to you because Great Britain today faces the possibility of entering on a tremendous struggle, confronted as she is with a challenge to her fundamental values and ideals. Moreover, I believe they are values which our two countries share in common, and I am convinced that if there is a certainty, it is that our two countries will never go to war with one another.

    Should the war which threatens break out, my advisers tell me that we would obtain a greater immediate increase in our effective power if we had the Norden bombsight at our disposal than by any other means we can foresee. Air power is, of course, a relatively new weapon which is so far untried on a large scale; there is the danger of unrestricted air attack which we for our part would never initiate. I am however most anxious to do all in my power to lessen the practical difficulties which may arise in operations even against legitimate military targets, and I feel that in air bombardment accuracy and humanity really go together. For this reason again I am certain that you would render the greatest service if you could enable us to make use of the magnificent apparatus which your Services have developed.

    I need hardly say how grateful I shall be if you can see your way to help us.

                    Yours sincerely,
                    Neville Chamberlain.

August 31, 1939.

My Dear Mr. Prime Minister;

I have been very glad indeed to receive your letter of August 25 and to hear from you directly with regard to the question set forth so clearly and so movingly therein.

The initial survey which I have had made of the situation and which I have myself studied very carefully leads me to the conclusion that under the existing legislation of this Government the request you make could not be granted unless the sight desired by the British Government were made available to all other governments at the same time it was made available to Great Britain. This clearly would not be in the interest of the United States, nor for that matter, I believe you will agree, in the interest of Great Britain.

I am sending you this preliminary reply to your letter because of my knowledge that you desire some indication from me in response to your request at the earliest moment. I may assure you, however, that I shall give continued consideration to the request you have made and to the possibility of acceding to it either under present conditions or under such new conditions as may arise, and that I shall write you again upon this subject as soon as a final decision can be reached.

Believe me

    Yours very sincerely,

The Right Honorable
    Neville Chamberlain,
        P.C., F.R.S., M.P.,
            Prime Minister,
                London.

XXIV

Compromise

Over the years the NBS had acquired a high profile. It was well known for its bombing accuracy and was fair game for espionage. Although it had been classified secret, the AAF was prone to advertise its capabilities. Of course, it was never placed on display, nor were any photos, performance data, statistics, specifications, etc., ever released until early 1944, when so many of our bombers had been shot down. There was no question that the Nazis had acquired many intact sights.

When C.L. Norden started his company, he hired many old world mechanics. Some of the employees where the author worked attended the Norden Bombsight school and can verify that many of the personnel were of German extraction. At that time many German craftsman had migrated to the United States for work and were quickly hired by industry for their expertise. One of these was Hermann Lang. Lang went to work for Norden in the middle 1930s and worked up to a position of trust. He was a draftsman and inspector and knew the bombsight well. The U.S. Government was aware of the espionage attempts being made by the Nazis on American defense industry secrets. The Federal Bureau of Investigation (FBI) had successfully penetrated the ranks of the major Nazi espionage ring in the U.S. On June 30, 1941, the FBI seized 29 persons on charges of being spies for a foreign government (Plate #1-14), of which Hermann Lang was one of them. This was later increased to 33 co-conspirators. Lang had become a naturalized American citizen and was given good jobs with the Norden Company. The spying activities started long before Lang was arrested. A Federal Grand Jury was impaneled in Brooklyn, New York, to hear the charges. The case was presented to them by government prosecutors beginning in late June and early July 1941 to hear testimony against the suspected spies. Within two weeks, the Federal Grand Jury returned indictments on all of the suspected 33 spies on charges of cooperating with the government of the

Third Reich, Germany, to obtain defense military information. All were immediately arraigned in Federal Court and charged with espionage against the United States.

The jury was selected in less than a day and were sworn in on the 3rd of September, 1941.[1] The government charged the defendants with conspiracy and a crime of high magnitude in secretly and covertly stealing defense information.[2] It was also brought out that government witnesses William G. Sebold and Walter Kipken, employed by Bureau investigators to assist in the investigation, were in the same position as Government agents themselves and were not accomplices in the conspiracy.[3] On October 16, 1941, the Secretary of War was quoted as having said to the press that he had no knowledge to believe that the Norden bombsight was in the possession of the Nazis. However, it was understood by the Government and defendant's counsel that this stipulation did not concede the materiality, relevancy, or competency of the foregoing quotation for the purpose of this case.[4] He was subpoenaed to testify, but there is no record that he ever appeared in court. There was some apprehension that Lang would not receive a fair trial because of his German extraction and the friction between the German Bund Organization and American citizens and other organizations.

Review of the court transcript shows that among the overt acts, mail drops were used, which are common with spy activities. Lang, in June 1938, met and conferred with Nikolaus Ritter (Master spy) in Hamburg, Germany.[5] Through radio messages that were intercepted by the FBI, it was alleged that 10,000 marks were deposited to the credit of Lang's wife in Germany.[6] Lang denied that he stole any part of the bombsight (Plate #2-15). Facts proved otherwise. It was felt that Lang did not steal the most vial parts of the bombsight, which was the rate end. The defense prepared a list of 155 questions and answers. On December 12, 1941, the Court charged the jury that war existed with Japan, a few days later with Germany, and that all testimony was concluded on December 2, 1941, that this case was to be decided as though these activities had not taken place, and the main question was that agents of the German Reich obtained and received information, of military and naval value, that could be used against military and civilian personnel. He also took to task the defense counsel about the removal of two questions from the list, thereby reducing it to 153.[7] The Government had requested that these two questions be deleted, as they were prejudicial to the government. It must be remembered that the Norden bombsight was to be handled with the "utmost secrecy" at that time. The defense made such a commotion about this deletion in such a very loud tone and impudence that was repeated throughout the trial that the judge spoke rather pointedly and with emphasis about these proceedings. Revealing these two items involving features of the bombsight would be advantageous to Germany, but prejudicial to the United States.[8]

The prosecution presented a well prepared and documented case. The Court's charge was a lengthy summary, too long to be repeated here. It brought out the damage done to the USA in the disclosing of secret data to a foreign government. It also took to task the grandstanding of defense counsel, which was conducted throughout the trial. On December 13, 1941, the jury found Lang guilty, as well as 14 other co-conspirators. Sentence was pronounced on January 2, 1942.[9] On count 1, Lang was convicted of criminal conspiracy to violate Section 233 of Title 22 of the United States Code, by acting as an agent of a foreign country without notice to the Secretary of the United States, and in the second count with conspiracy to violate Section 32 of Title 50 of the United States Code, by unlawfully disclosing information affecting national defense. The first count carried a term of imprisonment of two years, and eighteen years on count two. The sentences were to run concurrently without cost. On January 6, 1942, an appeal was filed by Lang with the United States Circuit Court of Appeals for the Second Circuit. The appeal was dismissed, as the defendant failed to file the record of appeal within the time required by law on July 20, 1942.[10] There is no information as to where or how Lang served his term. The case was officially closed in 1952.[11]

This was a high profile case, and the first of its kind resulting in a victory for the government. Due to the secrecy surrounding the bombsight, it was never revealed just what or how much Lang was able to steal. Thousands of pages of recordings and transcripts were made of the case. Due to the astronomical cost of reviewing and purchasing these files, only selected court transcripts were made of the case. Due to the cost of reviewing and purchasing these files, only selected court transcripts were obtained for research. In his book, "The Game of The Foxes," Ladislas Farago details the Abwehr's (German Intelligence Agency) actions in spying operations in North America. One chapter is devoted to the theft of the Norden bombsight. Anyone interested should read this book, as it is excellent. It can be acquired through any book research facility.

THE MELILLA INCIDENT
In February 1943, a Liberator Bomber (B-24) made a forced landing near Melilla, Spain. The airplane contained the secret Norden bombsight, and secret radio gear known as Identification Friend or Foe (IFF). Since Spain was a neutral country and not at war with either the U.S. or Germany, the airplane was detained and placed in the custody of the military. It was necessary for the U.S. State Department to act as liaison with the Spanish Government and the War Department. Much telegraphic correspondence concerning efforts in recovering this secret gear was made. This was a delicate matter, and it was necessary to work with the Secretary of Foreign Affairs (Spanish), requesting their cooperation in obtaining this gear as rapidly and as quickly as possible, and the return intact and uncompromised of the bombsight and IFF radio equipment reportedly to have been left in the bomber. The Spanish

officials were very cooperative. The airplane was flown to Barajas Airport. American technical personnel were advised that the Spanish technical people would deal directly with them in this delicate matter. Spanish officials felt that the secret equipment that was in the airplane at the Barajas Airport had never been touched.

American Officers from the Naval Attache's Office, accompanied by Spanish Air Ministry personnel, entered the Liberator bomber on March 13, 1943. The airplane was locked and was under heavy military guard. Upon inspection, the bombsight was found to be intact, in a locked compartment where it was customarily kept. The lock had to be broken to permit access. American Naval personnel thoroughly checked the bombsight and felt that it could not possibly have been compromised. The bombsight was in the hands of the American Embassy personnel the same day. However, the IFF was a different story. The IFF radio equipment could not be found. Spanish military personnel who flew the airplane from Melilla to Barajas thought the secret equipment had been blown up. No remains of the equipment were found. To ascertain if the gear had actually been destroyed, personnel from the Military Attache's Office interviewed the bomber crew at Alhama de Aragon where the crew was interred. The crew stated that it had not been destroyed, and a decision was made to mount another effort to locate and recover this equipment. Naval personnel ascertained from the planes' crew at Alhama that the IFF equipment had not been destroyed. Naval personnel obtained a diagram from the crew of the airplane, showing exactly where the equipment was installed. The Embassy personnel stated that their contact in the Spanish Air Ministry had promised to recover the equipment, if it was still in the airplane. On March 17, 1943, Naval personnel delivered the IFF to the U.S. Embassy, apparently in perfect condition and uncompromised. Like the bombsight, it was covered with a layer of dust which had been removed only in places where it had been handled.[12] The State Department handled this occasion in an excellent manner. The War Department and the cooperation of the Spanish Foreign Office and the Spanish Air Ministry made possible the recovery of this important equipment.[13]

The National Archives-Northeast Region has among its holding criminal case records of the United States District Court for the Eastern District of New York, including the Hermann Lang case Number 38425. This information was obtained from them in a letter to the author dated September 11, 1989.

29 SEIZED AS SPIES IN SWIFT ROUND-UP; MOST ARE GERMANS

TRAILED TWO YEARS

Suspects in Four States Got Defense Secrets, U. S. Charges

BROOKLYN CHIEF CENTER

Short-Wave Radio Set Found —Veteran Spy of Boer War Called Leader of Ring

The arrest of twenty-six men and three women, twenty-two of them German, on charges of conspiracy to undermine national defense by acting as spies, was announced here last night by J. Edgar Hoover, director of the Federal Bureau of Investigation.

The round-up, described by Mr. Hoover as "the greatest of its kind in the nation's history," was made on the instructions of Francis Biddle, Acting United States Attorney General. Eighteen were arrested in the New York area and will be arraigned this morning before a United States Commissioner in Brooklyn, as most of the activities of the ring were said to be concentrated in that section. Four were taken into custody in New Jersey, one in Michigan and one in Wisconsin. All will be brought to Brooklyn for arraignment. The remaining five are already in prison for other crimes.

The arrests, made by government agents in quick raids Saturday and yesterday, culminated more than two years of work on the case.

..rman Lang, 74-36 64th Place, Glendale, Queens, born Aug. 11, 1901, at Schwarzenbach am Wald, Germany. Naturalized American citizen. Machinist and draftsman.

U. S. BOMB SIGHT SOLD TO GERMANY, SPY JURY IS TOLD

Former Inspector in Norden Plant Disposed of Data in 1938, Prosecutor Says

SECRET CODE DESCRIBED

Counter-Espionage Agent Tells How Novels Were Used—Got $1,000 From the Gestapo

The secret of the Norden bombsight, this country's most jealously guarded air defense weapon, has been in the hands of the German Government since 1938, United States Attorney Harold M. Kennedy charged yesterday in his opening statement at the trial of sixteen persons accused of espionage and failure to register as agents of a foreign government, specifically Germany.

The trial is being conducted before Federal Judge Mortimer W. Byers and a jury in the Federal court in Brooklyn.

Data Sold to Germany

Mr. Kennedy, in his opening remarks, flatly accused one of the defendants, Herman Lang, former inspector in the Carl Norden plant, where the bombsight is manufactured, of having gone to Germany in 1938 and sold the information about the secret mechanism. Mr. Kennedy said another Nazi agent, not now on trial, was sent to this country with $1,500 as part payment for the information.

"We will prove," the Federal attorney added, "that on deposit right now in Germany for Lang are 10,000 marks for his work here."

Lang, along with twelve of the other defendants, is a fully naturalized American citizen, Mr. Kennedy pointed out. Two of the others have taken out their first papers, and only one, Paul Scholz,

Mr. Kennedy's allegation, together with a vivid word picture by a government witness of the counter-espionage activities undertaken by the United States Government to beat the Nazis at their own game, provided the highlights of the first day's testimony. A nearly filled courtroom alternately giggled and looked solemn as the espionage and counter-espionage activities of both governments were unfolded under Mr. Kennedy's questioning.

Notes:

[1] The Courts Charge to the Jury, Brooklyn, New York, December 12, 1942, page 147.
[2] United States of America, Plaintiff Against Hermann Lang, Defendant, et. al. Government's Request to Charge, no date, page 66.
[3] United States of America, Plaintiff, Against Hermann Lang, Defendant, et. al., Government's Requests to Charge, no date, page 69.
[4] United States of America, Plaintiff, Against Hermann Lang, Defendant, et. al., Stipulation circa October 16, 1941, page 134.
[5] The Courts Charge to the Jury, Brooklyn, New York, December 12, 1942, page 150 (Transcript page No. 8380).
[6] United States of America, Plaintiff, Against Hermann Lang, Defendant, et. al., Request to Charge, no date, page 111, submitted by Attorney for Defendant.
[7] United States of America, Plaintiff, Against Hermann Lang, Defendant, et. al., Summation, no date, page 158 (Transcript page no. 8425).
[8] Ibid, page 157 (Transcript page no. 8424).
[9] The United States District Court pronounced the sentence for Hermann Lang, on January 2, 1941 (page 2 court proceedings).
[10] Order of Mandate Criminal Case #38425 from United States Circuit Court of Appeals for the Second District, United States Appellee, Against Hermann Lang and all 16 other defendants, dated April 30, 1942, page 44A, and Order of Mandate dismissing the appeal on June 4, 1942, page 45 and recorded July 20, 1942 (page 35 of court proceedings).
[11] Letter, National Archives — Northeast Region, Bayonne, New Jersey from Archivist, dated September 11, 1989, National Archives.
[12] Letter from American Embassy, Madrid, Spain, Subject: Recovery of secret bombsight and secret IFF radio equipment from U.S. Army Liberator bomber forced landing near Melilla on February 25, 1943, National Archives.
[13] Enclosure No. 1 to dispatch No. 713 of March 18, 1943, from American Embassy.

Plate #1, 29 Seized as Spies in Swift Roundup; most are Germans, Copyright 1941 by the New York Times Company, Reprinted by permission.

Plate #2, U.S. Bombsight sold to Germany Spy Jury is Told, Copyright 1941, by the New York Times Company, Reprinted by permission.

XXV

Counting the Cost

The expenditure of funds for the production of the NBS system from 1941 to 1945 is not known. Accounting for the total cost of this weapon will never be accurately compiled, as there were so many contradictory quantities and allocation of funds for bombsights, associated equipment, attachments, and numerous contracts. Consideration must also be given to research and development and the many modifications that the various components, including the bombsight, went through during the war years. Many of the design and engineering changes were underwritten in the production contracts with the various prime and sub-contractors, which could not be separated. New factories, including the cost of very expensive machine tools, required an outlay of many millions, and heavy disbursements were made for new subsidiary facilities that must be included in the overall costs. In addition, terminations of contracts also played a major role in the cost of producing bombsights and related equipment. One of the most important factors in stabilizing costs was the introduction of mass production techniques on extremely precision parts by the various Norden bombsight contractors. Follow on funds allocated for the many bombsight modification and attachments added appreciably to the cost as the war progressed. Although physically the finished product was small, the various stages of fabrication, a large number of specialized skills, and expensive machine tools (lathes, mills, grinders, etc.) were mandatory.

To combat excessive costs the Vinson Act was passed (official name: The National Expediting Act). This also provided for the payment of advances of as much as 30% on the contract price to help contractors to get production started as soon as possible. These payments were authorized before completion of the contract to help the contractors and sub-contractors to meet these additional operation costs when undertaking projects far beyond their working capital. If this was not done, many of these concerns would have ended in bankruptcy. Unfortunately, there were no records or data on this phase on the procurement of this material in the National Archives files. Another element that must be considered and cannot be ignored was the direct and overhead wages of the huge number of civilian employees in and out of government in maintenance, supply, procurement, etc., employed by the Air Materiel Command depots (AAF) and Naval Materiel Command at the various bombsight repair buildings, AAF and Navy personnel assigned at the various bombsight manufacturer plants (i.e Resident Inspectors, engineers, etc), and also Technical Representatives from the various contractors assigned at field installations (world wide) that supported the NBS must be accounted for in arriving at a total cost. These costs would have been in the millions of dollars. All costs of materiel that was produced for the NBS are based on 1941-1945 dollars. These costs, if converted to current day prices, would increase the original costs many times.

The introduction of mass production methods did not substantially lower the cost. If costs were lowered, the surplus was used to cover increases in wages for employees and scarce material. The NBS did not materially increase in cost. Each successive year the sight became more complex and sophisticated due to design and engineering changes. Government auditors kept a tight reign on production and expenditures for NBS progress. However, in a highly fluid condition such as gearing up for war, it was inevitable that many unforeseen contingencies would occur during the conversion period, even with all the government super civilian agencies such as the War Production Board, Office of Price Administration, Joint Aircraft Committee, etc. There were few, if any, cost overruns. These excess costs were partly absorbed by the manufacturer. No comparison could be made to today's weapons procurement, where huge cost overruns seem to be the rule rather than the exception. It was remarkable that the cost did not skyrocket after 1941. If it did, the Vinson Act would have been applied (limiting profits on Army and Navy contracts). The establishment of newer and more efficient production lines increased the flow of materiel to the combat areas. The manufacturing cycle from raw materials to finished product was over seven months, and in some instances up to a year. No longer would one person be responsible for the completion for one unit. Now, it was one or more persons doing a certain specific part, and then it would move to another station and the next assembler/mechanic would do their operation, and so on until the unit was completed. The contribution of industry cannot be measured in terms of units delivered because so much time and money was consumed in converting from peace to wartime production. The building of new plants required years to complete and begin production before one unit was produced.

The total number of sight heads reported by the Navy Bureau of Ordnance from 1940-1945 was 43,826 for Navy controlled Ordnance plants and contractors.[1] This did not include the production of the Victor Co., who was the AAF prime contractor. Victor was awarded a contract between $50-60 million for the manufacture of the NBS and related items, including spare parts. Extensive research of the archive files did not reveal the quantities of sights produced. It is definitely known that Victor produced 4,982 sight heads. Since they made the M-9, 9A, and 9B series it can be reasonably assumed that they fulfilled their contractual agreement. In late 1944 and early 1945 the AAF required several thousand more sights. The weighted average in 1944 for a M-9 bombsight produced by Norden and Victor was $7,560 (M-9 sight is explained in another chapter). Using this average, the Victor Company production would equate to about 7,000 units.[2] The assumption of the above quantity is in the "ball park" in comparing this total with other NBS contractors. Another problem is that the Lukas-Harold Co. showed a total of over 14,000 sights produced. However, Navy sources show 12,743, or a difference of 1,257 sights. The question is "which is correct." The quantity of 14,000 sights was mentioned in the history of the Lukas-Harold Company.[3] The author is inclined to accept the figures of Lukas-Harold since they had all of the records and had to account for every sight. Hence, in using the above computations it would bring the estimated grand total of Navy and AAF sights produced from 1940-1945 to 52,083, which is consistent with bomber production (very heavy, heavy, medium, and light). Using the above weighted average the total cost of the sight heads would have amounted to $393.7 million.

Minneapolis-Honeywell Regulator Co. (now Honeywell, Inc.) in various correspondence stated that they made 35,000 C-1 automatic pilot sets, and the Bureau of Ordnance had 39,600 SBAE sets under contract for a total of 65,660 sets. However, this quantity appears to be a duplication. The Navy was the first to manufacture SBAE units. In the beginning, the Navy forwarded "skeletonized" units to Honeywell for installation of the new electronic mechanism until the Navy could catch up.[4] Various correspondence from July though Oct 1942 shows that in one order the AAF requested 9,000 sets with 11,000 sets yet to be delivered from Honeywell. It is now uknown, and there is no way to determine, if these numbers were part of those reported by Norden and Honeywell. Headquarters AAF directed that all types of bombardment airplanes equipped with bombardier noses would have the Norden sights installed including the automatic pilot.[5] The AAF obtained the auto pilot from the Navy and Honeywell Co. It would be safe to assume that SBAE/AFCE production would be in the neighborhood of at least 55,000 complete sets, as this would approximate bomber production. The cost per complete set was estimated at approximately $3,097.[6] Using this assumption, this program would have amounted to $170.3 million. The directional stabilizer was made by several Navy and AAF contractors. The examined files indicated that this unit was a separate item and was required with the Norden bombsight. The cost in Sep 1942 was $1,662.50.[7] Further research did not reveal any other costs, however, in the intervening years the price would have gone up due to increases in material and wages.

Using the same quantities of SBAE/AFCE, the cost would have amounted to approximately $91.4 million.

The next most expensive attachment was the Glide Bombing Attachment (GBA). The function of the GBA is discussed in another chapter. Navy and AAF records did not match the production quantities. These projections showed requirements and existing production of 18,108 units (scheduled cutback to the end of the war in Europe). Assuming that these projection sheets were correct and the unit cost of $2,836 each, the GBA program would have amounted to $51.4 million,[8] exclusive of design, engineering, and testing. Requirements for Low Altitude Bombing Attachment on all contracts was 21,440 units at a cost of $856.00 (1942 cost was $310), including 7% profit for a total of $12.3 million.[9] The total Anti-Glare Lens requirement is not known. The first contract was for 7,000 sets, and the second was for 1,000 per month until completed, however, the length of the contract is not known. There was one extremely urgent requirement in the European and Pacific area, and it is assumed that the contractual requirements were completed on schedule. Contracts were typically awarded from "at once," three to six months to over a year depending on the quantity and urgency. Since over 50,000 bomb sights were produced it would be safe to assume that at least half of the sights in service would be provided for. The 7,000 sets definitely known, plus the estimated 15,000 remaining on the contract would be approximately 22,000 units at $20.50 per set of six, including case would be $451,000.[10] The Bombardiers Information File describes these sets in detail. On occasion this item is available from surplus military outlets. An initial contract for Bombsight Carrying Case was awarded for 400 units at a cost of $21.60 each. Again the total quantities are not known. Thousands of these cases were required to carry the bombsight from the airplane to the vault, vice versa, or to the various repair shops. At least 25,000 were made available to the bombardiers. Some correspondence indicated that in production quantities, the price dropped to $13.80 per unit, amounting to $345,000.[11] The author has one on display. The first contract for bombing trainers for use in training bombardiers and bombsight maintenance students was for 50. Thirty-four additional units of this type were procured. The Link Belt Company produced 321 for a total of 405 units. The initial cost was $2,500 each, resulting in a cost of $1.01 million.[12]

The initial order for bombsight mounts type MK-l was for 120 units. Correspondence shows that a total of 60,000 were ordered from Lord Manufacturing Co., and by January 1, 1944, over 45,000 had been delivered. The entire contract was completed as scheduled. Exact contract costs for these mounts is not known, but correspondence shows that in 1943 they cost $64.50 per set. This would amount to $3.87 million.[13] The first Bombsight Super High Altitude Attachment cost $3,500, and seven additional units at $1,500 each for a total of $14,000.[14] Repair, remodel, and an addition to the Stratosphere Pressure Chamber cost $70,218.[15] Five Hundred Standard Sector Boxes on a cost plus fixed fee basis at $301 each amounted to

$150,870.[16] Rotor balancing sets used in balancing gyro rotors, 200 sets at an approximate cost of $621 each for $124,200.[17] Procurement of 11,000 Illustrated Manuals and Pamphlets for $112.000.[18] Modification of 80 Visual Link Trainers at $1,000 each for $80,000.[19] Fifteen bombsight top and rate end covers made from plexiglass for training bombardiers at a cost of $51,060[20] (direct labor only). Six thousand thirty-eight boxes complete with one bombsight metal box, one bombsight tray, and one stabilizer tray at $33.20 each for a total of $200,462.[21] The author has them on display. A total of 45,000 tachometers were ordered from the Jeager Watch Co. and Technical Oil Tool Co., and by Dec 1943, 24,000 were delivered. The contract was completed as scheduled at an estimated cost of $1.13 million.[22] No cost was found for the following support items: 8,500 Heating Blankets Mark I[23]; Delayed Action Automatic Release Mechanism to Dec 1943 was 14,324 units that were delivered or undelivered to the AAF for M-9, 26 volt bombsight, and an additional 4,000 were procured for sights already in service[24]; 500 each Wing Leveling Device[25]; 11,000 Data Card Holders for Bombsights, Stabilizers, and AFCE/SBAE sets[26]; 726 Calibration Outfits procured for squadron and depot overhaul shops; and 294 Motor Generators sets for power supply for squadron and depot repair shops.[27]

Funds for building new and expanding existing facilities cannot be accurately determined. Various letters, memos, and directives showed that about $33.9 million was spent or obligated in this phase between 1940-1945. It is not known if the above figure is correct, however, according to correspondence in the archives it appears fairly accurate. A word of caution here is that this amount is not to be construed as part of the contracts for production of bombsights and associated equipment. They were follow-on contracts required for expansion, acquisition of machine tools, land, facilities, training, etc. With tens of thousands of bombsights in the field, vast quantities of spare parts and supply items were desperately needed. Much correspondence was found in the files pertaining to the spare parts program. The range of support was from a low of 5% to a high of 20%, depending on the type of equipment. Using an average of 10% of the basic contracts it can be assumed that this program would approximate $77.5 million. Some data showed research and development in adapting the NBS to guided missiles (AZON/RAZON) used in the latter part of 1944 and 1945 amounted to about $15.2 million. Several contracts were awarded to Gulf Research and Development Co. In addition, the above amounts do not include the costs obligated for contracts canceled due to completion of a contract or the end of the war. A standard practice was to include a termination clause to get rid of unwanted military inventory upon completion or termination and to reestablish a factory to its original configuration for peacetime needs. These costs for various NBS contractors and sub-contractors would have been significant and would be in the neighborhood of tens of millions of dollars. No information was found in the examined files on this phase. It has been stated that over 3 billion was spent on developing radar in WWII. If only 5 or 10% of this effort was used to adapt the Norden sight with radar it would amount anywhere from $150-300 million. This was a "crash program" with a very high priority towards the latter part of the war. H2X (radar) was being developed as fast as possible for the European and Pacific areas. For example, from Apr 1944 to the end of the war in bombing Berlin, Germany, a total of 21,722 tons of bombs were dropped by the Mighty Eighth Air Force using a combination of the Norden sight and radar about 60% of the time, and visual the remaining 40%, consisting of 9,005 sorties. This gives the scope of the use of radar assisted bombing. The AAF expended for modification, research, and development $1.3 billion from 1942-1945. There is no breakdown of where or how it was spent. This does not count the Navy portion. It is not known how much was spent on the NBS. There are records at the MIT Radiation Laboratory. The author did not pursue research in this area. Another overlooked item was the purchase of precision anti-friction bearings from the various bearing manufacturers. No cost data was found for SKF Industries, Inc., New Departure (General Motors), Fafnir Bearing Co., Norman-Hoffman Bearing Co., etc. Considerable funds were allocated in converting these plants to produce acceptable bearings for the NBS for wartime use.

It was the same story with the following equipment: Reflex Open Optic Sight, requirements and production was 22,011; Formation Stick (C-1 Auto Pilot) requirements and production 11,468.[28] No data was found on the procurement of vast quantities of navigation and hand held bombing computers; Technical Orders for NBS Maintenance, Repair, and Calibration Manuals; Operation and Maintenance for the various bombsight attachments; Operation and Maintenance for the C-1 Automatic Pilot; AAF and Navy Training Films (NBS); Contracts to determine the effect on the Mark 15 bombsight astigmatism during warmup and precession. The development of the Beech AT-11 "Kansan" bombing trainer for the NBS. Land acquisition, construction, base support, and maintenance of 17 training bases for use in training bombardiers probably in the hundreds of millions.

No information or data was found in the files for the following ancillary equipment: the 7A-3 Bombing Trainer development and application by IBM—a very sophisticated advance trainer; super bombsights five feet high, eight feet long, and five feet wide used for explaining the functions of the NBS; development and production of VC-80 metal guard for SBAE/AFCE; modification of the A-6 bombing trainer for RAZON bombing; design and production by the AAF of an attachment called "crab" to apply to the standard NBS for effective RAZON bombing; contract to Columbia University for an electronic simulator for guided missiles; contract to University of Michigan to find the true vertical at a cost of about $800,000[29]; design and production by IBM for the A-5 bombing trainer; production and modification of the A-2 bombing trainer; Columbia University for Intervolometer Spacing; and the manufacture and assembly of Field Repair Shop (Airborne) for AAF, Navy, and United Kingdom; research, development, and production of H2X (VISAR, NOSMO, EAGLE) AN/APQ-7,13, etc. All of these were to be adapted to airplanes, and some correspondence estimated that about 20,000 sets were made. There were also the following: awarding of many contracts to Educational Institutions and commercial concerns for research, design, and development of new weapons; AAF Aviation Psychology Program, psychological research on bombardier training using the Norden sight[30]; procurement of tens of thousands of bombsight shipping trays (the author has one).[31] There were many more contracts awarded on the bombsights and auto pilot too numerous to mention, and were of a critical nature, such as the contract to Minneapolis Honeywell for the development of the Automatic Gyro Leveling Device. It would be impossible to list all of them.

The Norden bombsight program using known and assumed costs would have amounted to approximately $895.0 million, and with expenditure of radar would increase it to $1.05 billion. All costs are based on 1941-45 dollars. Present day prices would increase the total cost by many times. It appears that the U.S. Government received a bargain on this weapon in comparison with the spectacular overruns for present day purchases. There were no known overruns. Any losses were absorbed by the Norden Company. The addition of the other tangible equipment that no cost was found would increase this between $1 billion and 2 billion, almost approximating the Manhattan Project (Atomic Bomb). Military surplus outlets nationwide have for sale many of the items that no cost or quantities could be found. This in itself proves that these items were procured, used in combat, or declared surplus after the war. These assumed costs have a basis of fact, as the author has many of the items on display in a local museum. A bombsight heating blanket in 1989-1992 retailed for $75-$100; a Bombardier's Carrying Case—Type E-l, $85.00; an original Bombardier's Information File $275-$350, if available, and they are many times their original cost.[32] The author has all of the above and many more NBS artifacts on display. Since no cost or quantities could be found, it does not mean that they were not produced.

The author is well aware of the danger and pitfalls in formulating or extrapolating some of the computations by assuming various costs. Therefore, these items were computed on a best estimate basis in order to arrive at some overall production and associated funds for the NBS program. There is no question that this equipment was produced and used by the AAF, Allies, and the Navy, regardless of whether there is information in the National Archives. The author has seen, used, or repaired items where no quantity or cost could be found. The mass of facts and figures shown in this and other chapters may be subject to controversy or criticism. There was no simple rule to follow in the successful research of the NBS program and the procurement of the critically needed NBS and associated equipment, so it had to be done one step at a time.

Manufacture of the many accessory and ancillary items—sub-contractors, commercial, semi-commercial, and research centers—ranged from a small job shop to great industrial corporations with branches nationwide employing hundreds/thousands of employees. Many of these activities did not limit their activities to the bombsight field, however, their contributions to the war effort in producing other war materiel was outstanding. In the final analysis, the best measure of industry's impact can be seen by their reaction to do their utmost in providing the critical weapons that they had contracted to do, and was a true measure of industry's successful achievement in participation in the war effort in bringing the war to final victory.

Chapter XXV: Counting the Cost

Notes:

[1] The Navy's Mark XV (Norden) Bombsight. Its Development and Procurement 1920 - 1945, page 321, National Archives.
[2] Memo to Aero Equipment Branch, Production Division, Wright Field, AAF, to Budget Office, dated August 17, 1944, MS number 217.
[3] History of the Lukas-Harold Company was written by Mr. Gene Bowles of Indianapolis, Indiana. The author wrote to him but the letter was returned unopened with no forwarding address. Continuing research, a letter was written to the publishing company. The letter was forwarded to another firm with a similar name and they stated that this company was no longer in business. The author then wrote to the Naval Air Warfare Center, Naval Avionics which was the former Lukas-Harold manufacturing plant during WWII taken over by the Navy after the war. The author received a telephone call from an employee in the Aircraft Division and he stated that Mr. Bowles had passed away. He explained that he would try to obtain permission to use the material in the book either from his wife or Naval records. Two more letters were written but no answer was received. For some reason, they did not want to get involved or spend time to assist the author. The book contains rare photos of manufacturing and assembly shops on the Norden bombsight. This information was not used.
[4] From Honeywell, Inc. Celebration of their Centennial as stated in their newspaper concerning the C-1 Auto Pilot, No date.
[5] Letter from Production Engineering to Commanding General, AMC, Subject: Proposed Manufacture of Electronically Modified AFCE by C.L. Norden, Inc., dated October 12, 1942, Page 2, MS, National Archives.
[6] The Navy's Mark XV (Norden Bombsight). Its Development and Procurement 1920 - 1945, Page 321, National Archives.
[7] Letter from Bureau of Ordnance to C.L. Norden, Subject: Bombsight and SBAE Parts List A.F.P. 238456, dated September 2, 1942, page 10, National Archives.
[8] The Navy's Mark XV (Norden Bombsight). Its Development and Procurement 1920 - 1945, Page 321, National Archives.
[9] Letter from Bureau of Ordnance to Naval Inspector of Ordnance at the Norden plant, Subject: Low Altitude Bombing Attachments - Price Quotation for, dated August 24, 1944, National Archives.
[10] Polaroid Corporation to Bureau of Ordnance, Subject: Proposal for furnishing Ray Filter Kits, dated August 10, 1942 and Letter from Polaroid Corporation to Bureau of Ordnance, dated February 16, 1943, National Archives.
[11] Letter from Kamen Products Company, Inc. to Bureau of Ordnance, Subject: Procurement of Bombsight Carrying Case, dated April 30, 1943, National Archives.
[12] Letter, Naval Aircraft Factory to Chief of Bureau of Ordnance, Subject: Manufactur of 50 Bombing Trainers, dated September 12, 1940, Information from the former Ma Section, Subject: Bombsights and Related Equipment, No date, page 54, book #46, National Archives.
[13] Letter from L.P. Stuart, Inc. to Bureau of Ordnance, Subject: Bombsight Mounts MK-1, dated June 15, 1943 and Information from the former Ma Section, Subject: Bombsights and Related Equipment, no date, page 53, book #46, National Archives.
[14] Letter, C.L. Norden, Inc. to Naval Inspector of Ordnance at Norden Plant, Subject: Mark 15 Bombsight Super Altitude Attachment, dated July 21, 1943, National Archives.
[15] Authority for Expenditure, C.L. Norden Company to NIO, Subject: Addition to Stratosphere Chamber, dated March 27, 1944, National Archives.
[16] Price Analysis for Navy Contract Negotiations from C.L. Norden, Inc., Subject: Standard Sector Boxes, dated March 27, 1944, National Archives.
[17] Price Analysis for Navy Contract Negotiations from C.L. Norden, Inc., Subject: Rotor Balance Set, dated February 16, 1943, National Archives.
[18] Letter from Naval Proving Ground to Bureau of Ordnance, Subject: Preparation of Manuscripts and Illustrations for Bombsight Mark XV and SBAE, Mark I, dated April 29, 1943 and Letter from Naval Proving Ground to Bureau of Ordnance, same subject, dated March 9, 1943, National Archives.

[19] Letter from Link Aviation Devices, Inc. to Bureau of Ordnance, Subject: Conference on AFC Link, dated November 4, 1942, National Archives.
[20] From the authors files, recollections and news articles on the manufacturer of the plexiglass bombsight and news release March 1945.
[21] Letter from Naval Inspector of Ordnance to Bureau of Ordnance, Subject: Information on Metal Shipping Boxes, dated August 30, 1944, National Archives.
[22] Information on Bombsight and Related Equipment from former Ma Section, page 54, no date, book #46, National Archives. From the files of the author concerning the cost of tachometers. Calibrating instruments cost $900 each. These were not ordinary, off the shelf instruments but highly accurate and precision made. These units were selling in 1993 from $50 to $75 with case in excellent condition, and poor to average $37.50 to $50 each with no case or attachments. The author has one in excellent condition with case and attachments. On a "best guesstimate" basis it is assumed that they would have cost in the neighborhood of $25 - $50. This compares with the cost of a good railroad watch.
[23] Information on Bombsight and Related Equipment from former Ma Section, page 52, no date. These blankets in 1992 sold in retail military outlets for $85 - $100 each. No cost can be assumed since there is nothing similar for comparison.
[24] Letter from Bureau of Ordnance to Commanding General, AAF, Subject: Modification of M-9 Bombsights dated June 26, 1943, National Archives.
[25] Letter from Bureau of Ordnance to Naval Inspector of Ordnance, C.L. Norden Plant, Subject: SBAE Wing Leveling Device — Request for Quotations dated February 15, 1944, National Archives.
[26] Letter from Bureau of Ordnance to C.L. Norden, Subject: Bombsight,. Stabilizer and SBAE Data Card Holders, dated August 1, 1944, National Archives.
[27] Information on Bombsights and Related Equipment from former Ma Section, No date, page 52, book no. 46, National Archives.
[28] Army Air Forces Statistical Digest, 1942 - 1945, Target Summary, 8th Air Force from August 17, 1942 to May 8, 1945, National Archives.
[29] Memo for the Record from Chief Armament Section to Hq AAF, Subject: Present Bombsight Program, dated October 7, 1944, National Archives.
[30] University of Michigan, Division of Research, Development and Administration. Report consisting of 684 pages. Authors' letters dated June 26, 1991, Subject: Manual Stabilizer Error in the NBS, Project M-699A, B/S Testing, USAF Contract Number W-37-078-AC45318. Letter from DTIC dated December 5, 1990. The study was donated to the USAF Museum at Dayton, Ohio.
[31] Army Air Forces Aviation Psychology Program Research Report, Report No. 9, Psychological Research on Bombardier Training, Midland, Texas, AAF-AS-Sp 40, January 10, 1946, Library of Congress.
[32] Letter from C.L. Norden, Inc. to Naval Inspector of Ordnance, Subject: Modification of Mark XV Boimbsight Metal Shipping Tray, dated July 7, 1944, National Archives.
[33] From the files of the author concerning the cost of Norden Bombsight Artifacts from various surplus military outlets circa 1989 - 1994.

XXVI

A Bomb Drops on the Norden Company

It was a big surprise that charges of protecting the monopoly of bombsights were brought against the C.L. Norden Company and some of its officers. Mr. Carl L. Norden was not affected, as he was the consulting engineer and did not have any control of the business or its administration and was no longer affiliated with the company. The administration and the conduct of the daily business was left to Mr. Theodore Barth, President, and other officers of the company.[1]

A complaint was lodged by Remington Rand, Inc., concerning the Elmira, N.Y., plant, which had been given a large contract to build Norden bombsights for the Navy. Also, several other companies doing war work for the Navy but not manufacturing bombsights protested the Navy policy of using management teams in the production process. The United States Senate set up a Committee for the Investigation of the National Defense Program, commonly known as the Truman Committee. A special sub-committee was established to investigate the engineering service firm of Corrigan, Osborne and Wells. Hearings and executive sessions were conducted on May 16, 17, and 18, 1944 in the Senate Office Building, with Senator Harley M. Kilgore, West Virginia; Senator Carl Hatch, New Mexico; and Senator Homer Ferguson, Michigan. Chief Counsel was Hugh A. Fulton, and Rudolf Halley, Assistant Counsel. Witnesses for the plaintiffs were James H. Rand, Jr., Chairman of the Board and President of Remington Rand, Inc., and Albert M. Ross, Vice President, Remington Rand, Inc. For brevity, this company will be identified as RRI. RRI made a complaint to this committee charging that undue pressure was brought on the company by the Navy, namely Corrigan, Osborne and Wells, in the manufacture of bombsights for the Navy. RRI, before, during, and after WWII, was one of the first in the business machine industry. They had many plants and thousands of employees.[2] RRI was cognizant that the start up costs would be high, but then, as production increased, the costs would go down.³ Again, the specter of training employees on this type of work loomed very large. It was estimated that at least five months or more would be required to train personnel to the point of being proficient. During the training period, there would be low production until personnel became accustomed to close tolerance work. The Norden Company was to provide a complete set of engineering drawings. RRI engineers estimated that it would take from four to six months to complete the job.⁴ Actually, it took until September, 1942—over six months. The assembly drawings were not ready at that time, and it would be some months later before the parts drawings would be completed. It was not until January 1943 that they were finished. It took almost a year to complete this job with Norden and RRI personnel working steadily. It was also necessary to make drawings for tools that would be needed to support the production lines. Assembly drawings were vitally needed to get inter-relationship between the assembly and production lines.⁵

According to testimony at these hearings, it was claimed that there was very little cooperation from Norden personnel. One of the main reasons was the fear among Norden employees that RRI, being such a large organization, would take over the Norden Company, which was very much smaller, and they would then lose their jobs. It could not seem possible that RRI could be a failure.⁶ Their sales exceeded the sales of any other office machine manufacturer for the past five years.

Testimony revealed that Corrigan, Osborne and Wells was a large engineering company with offices in many large cities in the United States—a very powerful and capable company.⁷ In March 1943, the Bureau of Ordnance discussed the unsatisfactory status of production in the plant at Elmira, N.Y., for the manufacture of the Mark 15 under contract between C.L. Norden, Inc., and Remington Rand, Inc. It was necessary to train personnel in the production of parts and assembly of extremely close tolerances.⁸ It was explained that this delay was caused by failure of obtaining the delivery of critical machine tools. In addition, the heat in the new plant was not turned on until February, 1943. Personnel could not work in this environment without becoming ill. By September, 1943, RRI had gotten their people trained and their factory organization running and in production, and in the last week of August, 1943, they were producing bombsights at the rate of 240 sights per month.⁹ The RRI contract called for inspection and acceptance at the Elmira, N.Y., plant. This was disregarded by the Norden Company and the Navy. It seemed strange that the Naval Inspector of Ordnance complimented the RRI on the quality of the bombsights that they manufactured and stated they were as good or better than any of the sights being produced by other contractors. RRI had, by the end of August 1943, 304 bombsights that had been completed. Regardless of the fine work, the Navy was not satisfied. The Executive Order taking over the plant was signed by President Roosevelt on November 25, 1943.¹⁰ This Executive Order authorized the Secretary of the Navy to take possession of and operate the plant and facilities of

Chapter XXVI: A Bomb Drops on the Norden Company

the Elmira, N.Y., plant. The plant was physically taken over by the Navy Department on November 30, 1943. It was taken over on the premise of poor management and lack of production. They were given 10 days to take drastic action to revise management to a competent level.[11]

Another problem arose when the Navy (Bureau of Ordnance) requested that 304 bombsights at the Elmira, N.Y., plant, that had already been inspected, accepted, and approved by navy inspectors, be shipped to the Norden plant in New York for further review.[12] One hundred and fifty-four of the 304 bombsights were returned to RRI as not acceptable and would require rework. The Norden people had taken a sample of 20 sights, and their findings indicated that the rest of the group was not acceptable. To rework such a large quantity would require almost the complete shutdown of all production lines, idling many workers. All sights would have to be torn down, cleaned, reassembled, tested, calibrated, and inspected. The taking apart and putting back together of 154 delicate instruments was the equivalent of producing four times as many new bombsights. In reworking so many sights, it got down to the point where the assembly plant was doing nothing but disassembly. In the middle of September 1943, 154 of the 304 rejected sights were returned to RRI for rework. This threw the production lines into a tailspin, not only in morale, but also in the entire organization.[13] This shipment was in direct violation of the RRI contract.[14] The contract with RRI was for 6,930 bombsights, the number specified on the contract, for $54 million.[15] The Norden Company was to be the prime contractor. In order to accomplish this contract, $8 million was spent in converting from peacetime to war production.[16] RRI requested a report on the disposition of the 20 sights, but none was ever furnished due to the takeover.

It was later determined that 19 of 20 sights used in the sample test had been sent to the Services without rework. No information was available on the remaining sight.[17] At the time of the takeover by the Navy, RRI had reworked 77 of the rejected sights, and they had been forwarded to the Naval Inspectors.[18] In the meantime, the remainder of the sights were kept at the Norden plant in New York City pending resolution of the problem. RRI requested a report on the disposition of the 20 sights, but none was ever furnished.

The testimony and testifying was finished on May 18, 1943, concerning RRI and the Norden bombsight. Over 200 pages of testimony were given in this particular hearing. Other manufacturers' testimonies producing war material concerning this engineering firm are not shown in this write-up. However, those interested can obtain copies at a nominal cost. Thousands of pages of transcripts of the case needed to be reviewed in order to find those that affected the Norden bombsight. Only a very brief synopsis is shown about this case.

Because of the hearing held by the Truman Committee and the information that was revealed concerning these allegations, it was felt there was sufficient justification to present this to the Federal Grand Jury in Manhattan. It was alleged that the

Norden Company withheld engineering assistance, incomplete assembly drawings, and critical machine tools. In December 1944, the Grand Jury returned indictments against Commander Corrigan, Robert H. Wells, Theodore H. Barth, President of Norden Company, and other officers of the company. Carl L. Norden was not involved in this matter and was not even called to testify. It was turned over to the Justice Department. The case went to trial, but was dismissed for lack of evidence.

Notes:

[1] Investigation of The National Defense Program, Engineering Service Firms, Executive Session, Thursday, May 18, 1944, United States Senate, Special Committee Investigation of the National Defense Program, Washington, D.C., page 3.
[2] Ibid, page 4.
[3] Ibid, page 9.
[4] Ibid, page 17.
[5] Ibid, page 108.
[6] Ibid, page 21.
[7] Ibid, page 53.
[8] Ibid, page 55.
[9] Ibid, page 34.
[10] Ibid, page 43.
[11] Ibid, page 59.
[12] Ibid, page 40.
[13] Ibid, page 74.
[14] Ibid, page 70.
[15] ibid, page 108.
[16] Ibid, page 121.
[17] Ibid, page 71.
[18] Ibid, page 75.

XXVII

Norden Mark XV Bombsight
Bombing Accuracy - The Results

Uppermost in the minds of many civilian and military experts is one question—was the NBS of sufficient accuracy? One national journalist made a statement that the sight could not hit a target even with a good bombardier. This was later amended after a storm of protest. This and other disparaging remarks have been quite common over the years. A review of the European phase of bombing will show a great difference. Many years of testing were conducted before, during, and after Pearl Harbor. There is no question that a well trained bombardier could hit a target within acceptable limits at altitudes ranging from 15,000 to 30,000 feet. General Arnold mentioned the use of the NBS and its ability to hit targets after intense training. Recently, the bombing effort conducted by the AAF during WWII effectiveness was further intensified when a former Defense Department Secretary stated that several thousand bombers were required to hit a target, apparently comparing the NBS with present sophisticated weapons. Sometime later another military person added to the fray by stating that it took 6,000 bombs to hit a hangar. It is not known if these statements were in jest or serious. Regardless, these were unwarranted comments, especially for those who died in the line of duty.

There are many photos in the Smithsonian Institution Collection of WWII aerial bombing of Europe and Japan showing complete destruction of the intended target. The AAF conducted a study showing reduction in Radial Error of a Bombing Team as a result of practice, at an altitude of 15,000 feet. Over a seven day practice interval the error was 480 feet; 20 day practice interval, it was 300 feet; and with an additional 14 day practice interval, it was 164 feet. Date of the study is not known.[1] Some of the photos are spectacular. One is the sequence showing the bombs dropping on the Italian heavy cruiser "La Madallena" in the Mediterranean Sea, and its complete destruction from an altitude of 19,500 feet.[2] Another more dramatic series is the destruction of a Nazi ammunition convoy off Bizerte, North Africa.[3] Other photos show complete destruction of bridges, aircraft, and airframe factories, marshaling yards, etc. This remarkable bombing was accomplished under heavy attacks from "flak" and enemy fighters.

The Allies were greatly concerned about the aerial bombing conducted in Europe. To determine what the results were, in November 1944 the Secretary of War established the "United States Strategic Bombing Survey," directed by President Roosevelt. The survey team operated from London, England, and consisted of 300 civilians, 350 officers, and 500 enlisted men.[4] They established forward headquarters in Germany following the advance of the Allied Armies. They made close examination and inspection of several hundred German plants, cities, and areas, gathered volumes of documentary material, including top German government documents, and conducted interviews and interrogations of thousands of Germans, including virtually all of the surviving political and military leaders. While the European war was going on, it was necessary in some cases to follow closely behind the front, otherwise vital records would be irretrievably lost.[5] The results of this study are so large and extensive that it covers several volumes. The following statistics show some of the highlights of this massive aerial war. The American concept of precision daylight bombing was soundly criticized by many military experts. The idea that bombers could defend themselves against hostile airplanes did not prove itself in actual combat. No one had ever attempted this type of bombing in the past, and there was no precedent. The Nazis started night bombing, and the British retaliated. They felt saturation bombing would hit industrial targets, but it also resulted in the destruction of residential areas that were not devoted to the war effort. General Douhet and General Mitchell were early day advocates of strategic daylight bombing, but it was largely ignored. This controversy continued until late 1943 and 1944. Then the full weight of the air onslaught over Nazi Germany and Japan began to take its toll. The following statistics speak for themselves. In 1943, 5,092 tons of bombs were dropped on 14 plants, primarily airframe factories. Records show that Me 109 airplane production dropped from 725 airplanes in July to 357 in December 1943. Likewise, Focke-Wulf 190 airplane production dropped from 325 airplanes in July to 203 in December 1943. The culminating attacks on German aircraft industry began in the last week of Feb 1944. With the protection of long range fighter escort, 3,636 tons of bombs were dropped on German aircraft plants (airframes rather than engine plants) during that week. In that and succeeding weeks, every known aircraft plant in Germany was hit.[6] Detailed production data was gathered from Herr Messerschmitt and Herr Tank (of Focke-Wulf), who were interrogated at length about production. The deployment of long range fighter escort caused the reduction of the Luftwaffe and left the bombers to concentrate on the primary targets. The next priority targets were oil producing plants. Attacks resumed again in April 1944 on oil and oil products. Virtually complete records of German oil industry were taken by the Survey Team. Major plants were subject to attack, and their records were studied in detail. Production from these plants was on an average of 316,000 tons per month when the attacks began. Production fell to 107,000 tons in June to less than 17,000 tons in September. Output of aviation gasoline from synthetic plants dropped from 175,000 tons in April 1944 to 30,000 tons in July, and 5,000 tons in September. Production recovered somewhat, but for the rest of the war it was but a fraction of pre-attack output.[7]

By December 1944, according to Speer (Nazi head of production), fuel shortages were viewed by the Germans as catastrophic. Only through speedy recovery of damaged plants was it possible to repair partially some of the terrible losses. Leuna was the largest of the synthetic plants and protected by a highly effective smoke screen, as well as the heaviest "flak" concentration in Europe. Air crews viewed this as the most dangerous and difficult assignment of the war. Leuna was hit on May 12 and put out of production. The plant was hit many times from May to December, and production was stopped several times but recovered. From the first attack to the end, production at Leuna averaged 9% of capacity. There were 22 attacks by the 8th Air Force and 2 by the RAF. A total of 18,328 tons of bombs were dropped, and an entire year was required to demolish this plant. In Feb and Mar 1945, the Germans massed 1,200 tanks on the Baranov bridgehead at the Vistula to check the Russian Army. They were immobilized for lack of gasoline and overrun.

Albert Speer (Nazi Minister of Production) wrote that losses of aviation gasoline up to 90 percent was evident by June 1944.[8] The recovery of the damaged plants was possible by the use of as many 350,000 in a slave labor force for the repair, rebuilding, and dispersal of bombed out plants and new underground construction. According to Speer, when the Germans launched their counter offensive on Dec 16, 1944, their reserves of fuel were insufficient to support the operation. They expected to obtain fuel from captured American stocks. Nitrogen was indispensable for explosives, and the lack of this important item caused a general shortage on all fronts at the end of the war. The monthly output of synthetic nitrogen in early 1944, before synthetic plants were attacked, was about 75,000 tons. By the end of the year it had been reduced to 20,000 tons. Methanol production, necessary for TNT, hexagon, and high explosives was severely affected as the nitrogen production and allocations were dramatically and heavily cut. The loss of these two items caused a precipitous decline in production of explosives. Synthetic rubber suffered from the attacks on oil. By the end of 1944 overall statistics for the industry showed that production of synthetic rubber was reduced to 2,000 tons a month, or about one-sixth of wartime peak.[9] Had the war continued, Germany's rubber production would have been critical.

The next target was the production of steel in the Ruhr. Extremely heavy attacks on steel producing mills reduced production by 80 percent between June 1944 and the end of the year. Loss of high-grade steel in the Ruhr was a great loss, and became a bottleneck by the middle of 1944. Steel production had been reduced to critical levels by the end of 1944, and continued to fall until the end of the war. Three plants produced most of Germany's truck supply. One of these, Opel at Brandenburg, was knocked out completely in one raid on Aug 6, 1944, and did not recover. Daimler Benz was eliminated by attacks in Sep and Oct 1944. The Ford plant at Cologne, the third largest producer, was not attacked, but records show that production was sharply curtailed due to the destruction of component suppliers and the bombing of its power supply. By Dec 1944, production of trucks was only about 35% of the average of the first half of 1944. Heavy attacks during the late winter and early spring of 1945 closed all five of the major submarine plants, including the great Blohm and Voss plant at Hamberg.[10]

The attacks on transportation were the decisive blow that completely disorganized the German economy. It reduced war production in all categories and made it difficult to move what was produced to the front. These attacks limited the tactical mobility of the German Army. Prior to Sep 1944 there was no heavy bombing of German transportation. Beginning in Sep 1944, the attacks on marshaling yards, bridges, etc., was so great that by Mar 1945, the disorganization was at a point that no useful statistics were kept. In other words, their transportation system was in shambles.[11] Another item that did not go unnoticed were the major waterways, which did not go unscathed. The attacks on the waterways paralleled those on the railways, and were even more successful. On Sep 1943, attacks on the Dortmund -Ens and Mitelland canals completely stopped all water traffic between the Ruhr and points to the north coast and central Germany. By Oct 1944, traffic on the Rhine had been interdicted to a point where traffic on the Ruhr dropped sharply, and all water movement of coal to south Germany ceased. Finished and semi-finished components, commodities, and consumer goods had to be handled through the marshaling yards, and after the Oct 1944 attacks, moving this material was difficult, if not impossible. By Feb 1945, the complete interdiction of the Ruhr district was achieved.

The German anti-friction bearing industry was heavily concentrated. Approximately half of the output came from plants in the vicinity of Schweinfurt. The peak accuracy of 70% of the bombs that fell in the target area was reached in the month of Feb, 1945. In a series of raids beginning in Aug 1943, the bombing of the precision anti-friction ball bearing factories at Schweinfurt, 12,000 tons of bombs resulted in the plants being severely damaged. By Sep 1943 production was 35% of the pre-raid levels. In the second famous attack on Oct 14, 1943, the plants were again severely damaged, and one of the decisive air battles of the war took place. The bombers were escorted by fighter aircraft to a certain point, then the bombers were on their own and strongly attacked by German fighters. The factories were again in limited operation three months after the attacks by the use of tens of thousands of slave laborers and unlimited priority for men and materials. Also, Schweinfurt had one of the heaviest concentrations of anti-aircraft artillery and fighter airplanes of any city in Germany, and still the American bombers were able to get through to the target and destroyed 65% of the bearing production. It was not the fault of the bombardier or the NBS, but the lack of fighter escort that caused catastrophic losses of airplanes and personnel. Losses of this magnitude could not continue until long range fighter airplanes became available. In spite of these terrible handicaps, American airmen conducted themselves in heroic measures.

Official German records on raw materials and supplies showed that stockpiles of rubber were small at the beginning of the war—at most sufficient for only 2-3 months consumption. One major plant at Heuls was attacked as a primary target by the 8th Air Force in Jun 1943 and was closed for a month, requiring seven months to get back into full production. It operated on gas from the synthetic plant in the Ruhr, and when these were knocked out in the summer of 1944 production was again substantially reduced. Production at Schkopau, the largest of the synthetic rubber plants, was lost because it was dependent on hydrogen from the Leuna plants.[12]

The scope of the bombing is vividly shown in the Target Summary compiled by the 8th Air Force in Europe. During the period 17 Aug 1942 through 8 May 1945, the 8th Air Force carried out bombing operations on 459 days, dropping an average of 1,506 tons of bombs per day. Bombers dropped a total of 726,923 tons of bombs on combat operations, of which 691,470 tons were classified as dropped on the target, and 35,453 tons were jettisoned because of airplanes shot down or unaccounted for. Targets were listed by date, country, amount of tonnage dropped, type (high explosive, incendiary, fragmentation), number of airplanes taking part in the raid, and whether visual (Norden or Sperry) or radar (H2X) sighting was used. Of the 691,470 tons dropped on targets, 687,150 (99.4%) tons were by heavy bombers (B-17 and B-24), 4,285 tons by medium bombers (B-26s in 1943), and 54.0 tons by light bombers (DB-7s in 1942). This summary covers the bombs dropped by the 8th Air Force bomber airplanes' operation from bases in the United Kingdom, Italy, and Russia on "shuttle" missions. The Sperry bombsight was canceled and the Norden bombsight was used in its place. These statistics portray the effectiveness of the Norden sight. Visual bombing was used the most, with Radar (H2X) a close second. This was probably due to the newness of radar, which was still in its infancy. The H2X was being deployed as fast as possible. For example, in bombing Berlin, Germany, a total of 21,722 tons of bombs were dropped using both the Norden and radar about 60% of the time and 40% by visual bombing (Norden and Sperry). The tonnages listed are only those dropped by the 8th Air Force and do not include the tonnages dropped by the 9th, 12th, and 15th Air Forces. In some instances, the above Air Forces bombed the same targets as did the RAF. The bombing raids on Peenemunde, Germany, V- weapons, and experimental stations were not included in the totals.[13]

There is no question that Allied air power was decisive in the war in Western Europe. The results of the study suggest that even a first class military power cannot survive for long under full scale use of air weapons over the heart of its territory. It seems that the NBS was able to live up to its name as a precision high altitude bombsight.[14] The only method available was to penetrate and destroy the enemy's industrial capability. This was accomplished over Germany at a staggering cost. This was the concept as envisioned by Gen. Mitchell. The following summarizes the scope of the German phase of the air war: "The cost of our air war against Germany (tactical and strategic), we flew 1,700,000 combat sorties, dropped 1,555,000 tons of bombs, shot down 30,000 enemy aircraft, lost 26,700 American planes, suffered nearly 93,000 casualties, of which between 35,000 and 45,000 are expected to be listed as dead when final tallies are completed. We sent 7,235,500 individual airmen over Germany, the equivalent of 482 divisions of 15,000 men each. Total cost was 25 billion dollars."[15]

Notes

[1] Reduction in Radial Error of a Bombing Team, Smithsonian Institute, Photo No. 8329 AC, no date.
[2] Photo of Bombing the 110 Heavy Cruiser "Trieste", La Maddalana by 301 Bomb Group, 352nd Squadron Ship 351, National Archives, at 19.
[3] Photo attacking convoy off Bizerte, North Africa, 301 Bomb Group, 352nd Squadron, 6-4- 43, 11,000 feet, National Archives.
[4] United States Strategic Bombing Survey, Summary Report, (European War), September 30, 1945, National Archives, page 5.
[5] Ibid, page 1.
[6] Ibid, page 6.
[7] Ibid, page 8.
[8] ibid, page 8.
[9] Ibid, page 10.
[10] Ibid, page 12.
[11] Ibid, page 12.
[12] Ibid, page 9.
[13] Target Summary, 8thAir Force Tonnage Dropped on German Targets, May 10, 1950. Air University Historical Liaison Office. Obtained from USAF Academy.
[14] United States Strategic Bombing Survey, Summary Report, (European War), September 30, 1945. National Archives, page 15.
[15] AF Special Staff School, Orlando, Florida, Subject: Strategic Air Operations in Europe, dated March 1946, National Archives.

XXVIII

The Navy's Mark XV Bombsight Last Combat Use

The Norden Bombsight had been around for many years starting with the Mark XI in 1924. From 1932 through WWII and the Korean War, the Mark XV was used. Many attempts were made to combine the Norden Bombsight with radar technology and guided bombs. With radar, bombing could be conducted regardless of weather. Despite this fact, another need arose for the NBS once more. This equipment was still in the U.S. Air Force inventory as late as 1967-68. According to Air Force Warner-Robbins AFB, Georgia, and the Naval Air Systems Command, a total of 18 bombsights and stabilizers were to be sent to Observation Squadron (VO-67). Actually, the quantity of complete and usable bombsights and stabilizers received was 13. This weapon was brought out again for combat use.

The OP2E "Neptune" was manned by a crew of nine, with a pilot, co-pilot, and bombardier. Modification of these airplanes started in December 1967. At that time, Squadron 67 was engaged in a project to qualify all third pilots as bombardiers on the NBS. From February 1967 to July 1968, Navy Observation Squadron VO-67 was activated. The squadron was commissioned at Naval Air Station (NAS) Alameda, California, on February 15, 1967. Due to the continuing aircraft modification program and test commitments, flight crews spent considerable time ferrying aircraft between NAS Alameda, Martin Baltimore, and Norfolk, Virginia, for various aircraft modifications and preparations for deployment of 10 aircraft, which was accomplished in early November 1967. Two airplanes remained at Martin Baltimore undergoing final modifications. Requirements for training developed when it became apparent that the Navy MK-8 Reflex Sight was not sufficiently satisfactory for the required altitudes. The instructor for the NBS was from U.S. Air Force 93 Bomb Wing, Castle AFB, California.

One of the major problems encountered with the project was the short time the squadron had available to train bombardiers to use the NBS effectively, since none of the third pilots in the squadron had any experience on this outdated equipment. After the bombsights and stabilizers were repaired and calibrated, it was necessary to obtain airplanes, install the equipment, and train personnel in its use. The Squadron was stationed at NAS Alameda and conducted training flights at the NAS Fallon Range, Nevada, Hunter Liggett Military Reservation (U.S. Army), Jolon, California, around the San Joaquin Valley, California, Sierra Nevada Mountains, and NAS China Lake, California. The flights to the Fallon Range allowed the pilots and bombardiers to practice and evaluate dropping bombs from different altitudes and airspeeds. The flights around the San Joaquin Valley and the Sierra Nevada Mountains were to train the pilots and bombardiers to fly and navigate at low altitudes, while the flights to Hunter Liggett and China Lake combined low level navigation with low level sensor drops over terrain similar to that in Southeast Asia. These training flights were conducted until the squadron deployed.

The unit was deployed to Nakhon Phanom Air Base (NKP) in Thailand. This squadron was composed of 12 extensively modified OP2E "Neptune" airplanes. NBS training was conducted, and 12 bombardiers were qualified combat ready. The missions were to provide aerial delivery of acoustic sensors that were to be dropped at precise locations along the Ho Chi Minh Trail, which was the invasion route. Sensor buoys of various types can detect sound, vibrations, and currently (1990) can even sense heat sources. Techniques exist that will permit sensors to be soft landed to deliver fragile instruments. Others can even penetrate the ground to be hidden and sense vibrations after they have landed. Accou Buoys and Spike Buoys were used during this operation. They worked much like Sono Buoys work in the ocean. The use of buoys can help to pinpoint specific target locations and contribute toward selection of targets that are currently active. A monograph of this squadron was to be developed by the Navy, but due to budget cuts it was delayed indefinitely.

It must have seemed strange to see propeller driven planes not only for U.S. personnel, but also to the enemy, and to see a bombardier hunched over the bombsight, with the crew manning machine guns. Although it did not state the model of the bombsights that were used, it is presumed that they were the M-Series, M-9, M-9A, or M-9B. These were the latest types to be manufactured towards the end of the war, and it is assumed that the Air Force would keep the latest model in the inventory. Note: The above information was obtained by a professional researcher hired by the author and a personal visit through the courtesy of the Naval Aviation News, Washington Navy Yard Annex.

It was 23 years after WWII that the bombsight was used again to serve its country. The U.S. Government received its money's worth from this versatile weapon. As late as October 29, 1951, a 26 page dissertation was developed by the

Chapter XXVIII: The Navy's Mark XV Bombsight Last Combat Use

Norden Laboratories Corporation. In this engineering report, the theory of bombing was fully explored again, and changes were made to the sight. The name was changed, and it was called the Norden Horizontal Bombsight. It went into detail as to how the cross trail assembly worked, trail settings, etc. The author has a copy of this report on file. Although there were many attempts to improve this venerable sight, it had finally reached its usefulness and was removed from the inventory.

In order to keep abreast in maintaining the few sights that were still in the inventory, Technical Order 11B41-2-2-1, Operation, Service and Overhaul Instructions Bombsight Type M-9 was revised August 15, 1958, and changes up to February 27, 1967, were still being made. This Technical Order (T.O.) covers types M-9, M-9A, and M-9B bombsights utilized in precision aerial bombing operations, and incorporates all of the changes of previous T.O.s. It is now used as a useful book of instruction for collectors who want to try their skill in repairing and calibrating their sights. It is very detailed, and comprises some 368 pages of technical data required to work on the sight.

Epilogue

There will never again be armadas of bombardment airplanes streaming towards a target. The concept of 1,000 or more airplane raid is history. There will be no more bomber stream extending for miles. Today one airplane can duplicate what hundreds of airplanes did in WWII. At the time it was the state of the art and the best in the world. The question arises many times-did the U. S. Government err in following a policy of trying to develop a high altitude precision bombsight for daylight bombing? Millions of dollars were spent from 1922 to 1945 to improve the sight. The question is, "what would we have used if the Norden or other sights had not been highly developed at the time of Pearl Harbor? The Nazi's showed very clearly what damage could be inflicted in the lowlands, Poland and England. This was further refined by the Japanese attack on Pearl Harbor. Air power was now the main focus in planning the defeat of the Axis powers. The design, calibration, establishment of bombing tables, ballistic coefficient of bombs, testing, etc. was completed and perfected many years before WWII. Before and during the war the bombsight was modified and many attachments were made to make it more accurate for bombing at higher altitudes, but the basic overall design never changed throughout the war and many years beyond. Radar technology was being developed but it was still in its infancy. It was used by the British and the Germans with good success. Great strides were being made in this weapon for war time use and by 1944 it was beginning to be deployed in airplanes in bombing Germany. Also, experimentation was being conducted on radio guided bombs called AZON/RAZON. It appears that our planners were on the right track.

The destruction of the enemy's war capability was of paramount importance to the Allies. A review of the study "Strategic Bombing-Europe" revealed the utter destruction of all phases of the Nazi war machine. On the basis of this study, it seems this aerial onslaught was the correct approach. However, the final determination will be made by history.

Much criticism has been leveled at the strategy employed in the administration of America's air war in Europe. It became apparent that unless the U. S. knocked out the German Luftwaffe, air supremacy would not have followed. In systematically destroying Germany's oil production we brought the highly mechanized Wehrmacht to a dead halt for lack of fuel. The air attacks completely shattered their great marshaling yards and the destruction of their transportation lines seriously disrupted their ability to wage war. The American concept of air strategy followed closely of the air power advocated by General Douhet and General Mitchell and many others. The use of daylight bombing dispersed the enemy's defenses. By eliminating the enemy's retaliatory striking power, the Allies had control of the air.

It was through the determination and perseverance of the Navy Bureau of Ordnance, Bureau of Aeronautics, Navy Proving Ground and the Army Air Forces through failure and success that a workable precision bombsight was available when it was critically needed. Although it took time to get into production of bombsights and bombardment airplanes, again American industry arose to the occasion. A debt of gratitude is owed to those early pioneers who had the vision and foresight to continue experimenting with the bombsight. They faced many obstacles as there were no precedents on the manufacture of certain critical parts of the sight. Everything was done by trial and error. There were no micro chips or high speed computers to solve problems. All were done by hand. Some of the best minds, universities, research centers, manufacturing facilities were called upon for assistance in this program. It was the beginning of producing highly precision equipment on a production basis. The lessons learned in this program would go on to the development of future sophisticated equipment that is being used at the present time. It truly was the "Legendary Norden Bombsight".

After WWII thousands of bombsights were sold to the public ranging from $10.00 to $50.00. Many more thousands were destroyed and sold as scrap. As time went by the scarcity of sights grew to a point were a matched set (sight head and stabilizer) now commands a price from $2,500 to $5,000 depending on condition and model. A sight head in excellent condition (or new) sells for $1,200 to $1,800 or more. There are many who own a bombsight and are used as a conversation piece. The author has had many requests for Technical Orders or Manuals, especially Technical Order No. 11B41-2-2-1, Operation, Service and Overhaul Instructions, Bombsight Type M-9, dated August 15, 1958, Change 1, February 27, 1967 comprising of 355 pages for repair, maintenance and calibration. Many of these collectors are able to repair and actually turn on and operate the sight providing they have the correct power source. It is still a thrill to see the gyro and telescope operate. As it was during WWII good bearings are almost impossible to find. Since they are for only static display good bearings are not needed. The directional stabilizer and other parts of the auto pilot can also be made to operate.

Bibliographical Note

During the more than nine years that have past since the author started work on this volume, they have consulted the contents of dozens of government publications, tens of thousands of letters, memorandums, engineering reports, drawings, etc. There were no published volumes available on the Norden bombsight. Apparently over the years no attempt was made to write a history, although many excellent articles have been written. Broadly speaking this volume rests upon four types of information: first and most important were the working papers from the Bureau of Ordnance, Bureau of Aeronautic, Army Air Forces, and other government agencies at different levels of echelons from the President down through Cabinet to General and Flag Officers; Second, the files located in the National Archives and Records Administration (NARA) consisting of letters, studies , memorandums, monographs and other date produced during the war; Third, material from museums, libraries and universities consisting of data ranging from primary to remote interest on various phases in the development of the sight; Fourth, from individuals for first hand accounts and the few companies who cared to participate. This book has tried to depict the struggles, failures and successes of the Norden Company in developing the bombsight to a point of perfection that was ready for combat use when the United States entered the war in December 1941.

The preponderance of records were located in Washington, D. C. and Suitland, Maryland the official repository of NARA. Some records were obtained from the Department of the Air Force, Department of Justice and the Smithsonian Institution. A few records were found at Wright-Patterson AFB, Ohio and Materiel Division—which went through several changes. The files were not extensive, but they did shed some light on the continuing production problems concerning the critical shortage of anti-friction ball bearings that threatened to bring the entire bombsight production to a complete halt. Further data was located in the Central Air Documents Office (CADO). CADO has custody of vast amounts of technical and engineering data. Some Norden bombsight data was found in their files, but again the find was nominal. However, to obtain this data the Freedom of Information Act was invoked. If there were any further records at Wright-Patterson AFB they were not investigated.

It is unfortunate that records on expansion of bombsight facilities and the funds for procurement of material for this weapon was not found in the examined files. In addition to the above sources, information was obtained from various satellite government offices. Although NARA provided most of the data for this book, some usable material was found in various repositories of the Department of the Air Force, Department of the Army, Air Force field activities and Massachusetts Institute of Technology were queried with little success. The University of Michigan provided a study on determining the true vertical in space, University of Rhode Island offered some data on the extraction of spider silk from spiders and Columbia University did contract work on Intervalometer Spacing. Some scattered periodicals and Technical Orders were available for scrutiny at the Library of Congress, unfortunately, the material was limited and in some instances incomplete. Photographic data maintained by the Smithsonian Institute was of excellent quality. Photos were clean and sharp. Training films were also in good condition. Their personnel assisted us in every way possible. Although the production of the Norden bombsight was referred to the Joint Aircraft Committee (JAC) on several occasions and also the War Production Board (WPB), the authors did not find any records pertaining to decisions rendered by the JAC or WPB in the resolution of production deficiencies and allocations or what course of action was directed. A determined effort was made to locate these files, but it was to no avail. There were a few instances where it was necessary for contractors and government to obtain legal opinions concerning patents, patent liability, license to manufacture and patent infringement. These were quickly resolved by the litigants within a minimum amount of time, in order not to hold up the critical need for bombsights.

We encountered a somewhat strange course of events during our research at NARA. These depressing observations are recorded here more in dismay rather than to frighten interested researchers from using the NARA files. The author visited NARA facilities in Washington, D. C. and Suitland, Maryland, consisting of many days in reviewing files on the Norden bombsight. On our second trip they could not explain the reason why we could not obtain some of the material consisting mainly of engineering data that we had been able to previously research on the first trip. The second trip some material that had been researched during the first trip could not be found. However, after several days new material was found for us to continue. On the third trip we were not able to order from the files the material that we had previously scrutinized the first two trips. In addition, several days were lost waiting for NARA personnel to locate the missing files. This is an unheard of situation. We discussed this problem with the Chief of the Division and he could not intelligently explain or give a plausible reason of what happened to the files. This was a very frustrating and upsetting event as it entailed almost a 6,000 mile round trip for each trip not counting the financial burden. Due to the extreme cost, the author could not continue to pursue this phase of the research. In other words, "we gave up".

Prior to these trips, arrangements were coordinated with NARA personnel by letter and confirmed by telephone assuring us the files were available. Anyone seriously interested in doing research at NARA, it would be of the utmost importance to consult with them as to the availability and quantity of records.

Another disturbing event was that the National Archives files did not have any correspondence, engineering data, etc. after December 1944. Archival personnel could not answer this question, nor did they know where to find this information. Was everything discarded after 1944? We will never know. However, some data was found dating from 1945 to 1950, but these were only small and isolated studies or reports, nothing on the scope of the years from 1922-1944. WWII ended in August 1945 and all production on bombsights was stopped. Disposition of contracts and inventories will never be known.

The Navy "E"

The famous Navy "E" award was given to World War II Navy Contractors who were producing war material doing an outstanding job in meeting or exceeding production goals. Not all industrial war producing companies were given this coveted trophy for excellence in the performance of fulfilling production quotas. Thais award was reserved for those who really earned it. Originally, this award was limited only to Naval units. It is an honor not lightly bestowed and one that is accepted with honor. Since 1906, the Navy "E" was a traditional Navy symbol for excellence — for a job "well done". It was a much sought after award.

This award was presented to the Carl L. Norden Company on April 10, 1942 at the Waldorf Astoria in New York City. The event was attended by almost 200 high level Naval and Army Air Forces personnel, Cabinet Officers, Senators, and those involved with the bombsight and invited guests. There is no question that the Norden Company did their utmost to provide the critically needed bombsight. The company went on to win additional awards including the Navy "E".

The Navy "E" for Excellence

In the tradition of the United States Navy, the highest of all honors is the Navy "E."

It is the coveted emblem of Excellence which the officers and crew of every ship and plane in the fleet hope to attain. Each year it is awarded to the units which have earned special recognition for excellence and efficiency in such work as engineering or gunnery.

When awarded for the highest rating in engineering, the "E" is displayed on one of the funnels of the ship. When awarded for gunnery, it appears on a mast of the bridge. When awarded to the men of an individual gun turret, it is shown on the turret. Enlisted men in these honored crews wear the "E" as an arm band on the left sleeve. the white "E" signifies top ranking; the red "E" signifies a second-grade award.

Until very recently, it has not been the custom to award the Navy "E" to anyone other than the personnel of the U.S. Navy. By special permission of Franklin D. Roosevelt, President of the United States, and Frank Know, Secretary of the Navy, the Navy Department now awards the Navy "E" to Carl L. Norden, Incorporated as a recognition of outstanding performance in the production of naval materiel.

With this honor, Carl L. Norden, Incorporated is now privileged to fly the Navy Department blue burgee displaying the fouled anchor and the white Navy "E" for a period of six months. To each member of its personnel is awarded the official insignia bearing a ship, the Navy anchor and eagle, the Navy "E", and the legend "For Production".

In the words of Secretary of the Navy Knox:

"It's our way of saying

"Well Done"

In Memorium

Mr. Carl L. Norden, inventor of the famed bombsight that bears his name, passed away on June 16, 1965, at the age of 85, in Zurich, Switzerland, where he had retired. He had invented one of the most powerful weapons used during WWII. He preferred a cloistered life and rarely appeared in public. One of the few times was to receive the Holley Medal in 1944 from the American Society of Mechanical Engineers. Mr. Norden was hardly known outside his profession and among a small group of military and civilian engineers. He turned all patents over to the United States of America as represented by the Secretary of the Navy for the sum of one dollar. In 1994, he was inducted into the National Aviation Hall of Fame at Dayton, Ohio. He never became an American citizen.

Mr. Theodore H. Barth, co-inventor of the Norden Bombsight, passed away on June 20, 1967, at the age of 75 at Wareham, Massachusetts where he retired after WWII. He was President of the C.L. Norden Company since its founding in 1928, until it was purchased by other concerns. He took care of the business of the company and all of its administration. He and Mr. Norden worked together many years in perfecting the bombsight. Patents were issued to him on the Mark XV in 1935. He also turned over all patents to the United States of America as represented by the Secretary of the Navy. Mr. Barth was an outgoing person, just the opposite of Mr. Norden.

Aviation Ordnance, Mark XV

Aviation Ordnance
Bomb Sight, Mark XV, Mods, 4 and 5
Equivalent Air Corps Designation M-4, 5, 6 and 7
October, 1941
Navy Type

Type   Mark XV
Mod 3   Same as Army Air Corps M-2
Mark XV Mod 4   Same as Army Air Corps M-5
Mark XV Mod 5   Same as Army Air Corps M-7

Army Air Forces

Type	Serial Numbers	Identifying Features
M-2	425 to 440 Incl 461 to 480 491 to 533	(a) Designed for 12-Volt system (b) No precession motor or provisions therefor (c) Cannot be interconnected with stabilized bombing approach equipment. (d) Small gyro; electrostatic shielding. (e) Total weight—approximately 50 lbs.
M-2	574 to 583 Incl	These instruments are identical to M-2 sights except they have an integral tachometer and provisions for setting in trail up to 105 mils.
M-4	1 to 124 incl	(a) Modified M-1, M-2 & M-3 sights using the original stabilizer unit but equipped with new sight units. (b) 12-Volt D.C. System. (c) 16-Pole plug Integral Precession motor and fixed eyepiece. Optical power increased to 2.2 power plus or minus 5% prism tracks. Telescope cross wires smaller. Improved leveling and caging knobs. Bomb sight gyro housing drilled for Auto Erection system and the locking stem changed to a pin instead of a nut. Trail and cross trail increased to 130 mils. Lamp assembly changed from housing to bubble assembly. Automatic cross trail. Differential gear for disc speed in high-low altitude bombing, Slide prism replaced with mirror tilting knob in rate and housing which increases the sighting angle from 70° to 77°. Improved dropping circuit points. Reticle beam changed to adjustable reflector. Rheostat changed in reticle lamp. New type switches. Sight stem standardized to permit interchangeability of units. Power plug from stabilizer to sight changed from 4 pole to 6 pole. Dropping release switch and lever improved. Provisions for later incorporation of inertia release. Ground Speed drum larger and improved. Navigational constant changed. Single wire circuit for electric power, total weight approximately 70 pounds.

M-4	125 to 272 Incl	(a) Manufactured as M-4 sights instead of being a modification of earlier series sights. 12-Volt D.C. System. 16 Volt plug Integral precession motor and fixed eyepiece. Optical power increased to 2.2 power plus or minus 5% prism tracks. Telescope crosswires smaller. Improved leveling and caging knobs. Bomb sight gyro housing drilled for Auto Erection System and locking stem changed to pin instead of nut. Trail and cross trail increased to 130 mils. Lamp assembly changed from housing to bubble assembly. Bubbles improved. Automatic cross trail. Differential gear for disc speed in high-low altitude bombing. Slide prism replaced with mirror tilting knob in rate and housing which increased the sighting angle from 70° to 77°. Improved dropping circuit points. Reticle beam changed to adjustable reflector. Rheostat changed in reticle lamp. New type switches. Sight stem standardized to permit interchange of units. Power plug from stabilizer to sight changed from 4 Pole to 6 Pole. Dropping release switch and lever improved. Provisions for later incorporation of Inertia Release. Ground Speed drum larger and improved. Navigational constant changed. Single wire circuit for electric power. Total weight approximately 70 pounds.
M-5	1	Identical to M-4 sights Ser. Nos. 125 to 272 incl., except for operation on 24 volt current.
M-6	1 to 533 Incl	(a) Designed for 12-volt current and otherwise identical to M-4 sights 125 to 273 incl., except these instruments have an automatic erection for erecting the bomb sight gyro.
M-7	1 to 88 Incl	Identical to M-6 sights except designed for operation on 24-volt currents.

Source: Ordnance Pamphlet No. 650, Register No. 73, Aviation Ordnance, U. S. Navy, October 1941, Volume III

Norden Bombsight Contractors

Norden Bombsight Contractors

Direct Contractors — Army Air Forces
    Minneapolis Honeywell Regulator Company, Minneapolis, Minnesota
    Victor Adding Machine Company, Chicago, Illinois
Direct Contractors — Navy
    Barden Company, Danbury, Conneticut
    Burroughs Adding Machine Company, Detroit, Michigan
    Cardanic Corporation, East Hampton, Massachuetts
    Lukas-Harold Corporation, Indianapolis, Indiana
    Norden, C. L.
        Lafayette St, New York, New York
        Varick St, New York, New York
    Remington Rand, Incorporated, Elmira, New York
Sub-Contractors — Navy
    Cine-Simplex Company, Syracuse, New York
    International Projector Company, New York, New York
    Kollmorgen Optical Company, New york, New York
    Manufacturers Machine and Tool Company, New York, New York
    Robbins & Myers, Incorporated, Springfield, Ohio
    Simpson Optical Manufacturing Company, Chicago, Illinois
    SKF Industries, Incorporated, Philadelphia, Pennsylvania

Partial List of Contractors, Sub Contractors, Commercial and Semi-Commercial Companies and Agencies Supporting the Norden Bombsight

Able Tool Company
Ainsworth Manufacturing Company
Aircraft Radio Laboratory
Aluminum Company of America
Aluminum Industries
Allison Company
American Aviation Company
American Cystocope Makers
American Optical Company
American Time Products
American Thread Company
Arma Corporation
Armando, George Company
Atlas Ball Bearing Company
Austin Corporation

Baker and Company
Barden Company
Bardwell & McAlister Company
Bass Gould Company
Bausch and Lomb Corporation
Beech Aircraft Company
Bendix Corporation

Binswanger & Company
Bonney Tool Company
Breeze Corporation
Burroughs Adding Machine Company

Cannon Electric Development Company
Cardanic Corporation
Carrier Corporation
Chase Brass & Copper Company
Cine Simplex Corporation
Clorostat Manufacturing Company
Columbia Steel and Shafting Company
Columbia University
Conklin Brass and Copper Company
Corborundum Corporation
Cornish Wire Company
Curtis Wright Airplane Company

Davenport Machine & Tool Company
Delco Products
Devers & Eds Company
Dodge Steel Company
Douglas Airplane Company

Eagle Signal Corporation
Eastman Camera Work-Trenton Division
Eastern Aircraft Company-Trenton Division
Eastman Kodak Company
Eclipse-Pioneer Division
Edgeworth Hurde & Supply Company
Electric Chemicals Company (LABA)
Electric Thermal Company
Euclid Lamp World, Incorporated

Fafnir Bearing Company
Fairchild Corporation
Felsenthal & Sons Company
Firestone Tire and Rubber Company
Ford Instrument Company
Ford Motor Company
Franklin Institute
Frasse, Peter A. Company
Frieze & Sons Corporation
Frigidaire Corporation

Gaertner International Industries
General Electric Company
General Electronics Industries
General Luminscent Corporation
Gibbs Thomas Engineering Company
Goldmark James Company
Gulf Research & Development Corporation
Grumman Aircraft Company
Guarantee Speciality Company

Harris & Thomas Drop Forge Company
Harvey Radio Laboratory, Incorporated
Hasler Manufacturing Company
Herman Body Works
Herzog Minature Lamp Works
Homelite Corporation

International Business Machine Corp.
International Projector Company

Jack & Heinze Company
Jaeger Watch Company

Kaman Products Company
Keufel & Esser
Knepper, Harry Jr., Inc.
Kollmorgan Optical Company

La Salle Steel Company
Lewyet Metal Products
Liberty Motors & Engineering Company
Link Aviation Devices
Link Belt Company
Lockheed Aircraft Company
Lord Manufacturing Company
Lukas-Harold Corporation

Mallory, P. R. Company
Martin Rockwell Corporation
Martin Glenn L. Company
Maxon, W. L. Company
Massachutes Institute of Technology
Menzer C & H Company
Merril Brothers Forge Company
Mitchell Camera Company
McClintock, O. B. Company
Moore Products Company
Motch and Merryweather, Incorporation
Moyer, W. A. and Sons

National Carbon Company
National Defense Research Council
National Research Corporation
National Screw Manufacturing Company
Naval Aircraft Factory
Naval Gun Factory
Naval Research Laboratory
New Departure Bearing Company
Norman-Hoffman Bearing Company
North American Aviation Company
Norton Abrasive Company

Office of Scientific Research Development

Patton MacGuyver Company
Philco Corporation
Pioneer Instrument Company
Polaroid Corporation
Pratt & Whitney Corporation
Prestone Corporation

Radiation Laboratory-MIT
Radio Corporation of America
Remington Rand, Incorporated
Richard Canman Insurance
Roebling and Sons, John A.

Robbins and Myers, Incorporated
Rockford Machine & Tool Company
Roller Smith Company
Ross Gould Company

San Francisco Grease Company
Schwein Engineering Company
Service Tool Company
Simpson Optical Manufacturing Company
SKF Industries, Incorporated
Socony-Vacuum Oil Company
Sparks Withington Company
Specialties Company
Sperry Gyroscope Company
Standard Aircraft Products, Incorporated
Standard Oil Company-Home Oil Division
Star Watch Company
Stich, H. H. Company

Strahs Aluminum Company
Stuart, L. P. Company
Synthane Corporation

Technical Oil Tool Company

U. S. Radium Corporation

Victor Addimg Machine Co.

Waltham Screw Company
Weidenback Brown Company
Western Electric Company
Westinghouse Electric Company
Weston Meter Company
White Rogers Electric Company
Wire Rope Corporation of America
Wollensak Optical Company

Norden Bombsight Production

Norden Bombsight Production
By Contractor 1940 - 1945

Contractor	Reported by Navy	
Bureau of Ordnance Additions		
Burroughs Adding Machine Company	6,041	
Lukas-Harold Corporation	12,743	1,257[1]
Naval Gun Factory	155	
Norden, Carl L., Inc.	21,437	
Remington Rand, Inc.	3,450[2]	
Victor Adding Machine Company	7,000[3]	
Sub-Total	43,826	8,257
Total Estimated Sights:	52,083	

Notes:

[1] The History of Lukas-Harold Company by Mr. Gene Bowles stated that over 14,000 sights had been manufactured. The author tried to contact Mr. Bowles but was informed that he had passed away. The publishing company of Mr. Bowles' book was contacted but the letter was returned. We were advised that the company was no longer in business. An attempt was made to contact the Naval Air Warfare Center, Indianapolis, Indiana (the former Lukas-Harold Company). To our several letters, only one telephone call was received and no information of any kind was furnished. No further attempts were made to contact them.

[2] Closed October 1, 1944.

[3] No firm quantity was found concerning this company. Please see Chapter XXV.

Definitions and Nomenclature of Norden Bombsight Terms

Automatic Release Mechanism — Located on quadrants in rate end. It provides automatic electrical release when the indices match and the release lever is up.

Autopilot Clutch — Located on the top of the stabilizer. It transmits stability of the directional gyro to the directional panel.

Autopilot Clutch Engaging Knob — Located on autopilot clutch and is used to engage autopilot clutch to the directional gyro.

Autopilot Connecting Rod — It connects the autopilot clutch to the drift gear clutch, allowing turns to be made from the bombsight through the directional panel.

Bombsight Clutch — Located on the top of the stabilizer. It transmits stability from the directional gyro to the sighthead.

Bombsight Connecting Rod — The link between bombsight clutch and stabilized sector.

Bombsight Switch — Located on right side of stabilizer. It completes or breaks the circuit to the sighthead and the vertical gyro.

Bubbles — Located on the top of the vertical gyro. They indicate the position of the vertical gyro's axis.

Bubble Light — Located at junction of bubble tubes. It lights bubbles for night bombing.

Caging Knob — Located on top of sighthead case. It locks the vertical gyro to the case.

Clevis Pin — The pin which fastens the bombsight connecting rod to the stabilized sector.

Coincidence Pointers — Two pointers, one on telescope cradle, the other on mirror sector, used in checking length of mirror drive cable.

Course Knobs — Two knobs located on lower right side of sighthead. They are used to set up the course of the airplane.

Crosshair Rheostat — Located on rear of sighthead case beneath eyepiece. it controls intensity of the light on the crosshairs.

Degree Scale — Seen through index window in top right side of sighthead case. It is used to measure the sighting angle.

Directional Gyro — Located inside the stabilizer. It is used to give the azimuth stabilization of the bombsight and the autopilot.

Disc Speed Drum — Located on rate end. It determines speed of rate motor by the spring tension holding breaker points.

Disc Speed Gear Shift — Located on the rate end. It is used to select the range of the disc speeds.

Displacement Knob — Located on the rate end. It is the outer of the range knobs, used to displace the lateral crosshair without changing range synchronization.

Dovetail Locking Pin — The pin which fastens the dovetail shaft to the dovetail locking bracket on stabilizer.

Drift Gear — Located on top, right forward corner of stabilizer. It transmits motion from drift worm to drift gear clutch.

Drift Gear Clutch — Located below the drift gear. Transmits motion or stabilization from the stabilizer to the PDI brush.

Drift Knob — Located on rate end. It is the inner course knob, used to displace PDI and direct airplane without changing the line of sight.

Drift Pointer & Scale — Pointer is located on rear lower art of sighthead. Drift scale is on stabilizer under pointer. They indicate amount of drift set into bombsight.

Drift Worm — Located on sighthead below turn worm and meshed with drift gear. Transmits motion from drift knob to drift gear.

Dropping Angle Index — Seen through index window on left side of tangent scale. It indicates the tangent of the dropping angle.

Extended Vision Knob — Located on rate end. It increases forward vision 20°.

Fore-and-Aft Bubble — Located on top and left of vertical gyro housing. It indicates the fore and aft position of vertical gyro axis.

Fore-and-Aft Crosshair — Located on the lens inside telescope tube. It serves as a reference to synchronize for course.

Lateral Bubble — Located on the top and rear of vertical gyro housing. It indicates lateral position of the axis of vertical.

Lateral Crosshair — Located on the lens inside the telescope tube. Used for movements of gyros' axis to the left and right.

Mirror Drive Clutch — Located in center of displacement knob. It engages the mirror drive by locking the lower traction gear.

Pilot Directional Indicator (PDI) — Located on pilot's instrument panel. An electrical meter that indicated to the pilot the direction to correct the airplane's flight.

PDI Brush & Coil — Located on top of stabilizer. Brush is attached to drift gear clutch collar and moves over coil. Brush moving over coil sends a signal to pilot's PDI.

PDI Switch — Located on rear of stabilizer. Switch for PDI circuit to pilot's PDI.

Range Knobs — Two knobs located on the rate end. They are used to determine and set up the dropping angle (range) at which the bomb is released.

Rate End — Located on the right side of sighthead. It solves the range problem by determining groundspeed and dropping angle.

Rate Knob — The inner of the two range knobs. It is used to determine the speed of closure and set up the dropping angle.

Rate Motor — Located inside the rate end of sighthead. It is used to rotate the disc.

Rate Motor Switch — Located on rate end. It completes the circuit to rate motor.

Release Lever — Located on tight rear of sighthead. It permits the automatic release points to close and completes the bomb release circuit.

Search Knob — Located on the lower part of the rate end. Allows you to make rapid displacement of the lateral crosshair.

Sighthead — The upper unit of the bombsight assembly. It stabilizes the optics in pitch and roll and solves the range problem.

Sighting Angle Index — Seen through index window on right side of degree scale. It indicates the sighting angle in degrees.

Sight Stem — Located on the bottom of the sighthead. A projection tube which fits into sight stem sleeve on the stabilizer.

Sight Stem Sleeve — Located on the front of the stabilizer. It is the bracket in which the sighthead is mounted.

Stabilized Gear Sector — Located on the underside of the sighthead. I t aids in transmitting stability to the sighthead and positioning the sighthead in azimuth.

Stabilizer — Lower units of bombsight assembly. It stabilizes sighthead in azimuth.

Stabilizer Switch — Located on the right side of stabilizer. Completes or breaks the circuit to stabilizer and directional gyro.

Tachometer Adapter — Located on rear of sighthead. It is connected to a shaft running from disc. A tachometer can be fitted into adapter to read the disc speed in rpm.

Tangent Scale — Seen through index window on top right of sighthead case. It is used to measure the dropping angle.

Telescope — Located inside the sighthead on the telescope cradle. The unit in the bombsight that magnifies the target image and projects the crosshairs on the mirror.

Torque Motor Switch (SERVO) — Located on the right side of the stabilizer. It completes or breaks the circuit to the torque unit.

Torque Unit — Located inside front of stabilizer. It keeps spin axis of directional gyro horizontal in relation to stabilizer case.

Trail Arm and Trail Plate — Located on top of rate end. It provides a method of putting desired trail into the bombsight.

Trail Arm Clamp Screw — Located on the end of the trail arm. It provides a method of locking the trail into the sight.

Trail Arm Pinion — Located on top and at the pivot point of trail arm. It transmits motion from the trail arm to trail rack.

Trail Bell Crank — Located on front of sighthead. it transmits motion from trail rack to push rod. This sets in potential crosstrail.

Trail Rack — Located on trail plate. Transmits motion from trail arm pinion to trail bell crank.

Trail Scale — It is marked on the trail plate in mils to allow the proper trail setting.

Turn Knob — Located on rate end. It is the outer course knob. It turns sighthead around stabilized gear sector, changing the line of sight and displacing PDI.

Turn Worm — It is mounted on the turn knob shaft and meshed with the stabilized gear sector. It transmits stability from stabilized gear sector to sighthead.

Vertical Gyro — Located inside left end of sighthead. Stabilizes optics in pitch and roll.

Glossary

AAC	Army Air Corps
AC	Air Corps
AAF	Army Air Forces
AES	Automatic Erection System
A-5	Sperry Auto Pilot
AFCE	Automatic Flight Control Equipment—AAF
AGLD	Automatic Gyro Leveling Device
AN	Army and Navy - Used by both services
AN/APS-3	Army-Navy, Airborne Radar Search Equipment N. 3 AN/APQ-13 is Army-Navy, Airborne Radar Special Equipment No. 13, etc.
AN/APQ-7	X-Band Radar Designed for high altitude bombing
AN/APQ-13	X-Band Radar designed for high altitude bombing
ANMB	Army Navy Munitions Board
ATSC	Air Technical Service Command
AZON	Azimuth only—Radio Guided Bomb
B-1	Automatic Pilot developed by Norden, Incorporated
Bomber	Short for Bombardier-Navy
BurAero	Bureau of Aeronautics-Navy
BurOrd	Bureau of Ordnance-Navy
BTO	Bombing Through Overcast
Burroughs	Burroughs Adding Machine Company
CADO	Central Air Documents Office
C-1	Automatic Pilot Developed by Minneapolis Honeywell Regulator Company
CBI	China, Burma, India Theater of War
C of AC	Chief of the Air Corps
CNO	Chief of Naval Operations
DS	Disc Speed. The speed in RPM at which the disc rotates in the M-Series Norden bombsight. The time factor is derived by dividing ATF into 5300=DS=5300. ATF
D-4-8	Bombsight made by Estoppey
FBI	Federal Bureau of Investigation
GBA	Glide Bombing Attachment
Honeywell	Minneapolis-Honeywell Regulator Company
Hqs	Headquarters
IFF	Identification Friend or Foe
JAC	Joint Aircraft Committee
JAG	Just Another gadget—Device used in RAZON bombing
LABA	Low Altitude Bombing Attachment
LOI	Letter of Instruction

347

Mark 15	Navy designation of Norden bombsight, same as Mark XV
Mil	In bombing, an angle which subtends a distance on the ground equal to 1/1000 of the bombing altitude
MOD	Model or modification
M-Series	AAF Designation of the Norden Bombsight
MIT	Massachusetts Institute of Technology
NARA	National Archives and Records Administration
NBS	Norden Bombsight
NOSMO	Norden Optical Sight Modification
NDRC	National Defense Research Council
Norden	C. L. Norden, Incorporated
NPG	Naval Proving Ground, Dahlgren, Virginia
NIO	Naval Inspector of Ordnance
OCAC	Office of the Chief of the Air Corps
PDI	Pilot Directional Indicator. An indicator in the pilot's instrument panel connected by electrical linkage to the bombsight. It indicates the direction, and to some extent the amount of correction desired by the bombardier.
RAF	Royal Air Force-British
RCAF	Royal Canadian Air Force
RADAR	Radio detection and ranging Army/Navy designation
R&R	Request and Routing
RRI	Remington Rand, Incorporated
RAZON	Range and Azimuth—Guided Bombs/Missiles
S-1	Sperry Bombsight
SBAE	Stabilized Bombing Approach Equipment-Navy
T.O.	Technical Order
Tech Rep	Technical Representative-Normally furnished by the contractor.
UK	United Kingdom
UR	Unsatisfactory Report on equipment
USC	United States Code
Victor	Victor Adding Machine Company
WD	War Department
WF	Wright Field-Army Air Forces
WPB	War Production Board
WWII	World War II
VISAR	Visual Radar
Vinson Act	Named after Congressman Carl Vinson, Georgia, Chairman of the House Naval Affairs Committee. Profit curbs were set at 7% and 8% for cost plus fixed fee contracts.

Government Publications

1. Army Air Forces, Case History of the Norden Bombsight and C-1 Automatic Pilot, Historical Office, Air Technical Service Command, Wright Field, Ohio January 1945.
2. Army Air Forces, Department of Armament, Bombsight Maintenance Division, Maintenance and Calibration, September 1943.
3. Army Air Forces, Handbook of Instructions for Bombsight Type M-9, dated June 5, 1945.
4. Army Air Forces, M-Series Bombsight Maintenance and Calibration, WTTC Manual 30-17, dated October 26, 1944.
5. Army Air Forces Proving Ground Command, Orlando, Florida, Comparative Bombing Test of the AN/APQ-13 Radar and Norden Equipment on B-29 Aircraft, September 1945.
6. Army Air Forces Statistical Digest, WWII, The Office of Statistical Control, December, 1945.
7. Army Air Forces, Tactical Center, Intelligence Manual Combat Intelligence Division, Orlando, Florida, February 15, 1944.
8. Armament Specifications for the Mark XI Bombsight Installation, U. S. Navy Bureau of Aeronautics, May, 15, 1931.
9. Army and Navy Publication, Handbook of Operation and Service Instruction, Type C-1 Automatic Pilot, June 25, 1943.
10. Aviation Psychology Program, Psychological Research on Bombardier Training, Army Air Forces, January 10, 1946.
11. B-1 and C-1 Automatic Pilot Preflight and Operational Procedures, Bombsight Maintenance School, Lowry Field, CO. 15 June 3, 1943.
12. Bombardiers' Information File, Army Air Forces, November 1945.
13. Bomb Release Interval Control (commonly called an Intervolmeter), Types B-3A, AN B-3, AN-B2A and B-2.
14. Department of Fire Control, M-Series Bombsight, Bombsight Maintenance Division, Maintenance and Calibration, Lowry Field, September 1944.
15. Experimental Investigation in Connection with High Angle Dirigible Bombs VB-3, Ballistic Data, September 1945, Office of Scienctic Research and Development.
16. Experimental Investigation in Connection with High Angle Dirigible Bombs, RAZON VB-3, Office of Scientific Research and Development, April 1943.
17. Investigations for the National Defense Program—U. S. Senate May 16, 17, and 18, 1944. Proceedings, Executive Sessions on Engineering Firms—Corrigen, Osborne & Wells.
18. Memo No. 1034-44, Tests of the Mark 17 Bombsight (Norden Aiming Angle and Dive Sight.).
19. Naval Proving Ground, Dahlgren, Virginia, Experimental Department, Memo No 1017-43. Theory of Glide and Climb Bombing with the Norden Glide Bombing Attachment, April 7, 1944.
20. Navy Bureau of Aeronautics, Precision Bombing, Naval Air Operational Training Command, NAV AER 30 7OR, June 1944.
21. Navy Bureau of Ordnance, Glide bombing, Mark 2, Mod 1, Parts Catalog and Testing Procedures, Pamphlet 1116-A, April 8, 1944.
22. Navy Department, Aviation Ordnance, bombsight Mark XV, Mods 4 and 5; Equivalent Air Corps Designation M-4, 5, 6 and 7.
23. Navy Department, Aviation Ordnance, Bombsight Mark XV, Maintenance, Ordnance, Pamphlet 649, July 1943. Bombsight Mark XV, Mods 4 and 5. Equivalent Air Corps Designation M-4, 5, 6, and 7.
24. Navy Bureau of Ordnance, Glide Bombing Attachment, Mark 2, Mod 1, Pamphlet Number 1116, March 25, 1944.
25. Navy's Mark 15 (Norden Bombsight). Its Procurement and Use 1920-1945, National Archives.
26. Navy Bureau of Ordnance, Telephone Messages, December 1941-July 1944, National Archives.
27. Norden Laboratories Corp., Mark XV Horizontal Bombsight, Engineering Report, November 26, 1957.

28. Students Manual-Bombing-Army Air Forces Training Command, Manual No. 51 340-1, September 1, 1949.
29. Students Manual-Bombing, Army Air Forces Training Command, no date.
30. Technical Order, Bombsight Series No. 11-5A and No. 11-30 Series (Individual T. O.'s not listed.).
31. Technical Order No. AN-60AA, Handbook-Operation and Service Instructions, Type C-1 Automatic Pilot, June 25 1943, Revised June 23, 1950.
32. War Department, Technical Manual, Precision Bombing Practice, TM 1-250, May 28, 1942.
33. Video Tapes and Laser Discs.
    Basic Electricity-Training Film, Honeywell, Incorporated (Formerly Minneapolis-Honeywell Regulator Company, 1942).
    The Norden Bombsight, Part I, Principles, N. 0871.
    The Norden Bombsight, Part II, Operations, No. 0872.
    The Norden Bombsight, Part III, Preflight Inspection, No. 0873.
    The Norden Bombsight, Part IV, Conduct of The Mission, No. 0874.
    The Norden Bombsight, Part V, The Leveling System, No 0875.
    Operation of The Norden Bombsight on Ground Training, No. 3365.
    National Air & Space Museum Archival Video, Disc 1 through 5, numbering some 500,000 images.
34. War Department T.M. 1-252, Technical Manual, Bombing Aids, July 20, 1943.

Sources of Information

Department of The Air Force
United States Air Force Museum, Technical Research Department, Wright Patterson AFB, Ohio.
United States Historical Research Center, Maxwell, AFB, Alabama.
Secretary of The Air force, Office of Public Affairs, The Pentagon, Washington, D. C.
United States Air Force Information and News Center, Kelly AFB, Texas.
Unites States Military Airlift Command, Audio Visual Services, March Air Force Base, San Bernadino, CA.
Hq, Air Force Logistics Command, Aeronautical Systems Division, Wright Patterson AFB, Ohio.
Hq, Air Force Systems Command, Historian's Office, Andrews AFB, Washington, D. C.
Air Force Development Test Center, Eglin AFB, Florida.
Eglin Armament Museum, Eglin AFB, Florida.
Hq Air Force Office of Air Force History, Bolling AFB, Washington, D. C.
The Air Force Historical Foundation, Andrews AFB, Maryland.
Department of Defense, Still Media Depository, Washington, D. C.

Department of Commerce
Patent and Trademark Office, Washington, D. C.

Library of Congress
Photo Duplication Division, Washington, D. C.
Science and Technical Division, Washington, D. C.
Research Services Department, Washington, D. C.

Defense Logistics Agency
Defense Technical Information Center, Cameron Station, Alexandria, Virginia.

Department of Justice
United States District Court, Eastern District, Brooklyn, New York, N. Y.

Department of The Army
Military Historical Institute, Carlisle Barracks, PA.

Presidential Libraries
Franklin D. Roosevelt Library, Foreign Affairs, January 1937-August 1939, Volume Sixteen.
Harry S. Truman Library, Independence, Missouri.

Service Organizations
Confederate Air Force, Midland, Texas
Bombardiers Incorporated, Daphne, Alabama
Pueblo Historical Aircraft Society, Pueblo, Colorado

Smithsonian Institution
Film Archives Branch
Archives Division
Records Management Division
Printing and Photo Services Division
Flight Material Aeronautics Department
Archival Support Center, Gerber Facility, Suitland, Maryland.

University and Public Libraries
Black Gold Information Center, Santa Barbara Public Library, Santa Barbara, California.
California State Library, Sacramento, California.
California Polytechnic State University, San Luis Obispo, California.
Case Western Reserve University, Cleveland, Ohio.
Columbia University, New York City, New York.

Massachusetts Institute of Technology Libraries, Cambridge, Massachusetts.
New York City Public Library, New York City, New York.
Public Records Office, London, England (British Archives).
Santa Maria Public Library, Santa Maria, California.
Stanford University Library, Palo Alto, California.
University of California at Santa Barbara, Santa Barbara, California.
University of California at San Diego, San Diego, California.
University of Michigan, Ann Arbor, Michigan.
University of Rhode Island, Providence, Rhode Island.

Federal Records Centers
San Francisco Branch, San Bruno, California.
Seattle Branch, Seattle, Washington.
New England Region, Waltham, Massachusetts.
Laguna Niguel Branch, Laguna Niguel, California.
Federal Information Center, 2100 Cottage Way, Sacramento, California.

Magazines
Popular Science, December, 1943
Popular Science, July, 1942
Popular Science, August, 1943
Popular Science, June, 1945
Flying and Popular Aviation, February 1941
Mechanics Illustrated, September 1944
True Magazine, July 1967
Science Newsletter, July 15 and December 9, 1944
Office of Scientific Research and Development,
New Weapons for Air - 1947

National Archives and Records Administration
Still Pictures Branch, Washington, D. C.
Suitland Reference Branch, Suitland, Maryland
Motion Picture, Sound and Video Branch, Special Archives Division, Alexandria, Virginia
Center for Legislative Archives, Reference Branch, Washington, D. C.
Criminal Case Legal Documents Repository, Northeast Region, Bayonne, New Jersey

Department of The Navy
Naval Air Warfare Center, Aircraft Division, Indianapolis, Indiana
U. S. Naval Surface Weapon Weapons Center, Dahlgren, Virginia
Naval Aviation News, Washington Navy Yard, Washington, D. C.
National Museum of Naval Ordnance, Pensacola, Florida
Office of The Chief of Naval Operation, Washington, D. C.
Naval Surface Warfare Center, White Oak, Silver Springs, Maryland
Naval Historical Center, Washington Navy Yard, Washington, D. C.
Naval Historical Foundation, Washington, D. C.
Secretary of The Navy, Bureau of Aeronautics/Ordnance, Military Reference Branch

Private Companies
Honeywell, Incorporated, Minneapolis, Minnesota
Barden Company, Danbury, Connecticut
United Technology Company, Norden Systems, Incorporated, Norwalk, Connecticut.
Victor Technologies, Addison, Illinois.

ISBN: 978-0-7643-0723-2